ISAC and ARIEL: The TRIUMF Radioactive Beam Facilities and the Scientific Program

Jens Dilling • Reiner Krücken • Lia Merminga
Editors

ISAC and ARIEL: The TRIUMF Radioactive Beam Facilities and the Scientific Program

Previously published in *Hyperfine Interactions*
Volume 225, 2014

Springer

Editors
Jens Dilling
TRIUMF
4004 Wesbrook Mall
Vancouver
BC V6T 2A3
Canada

Reiner Krücken
TRIUMF
4004 Wesbrook Mall
Vancouver
BC V6T 2A3
Canada

Lia Merminga
TRIUMF
4004 Wesbrook Mall
Vancouver
BC V6T 2A3
Canada

ISBN 978-94-024-0293-3 ISBN 978-94-007-7963-1 (eBook)
DOI 10.1007/978-94-007-7963-1
Springer Dordrecht Heidelberg New York London

Printed on acid-free paper

Springer is part of Springer Science+Business Media (www.springer.com)

Contents

DOI 10.1007/978-94-007-7963-1_1
Reprinted from *Hyperfine Interactions* Journal, DOI 10.1007/s10751-013-0877-7

ISAC overview

J. Dilling · R. Krücken · G. Ball

Published online: 10 October 2013

Abstract TRIUMF operates the rare ion beam facility ISAC, which provides intense beams of short lived isotopes in the energy range from 0.15 to 5 A MeV for A < 30 to a full suite of world class experimental apparatus, enabling a forefront research program focused on understanding the evolution of nuclear structure towards the limits of existence, the origin of the chemical elements in the universe, searching for physics beyond the standard model of particle physics, and the characterization of magnetic properties of new materials at surfaces and interfaces.

Keywords First keyword · Second keyword · More

1 Introduction

The quest for developing a unified theoretical description of all nuclei and nuclear matter, a full understanding of the origin of the heavy chemical elements, as well as the discovery of new particle and forces with precision experiments in atomic nuclei is driven by the study of nuclei far from stability. In order to provide the necessary variety and intensities of exotic nuclei major investments are made worldwide to build next generation rare isotope beam (RIB) facilities. The Isotope Separator and Accelerator (ISAC) facility at TRIUMF currently provides a wide variety of intense beams of exotic nuclei produced using the ISOL method via the bombardment of thick high-power targets [1] with up to 100 μA of 500 MeV protons. By coupling different ion sources, including surface, plasma (FEBIAD) [2], and laser (TRILIS) [3] sources, to target materials (e.g. Si, Ti, Nb, Zr, Ta, U) different varieties of exotic

ISAC and ARIEL: The TRIUMF Radioactive Beam Facilities and the Scientific Program.

J. Dilling (✉) · R. Krücken · G. Ball
TRIUMF, 4004 Wesbrook Mall, Vancouver, BC V6T2A3, Canada
e-mail: JDilling@triumf.ca

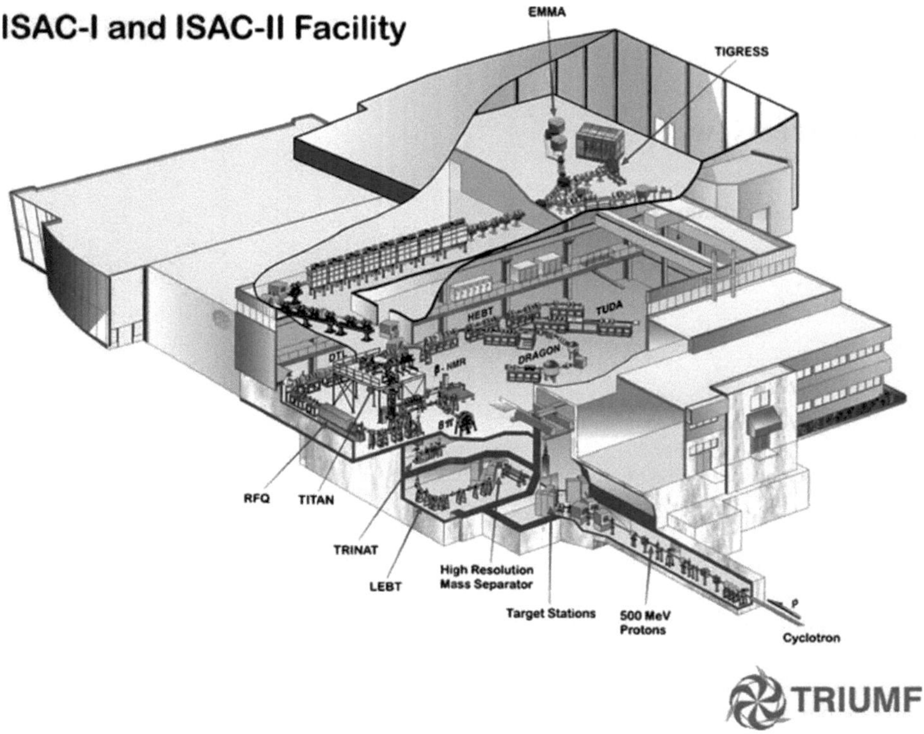

Fig. 1 Schematic view of the TRIUMF-ISAC facility

nuclei can be produced, extracted, and delivered to experiments. The ISAC facility has some of the world wide most intense beams available for certain isotopes such as 11Li, 11Be, 21Na, 18F, 23Mg and 26gAl. Only recently the variety of produced RIBs was significantly extended by the use of UC_x targets, which can currently be bombarded with up to 10 μA proton beam intensity. Plans exist to increase this to 40 μA in the future.

The short-lived isotopes can be delivered to experiments using a low-energy (<60 keV) electrostatic beam transport system, available since 1999, or re-accelerated beams. For re-accelerated beams two acceleration stages are available, the first for energies of 0.15 to 1.8 AMeV since 2001, designed for high intensity RIBs for studies of nuclear reactions relevant for explosive nucleosynthesis, and a second superconducting stage for energies of at least 6 AMeV for masses of A<150. A schematic view of the ISAC facility is shown in Fig. 1.

A wide range of experimental set-ups is available at the ISAC facility to investigate questions of current interest in nuclear astrophysics, nuclear structure and reactions, and electro-weak interaction studies. In addition to this main focus of the facility ISAC also hosts world-wide unique βNMR and βQMR set-ups for the study of magnetic properties in thin materials and at material interfaces. Hereafter we will provide a brief overview of the experimental program at ISAC. More details will be presented in the dedicated articles of this special issue of Hyperfine Interactions.

1.1 Nuclear structure and reactions

In recent years major theoretical advances were made in describing the structure of light nuclei using ab-initio methods and bare nuclear forces. In particular the use of forces based on chiral effective field theory (ChEFT) promise to provide the long thought connection between low-energy QCD and nuclear interactions. Ab-initio calculations using various methods, such as GFMC [4] and NCSM [5] have demonstrated the importance of three-body and tensor components in the bare nuclear force to correctly describe light nuclei. Further developments of the ChEFT interactions rely on precise experimental tests based not only on level energies but expanding to transition matrix elements and even reaction rates. An a area of particular focus has been the structure of halo nuclei, with their weak binding, extended wave-functions, and intricate correlations involving bound and continuum state. With its high intensity beams of ^{11}Li, ^{11}Be, and other light nuclei ISAC is ideally positioned to make important measurements in this area. The first ISAC experiment on halo nuclei was a study of the beta decay of ^{11}Li using the 8π spectrometer [6]. This was followed by a measurement of the charge radius of ^{11}Li was measured using resonant laser ionization particle spectroscopy [7] and the observed increase relative to ^{9}Li was found to be due to a combination of halo neutron correlation and intrinsic core excitation. A program of high precision Penning trap mass measurements on halo nuclei was carried out at the TITAN Penning trap facility including a measurement of ^{11}Li with its very short half-life of the 8.75 ms half-live [8–11]. The very first experiemnt at ISAC-II was using the MAYA active active target from GANIL in a pioneering experiment studying the p(^{11}Li,t)^{9}Li. It was carried out using the ISAC-II superconducting post-accelerator to accelerate ^{11}Li ions to 3 AMeV [12] and 4 AMeV [13]. Further reaction studies on the structure of halo nuclei were performed using ISAC-II re-accelerated beams, in particular the d(^{11}Be,p)^{12}Be transfer reaction [14] and a study of the E1 transition matrix element from the first excites 1^- state in ^{11}Be with the TIGRESS gamma-ray spectrometer using inelastic scattering on a nuclear target.

In heavier nuclei, not accessible to ab-initio theory yet, effective nuclear interactions are used with many-body models, such as shell model calculations or density functional theory are used to describe their structure. The shell structure in nuclei is a benchmark in the study of heavier nuclei and investigations in recent years have revealed that the magic numbers may change locally in exotic nuclei via the disappearance of classic shell gaps and the appearance of new magic numbers [15, 16]. The first region in which modifications of shell structure were observed is the *Island of Inversion* around ^{32}Mg, which has been investigated extensively over the years. In 2009 the transition matrix elements in ^{29}Ne obtained from a Coulomb excitation experiment [17] with the TIGRESS gamma-ray spectrometer established it to be on the transition line between normal and inverted shell structure.

Today it appears that these local changes to the shell structure are driven by the monopole components of the residual interaction between the valence nucleons and that in particular the tensor interaction[18–21] and three-body forces [22, 23] play an important role. In particular new magic shell closures for ^{24}O, which has been confirmed experimentally, and ^{54}Ca have been associated with these effects. With the TITAN penning trap facility, it has recently been possible to extend precision mass measurements to ^{52}Ca [24]. The extracted neutron separation energies indicate that experimental trends can be best described by using effective interactions derived from realistic nuclear forces including three-body components.

Aside from investigating the evolution of shell structure directly, collective motion in nuclei provides important clues on the working of the underlying nucleonic dynamics that is responsible for the emergence of collective degrees of freedom, such almost degenerate states with different competing shapes, as well as rotational and vibrational motion. Some of these studies are carried out with the 8π gamma-ray spectrometer, that is being used for the spectroscopy of excited states populated in the beta decay of the isotopes extracted from the ISAC target and implanted into a tape positioned in the center of the 8π vacuum chamber. Experiments at TIGRESS with re-accelerated beams, e.g. via Coulomb-excitation or Doppler-shift lifetime measurements will provide complementary information. Recently, a precision decay study of states in ^{112}Cd [25] revealed evidence that the structure in this test book example for a nucleus for vibrational motion shows some substantial deviations from the vibrational model. Using the new availability at ISAC of neutron-rich fission fragment beams produced in a UC_x target the 8π was used for the spectroscopy of neutron-rich Sr iosotopes following the beta decay of ^{96,98,100,102}Rb, a region of rapid transition from spherical to deformed ground states in the Sr and Zr isotopes around neutron number 60.

1.2 Nuclear astrophysics

The ISAC facility was specifically designed to address a wide range of questions related to the energy production and nucleosynthesis in various astrophysical environments. The nuclear astrophysics program at ISAC is best known for the direct measurements of hydrogen and helium radiative capture reactions with the DRAGON recoil separator. Explosive hydrogen and helium burning plays a key role in the energy generation and element production in the proton-rich outflow in x-ray bursts, nova explosions and supernova explosions. Recent work at DRAGON and TUDA made important contributions to constraint the production and destruction reactions in classical novae for radioisotopes that are either observed or searched for in gamma-ray satellite missions. An important result for Novae is the DRAGON measurement of the ^{23}Mg(p,γ)^{24}Al reaction [26]. The experiment drastically reduced the uncertainty for this critical reaction in the transition between the NeNa and MgAl cycles in nova nucleosynthesis, having a bearing on the ejected abundances of both ^{22}Na and ^{26g}Al. Recently important constraints on the production of 511keV annihilation gamma-rays from the ^{18}F decay were set by measuring the ^{18}F(p,α) reaction with TUDA [27] and constraining the gamma-width of the dominant resonance in a DRAGON experiment (Laird, private communications). It was subsequently shown that DRAGON can also be used for direct capture reactions of nuclei with mass up to A=80, enabling direct reaction rate studies for several reactions critical for the production of p-nuclei in this mass-range.

Long-lived radioisotopes produced at ISAC can be implanted in thin foils to produce targets of these radioisotopes for reaction studies at other facilities. The direct measurement of the ^{22}Na(p,γ) reaction at the University of Washington in Seattle was enabled using an implanted target produced at ISAC [28, 29]. The experiment revealed a much stronger destruction rate of ^{22}Na in novae, reducing the distance for novae from which ^{22}Na gamma-rays are observable by about a factor of two. Similary, implanted ^{26g}Al targets can now be produced for charged particle and neutron induced reactions on this key nucleus.

In addition to the direct reaction studies important properties of relevant resonances, such as energies and partial decay width, the nuclei involved in key astrophysical reactions can be studied with transfer reactions using TUDA, SHARC, EMMA, or level lifetimes can be measured using the DSL Facility. In addition, important nuclear properties governing the reaction flow in the large reaction networks such as Q-values (TITAN) [30] and half-lives (8π) can be determined.

The rapid neutron capture r-process is considered to be the source of about half of the heavy elements above iron. It occurs in a very neutron-rich, high entropy environment where seed nuclei from an alpha-rich freeze-out capture neutrons until an equilibrium abundance of isotopes for an element (Z) is reached between neutron capture and photo-dissociation in the hot environment. Subsequent beta-decay leads to the heavier elements where again neutron capture occurs and through successive steps all heavy elements above iron can be produced. The shell structure in these very neutron-rich nuclei is imprinted in the abundance pattern. However, only for a few nuclei along the possible r-process path experimental data exist. With the availability of very neutron-rich fission fragments produced with protons or electrons at the future ARIEL facility, TRIUMF will gain access to a large number of potential r-process nuclei for which important properties such as masses, half-lives, P_n values, and information on single-particle structure can be extracted.

1.3 Electroweak interaction studies

The precision measurements of the decay properties of superallowed $0^+ \rightarrow 0^+$ Fermi-decays provides unique access to test the conserved vector current (CVC) hypothesis and to extract the V_{ud} matrix element of the Cabibbo-Kobayashi-Maskawa (CKM) quark mixing matrix, enabling in turn the test of the unitarity of the CKM matrix [31]. At ISAC an experimental program to determine decay Q-values, from mass measurements as well as half-lives, and decay branching ratios from beta decay spectroscopy, has been carried out for many of the relevant isotopes. Most recently the half-life [32] and the decay branching ratio [33] of ^{26m}Al were measured while the decay Q-value was determined for ^{74}Rb [34], using for the first time TITAN's new capability to use highly charged ions. In a number of the relevant nuclei, the uncertainties are now dominated by the nuclear corrections that have to be applied to extract the corrected Ft values. In particular, the isospin breaking corrections are still not fully understood and more data on this is needed. A recent determination of the charge radius of ^{74}Rb [35] helped to substantially reduce the uncertainty of this correction for this nucleus. A further measurement of the decay branching ratio was performed with the 8π spectrometer. The data is still under analysis.

Precision measurements of the angular correlations between electron (positron) and anti-neutrino (neutrino) emitted in the beta-decay of spin aligned nuclei allows to search for new components of the electro-weak interaction, such as the exchange of new bosons not described in the Standard Model. Since the neutrino cannot be detected efficiently the correlation is determined from the measurement of the momenta of the electron and recoiling nucleus. The TRINAT experiment employs and magento-optical trap (MOT) for such experiments and currently has made the most sensitive experiment in the superallowed Fermi decay of ^{38m}K [36] providing the most stringent constraint for scalar components in the electro-weak interaction. Moreover, limits were set on the admixture of heavy neutrinos to the electron neutrino [37] With currently ongoing upgrades to the TRINAT set-up a further improvement of the achievable precision by a factor of five is anticipated in the near future.

The measurement of a nuclear anapole moment, a parity nonconserving (PNC), time reversal conserving moment that arises from weak interactions between the nucleons, provide unique access to neutral weak interactions inside the nucleus. Experimentally the anapole moment would reveal itself in a nuclear-spin dependent signal of a certain electric dipole transitions between hyperfine states of the atom that is parity forbidden but would be enabled through the PNC effect of the anapole moment in the electron-nucleus interaction. The parity violating components scale with Z^3 and hence heavy nuclei are favorable for such experiments. Francium atoms are particularly well suited and due to the high production yields of a broad range of Fr isotopes at ISAC, together with the prospects for substantial amounts of beam time due to multi-user capability of the facility once the ARIEL facility is operational, there is a unique opportunity for such a program at TRIUMF. Recently a fully electromagnetically shielded room with at Fr MOT was installed at ISAC and first trapping and hyperfine spectroscopy experiments have been carried out on $^{206-212}$Fr.

Heavy nuclei with large stable octupole deformation in the ground state are particularly well suited to search for experimental evidence of an atomic electric dipole moment (EDM), which would provide the most sensitive available probe of flavor non-changing, CP violating physics beyond the standard model. Heavy Rn isotopes around ^{223}Rn are expected to exhibit such octupole deformations and due to the high nuclear charge the enhancement of the effect would be much larger than in other systems under investigation, such as Hg or Xe. With the availability of Rn beams from UC_x targets a detailed investigation of the structure of heavy Rn isotopes is now possible, e.g. with the 8π and TIGRESS spectrometers. At the same time a set-up for an EDM experiment is under development in which the decay asymmetries of polarized Rn atoms will be measured.

In the simplest model the half-live for neutrinoless double beta decay is directly connected to the electron neutrino mass. This connection involves the effective nuclear matrix element for this transition, which can only be predicted theoretically. However, predictions for different candidate nuclei by different theoretical models vary by factors of 2–10. The predictions of these same models for double beta decay and for the beta decay and electron capture decay can be determined experimentally and provide sensitive tests of the models. The TITAN-EC project aims to measure the very weak EC branching ratios in the decays of the intermediate odd-odd nuclei in the converted EBIT trap of the TITAN facility. After a successful proof of principle demonstration the detector system has been completed and first measurements are imminent.

1.4 Material science

The beta-detected nuclear magnetic resonance (β–NMR) facility at ISAC is constructed specifically for using radio-isotopes for experimental studies in materials science, utilizing a probe beam of spin-polarized radioactive ions such as $^8Li^+$ (half-life = 848 ms) to monitor the local electromagnetic properties of the host material sensed at the atomic scale. A crucial capability of this technique is control of the average implantation depth of the probe, thus obtaining depth resolution on an interesting length scale (five 200 nanometers). The β–NMR facility at TRIUMF is unique in the world. This local probe technique provides important complementary tools to reciprocal space methods like polarized neutron reflectometry (PNR) and synchrotron techniques such as Resonant Inelastic Xray Scattering (RIXS) that also study depth-dependent electromagnetic properties of materials. As with conventional NMR, the nuclear spin senses its local environment, specifically the local magnetic field (which is determined by the surrounding electronic structure of the host), and the electric field gradient via the quadrupolar coupling, as well as fluctuations of these quantities that result in relaxation of the probe nuclear spins. The β–NMR facility is used for a number of experiments in magnetism and superconductivity as well as on novel ultra-thin heterostructures exhibiting properties that cannot occur in bulk materials (see [38] for a recent example). Recently the scope of research using this facility has expanded into soft condensed matter and ionic conductivity.

1.5 Summary

The brief overview presented above provides a flavor of the broad scientific program with RIBs accessible at TRIUMF today. With the completion of the ARIEL facility even higher intensity and more exotic neutron-rich beams will become available. At the same time the multi-user capability with up to three parallel RIB production targets delivering beams to the experimental set-up, tremendously enhancing the efficiency in exploiting these world-leading instrumentation and enabling the often month-long beam times necessary for the experiments in search for physics beyond the Standard Model using weak decays and measuring nuclear astrophysical reaction rates.

References

1. Bricault, P., et al.: Nucl. Instrum. Methods **B204**, 31924 (2003)
2. Bricault, P., et al.: Rev. Sci. Instrum. **79**, 02A908 (2008)
3. Lassen, J., et al.: AIP Conf. Proc. **1099**, 76973 (2009)
4. Pudliner, B.S., et al.: Phys. Rev. **C56**, 1720 (1997)
5. Navratil, P., et al.: Phys. Rev. Lett. **99**, 042501 (2007)
6. Sarazin, F., et al.: Phys. Rev. **C70**, 31302 (2004)
7. Sanchez, R., et al.: Phys. Rev. Lett. **96**, 033002 (2006)
8. Smith, M., et al.: Phys. Rev. Lett. **101**, 202501 (2008)
9. Ryjkov, V.L., et al.: Phys. Rev. Lett. **101**, 012501 (2008)
10. Ringle, R., et al.: Phys. Lett. **B675**, 170 (2009)
11. Ettenauer, S., et al.: Phys. Rev. **C81**, 024314 (2010)
12. Tanihata, T., et al.: Phys. Rev. Lett. **100**, 192502 (2008)
13. Roger, R.: Ph.D. thesis (2009)
14. Kanungo, R., et al.: Phys. Lett. **B682**, 391 (2010)
15. Sorlin, O., Porquet, M.-G.: Prog. Part. Nucl. Phys. **61**, 602 (2008)
16. Krücken, R.: Contemp. Phys. **52**, 101 (2011)
17. Hurst, A.M., et al.: Phys. Lett. **B674**, 168 (2009)
18. Otsuka, T., et al.: Phys. Rev. Lett. **87**, 082502 (2001)
19. Otsuka, T., et al.: Phys. Rev. Lett. **95**, 232502 (2005)
20. Otsuka, T., et al.: Phys. Rev. Lett. **97**, 162501 (2006)
21. Otsuka, T., et al.: Phys. Rev. Lett. **104**, 012501 (2010)
22. Otsuka, T., et al.: Phys. Rev. Lett. **105**, 032501 (2010)
23. Holt, J., et al.: J Phys. **G39**, 085111 (2012)

24. Gallant, A., et al.: Phys. Rev. Lett. **109**, 032506 (2012)
25. Garrett, P.E., et al.: Acta Phys. Polon. **B42**, 799 (2011)
26. Erikson, L., et al.: Phys. Rev. **C81**, 045808 (2010)
27. Beer, C.E., et al.: Phys. Rev. **C83**(R), 042801 (2011)
28. Sallaska, A.L., et al.: Phys. Rev. Lett. **105**, 152501 (2010)
29. Sallaska, A.L., et al.: Phys. Rev. **C83**, 034611 (2011)
30. Simon, V.V., et al.: Phys. Rev. **C85**, 64308 (2012)
31. Towner, I.S., Hardy, J.C.: Rep. Prog. Phys. **73**, 046301 (2010)
32. Finlay, P., et al.: Phys. Rev. Lett. **106**, 032501 (2011)
33. Finaly, P., et al.: Phys. Rev. **C85**, 55501 (2012)
34. Ettenauer, S., et al.: Phys. Rev. Lett. **107**, 212502 (2011)
35. Mane, E., et al.: Phys. Rev. Lett. **107**, 212502 (2011)
36. Gorelov, A., et al.: Phys. Rev. Lett. **94**, 142501 (2005)
37. Trinczek, M., et al.: Phy. Rev. Lett. **90**, 12501 (2001)
38. Salman, Z., et al.: Phys. Rev. Lett. **109**, 257207 (2012)

DOI 10.1007/978-94-007-7963-1_2
Reprinted from *Hyperfine Interactions* Journal, DOI 10.1007/s10751-013-0878-6

The TRIUMF 500 MeV cyclotron: the driver accelerator

I. Bylinskii · M. K. Craddock

Published online: 10 October 2013

Abstract ISAC's radioactive ion beams are currently produced in nuclear spallation reactions driven by an intense 480-MeV proton beam from the TRIUMF sector-focused cyclotron. In this article we describe the design and operation of the cyclotron, and also the extensive program of upgrading the equipment and beams that has enhanced its performance and reliability over recent years.

Keywords Cyclotron · Accelerator · Negative hydrogen ion

1 Introduction

TRIUMF's core accelerator is a sector-focused H^- cyclotron capable of delivering four independently controllable proton beams at energies from 70 to 520 MeV with a total current of up to 300 μA. The design was originally conceived by Richardson [1] in 1962 for a 700-MeV H^- meson factory at UCLA. But when the U.S. government decided to fund an 800MeV proton linac at Los Alamos instead, he agreed to allow a scaled-down version to be used by the TRIUMF group in Western Canada, led by Warren [2]. The Canadian proposal was funded in 1968, and the cyclotron completed and first beam extracted in 1974, under Richardson's direction [3].

Although this machine does not match its rivals (the LANL linac and PSI proton cyclotron) in raw beam power, its cw operation and the flexibility provided by its

ISAC and ARIEL: The TRIUMF Radioactive Beam Facilities and the Scientific Program.

I. Bylinskii (✉) · M. K. Craddock
TRIUMF, Vancouver, BC, Canada
e-mail: bylinsky@triumf.ca

M. K. Craddock
University of British Columbia, Vancouver, BC, Canada

Fig. 1 Staff pose on the lower magnet sectors, January 1972

multiple beams and variable energy have enabled it to maintain comparable scientific productivity. This flexibility is made possible by the use of negative hydrogen ions, which have a binding energy of only 0.75 eV, and so can be extracted in a simple and highly efficient way by stripping them to protons in thin (11 μm) pyrolytic graphite foils. (All but 3 parts in 10^{18} are fully stripped at 500 MeV, and all but 1 part in 10^{55} at 100 MeV.) The fragility of H^- ions, however, means that they can also be readily disrupted in strong electromagnetic fields or by collisions with gas molecules. Their use therefore incurs design penalties: to limit the beam power loss by electromagnetic stripping to 7 % by 500 MeV, the magnetic field strength must not exceed 0.58 T; to limit that by collisions to 3 %, the vacuum must be better than 10^{-7} Torr. The former requirement implies much larger orbits than in a proton accelerator, perhaps by a factor 3. Thus the TRIUMF cyclotron has the dubious distinction of having the largest diameter of any yet built: for the magnet poles, 17.2 m; for the yoke, 21.5 m. As the electromagnetic stripping is only significant above 450 MeV, it is possible to run higher currents for the same absolute loss by extracting the beam at energies below 500 MeV.

2 Cyclotron systems

The 4000-tonne magnet is composed of six separate sectors, radial near the centre, but increasingly spiraled at large radii to provide sufficient vertical focusing (Fig. 1). The space between them is left free of iron to maximize the magnetic flutter. The magnet is excited by a pair of circular coils 19 m in diameter, each consisting of 15 vertical sheets of aluminum (45.7 cm × 2.5 cm) with internal water cooling channels, and weighing 77 t. The power supply provides 18,400 A at 75 V with stability 10^{-6}.

The vacuum chamber is roughly circular (maximum diameter 17.9 m) and 46 cm in height, the lid being sealed by a pair of rubber O-rings. A vacuum better than 5×10^{-8} Torr is maintained by two long 4-K cryopanels and 6 cryopumps, backed up by turbopumps. Externally, the lid and base carry 54 circular and 78 harmonic trim coil pairs for fine adjustment of the magnetic field. The 2400-t atmospheric load on the base is supported by 332 steel tie rods anchored to the vault floor; that on the lid by another 332 tie rods bolted to a 109-t "spider" of steel I-beams above. The upper half of the magnet, the spider and the tank lid can all be raised 1.22 m by

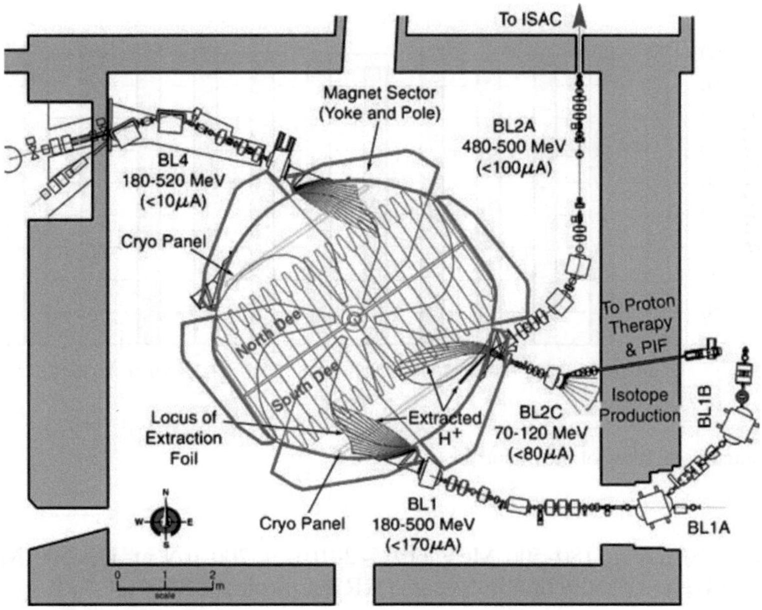

Fig. 2 The cyclotron and external proton beam lines

12 electrically-driven jacks to permit maintenance work inside the tank. Much of this work can be carried out remotely (including replacement of the rf resonator sections) using a 9-m-long bridge that can be inserted and rotated about a central pivot.

H^- ions are produced in an external cusp source mounted in a Faraday cage raised to 300 kV by a Cockcroft-Walton set. The 300-keV ion beam is then transported horizontally over the cyclotron vault roof and bent 90° downwards for axial injection. Electrostatic focusing and steering is used throughout the 46-m-long injection beam line. An electrostatic spiral inflector and horizontal deflector are then used to steer the beam into the median plane of the cyclotron for acceleration by the dees.

Acceleration is by 5th harmonic rf (23.055 MHz), the two 180° dees each being composed of 40 quarter-wave resonators 75-cm-wide, 3-m-long, half mounted on the vacuum tank base and half on the lid. The system is powered by eight 250-kW Eimac tetrodes, producing a dee voltage of 95 kV (with 10^{-4} stability) and maximum energy gain/turn $\Delta E = 380$ keV. There is also a 92-MHz booster cavity to increase ΔE by $\leq$ 200 keV above 450 MeV, significantly reducing the beam loss.

The primary beam diagnostic tools are two "low-energy" and three "high- energy" intercepting probes equipped with multi-finger heads that can be moved radially to cover the whole energy range. These provide data on total beam current, radial and vertical intensity distributions, and time of flight. Visual access is also available via CCTV camera, periscope and rotatable mirrors.

Extraction systems are provided for four external beam lines (see Fig. 2):

- BL1: $\leq$ 170 μA at 180–520 MeV for pion and muon production;
- BL2A: $\leq$ 100 μA at 472–500 MeV for radioactive ion production;
- BL2C: $\leq$ 80 μA at 70–120 MeV for radioisotope production, proton irradiations and cancer therapy;

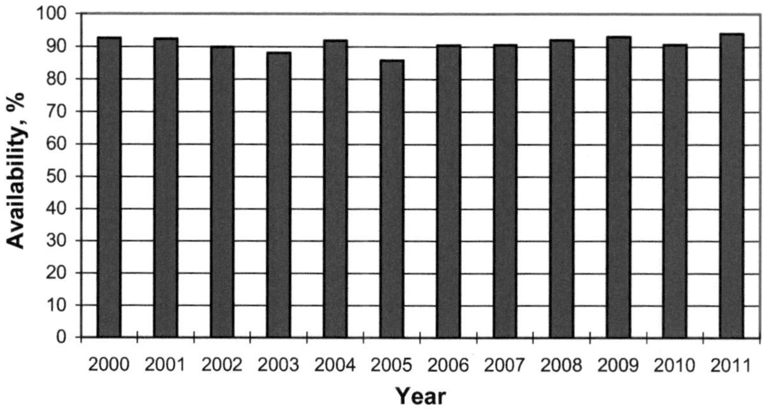

Fig. 3 Annual availability of the cyclotron

- BL4: $\leq$ 10 μA at 180–500 MeV (1975–2010); $\leq$ 200 μA at 470–500 MeV for radioactive ion production in 5 years (ARIEL project).

The stripping foils may be moved radially to change the extracted beam energy, and azimuthally to direct the beam to an external "combination" magnet that steers it down the beam line. Multiple foils are available in each extraction probe cartridge and they may be changed remotely, either in case of damage, or when a different foil shape is required (as for the lower energy beams, where only a fraction, ranging from 1/50,000 to 1/2, is to be extracted). The extracted beams typically have an energy spread of 1 MeV, radial and vertical emittances $4\varepsilon_{\rm rms} = 1 - 2\ \pi\,\mu$m, and a 4-ns, 23-MHz, bunch microstructure. When required, a chopper in the injection line can be used to create a pulse macrostructure with a duty cycle variable from 0.1 to 99 %.

3 Operation and performance

For decades the cyclotron operated in a fairly steady mode of 24/6 production with regular weekly 8- to 36-hour maintenance periods and two annual shutdowns: 7 days in September and 3–4 months in winter, depending on service and repair needs. Lately, aiming at higher scientific production and as a result of multiple machine upgrades, the number of maintenance days has been greatly reduced (by ~30 %).

Long maintenance periods are usually driven by ion-source filament replacements (every 3 weeks) and cryo-panel preventative defrosts (every 6 weeks). Over the last decade the machine has demonstrated fairly reliable operation with annual availability around 90 % (see Fig. 3).

Both peak and integrated beam production are driven by the beam delivery schedule determined by the users' needs. With the recent deployment of actinide targets in ISAC, beam demand down beam line 2A has dropped from 70–100 μA (high power targets) to 10 μA for a significant fraction of experiments. A histogram of beam charge delivered over the last decade is presented in Fig. 4. The reduction in BL1 charge in 2011 is associated with an extended shutdown of this beam line for

Fig. 4 The cyclotron's annual production of beam charge and its distribution among the beam lines

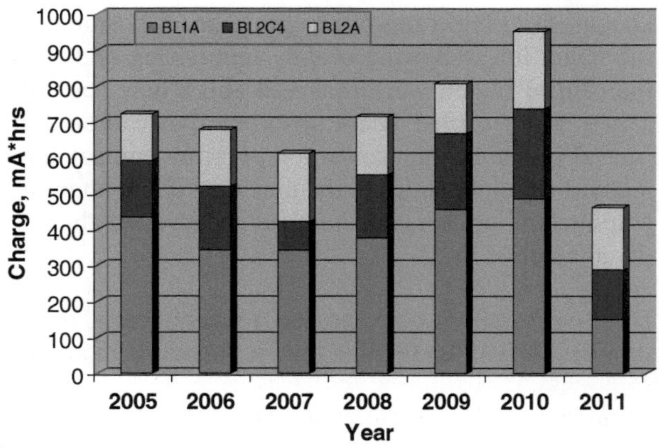

installation of a new M20 secondary channel and repair of a vacuum leak at the T2 target extraction port; the reduction in BL2C4 charge is due to an FDA investigation of an isotopes breach in the USA and the subsequent interruption of ^{82}Sr production.

The most significant issues impeding operations and requiring extended effort during shutdowns are usually associated with water leaks (cyclotron resonator panels, Meson Hall magnets) or vacuum leaks (distorted joints in high radiation areas).

Under optimal conditions transmission up to 70 % has been measured between injection and extraction to all beam lines. The extracted beams have good spot sizes (3–7 mm) at the targets and a 4-ns-long time structure.

4 Hardware upgrades

Since 2000, when ISAC-I construction was substantially completed, the cyclotron has received government funding for its refurbishing and upgrade of ~400 k$ annually. Within this program many subsystems and components have been upgraded [4]:

RF system [5] Power amplifier (PA) resonators upgrade, new high-power rf switch, transmission-line matching section upgrade, dee-voltage monitoring upgrade, new monitoring systems for the PA and power distribution dc and rf signals, hv crowbar and high reverse-power protection upgrade, rf coupler upgrade, new 12kV ac switch gear.

Vacuum system [6] New helium refrigerator deployment for the cyclotron cryopumping provided higher reliability, reduced maintenance and better vacuum.

Injection system [7] A new 12m-long vertical section of the electrostatic beam line employing low-maintenance reliable design dramatically extended the diagnostics and tuning capabilities, and supports high-intensity (up to 5 mA) beam transport; an Alison type emittance scanner has been installed downstream of the ion source, greatly expanding beam characterization and tuning capabilities.

Diagnostics New non-intercepting beam position monitors have been installed in the beam lines allowing on-line monitoring and tuning; old leak-prone devices in the central region were removed and a new vertical flag developed and installed; a new water-cooled beam absorber that captures un-accelerated particles in the central region has improved temperature control and reduced rf sparking, allowing higher beam intensities in the machine; deployment of new highly-oriented pyrolitic graphite material has dramatically improved the life time of the extraction foils (by a factor of 3).

DC power supplies A project to completely replace all of the 310 kW power supplies (120 units) feeding the cyclotron's trim and harmonic coils is half complete.

Cyclotron vault cable infrastructure Wiring in the vault is exposed to harsh conditions of radiation, humidity and temperature. This leads to premature failures and destructive damage. To address this issue TRIUMF has embarked on replacement of all the cables (~1500 units) ending in the cyclotron vault. A complementary parallel cable tray infrastructure has been created and half of the cables replaced.

5 Cyclotron beam development

Since 1995, when the ISAC project got under way, new requirements for beam quantity and quality have been established. First of all, the total beam intensity demanded has grown from 200 to 300 μA. However, this growth in production had to be achieved without increasing facility activation due to beam loss. Also, due to the vulnerability of the ISAC targets and to the high sensitivity of their yields, stringent limits have had to be set on fluctuations in intensity and beam position on the BL2A target, setting high demands on machine stability and reproducibility. To address these issues a number of beam developments have taken place and several machine improvements have been implemented.

The cyclotron design does not support operation much beyond 500 MeV because above ~450 MeV there is a rapid rise in electromagnetic stripping losses (see Fig. 5), due to the Lorentz force tearing one of the weakly coupled electrons off the hydrogen ion. In 2009, after careful evaluation of the impact on the experimental program, the extraction energy was reduced from 500 MeV to 480 MeV. This has led to ~30 % reduction in both prompt and residual activation for the same beam intensity [8]. Alternatively, it allows an equivalent intensity boost within the traditional activation dose budget.

Several improvements were made to get the required higher stability for the ISAC primary beam. First, the BL2A beam intensity was stabilized to ±1 % by introducing a feedback loop between the electron current caught on the stripper and the pulser at injection, regulating the beam's duty cycle. Decreases in the stripper current are compensated by increases in duty cycle and vice versa [9]. Second, automatic beam steering was implemented to keep the beam centred on the production target. Thirdly, the beam line 2A tunes were developed to form an image at the target of the spot on the stripping foil, thereby minimizing the beam halo on the target. Automatic beam steering was also implemented for all targets in BL1A.

Stabilizing the BL2A intensity was found to produce a side effect—magnified intensity fluctuation in the other primary beam lines. The root cause of the instability

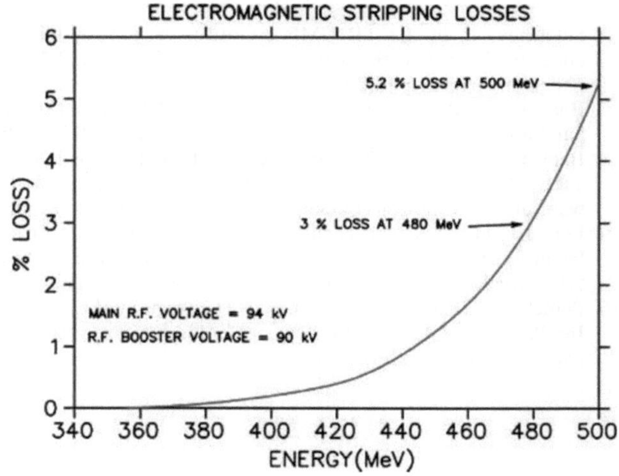

Fig. 5 Electromagnetic stripping loss as a function of energy

was traced to the cyclotron's $\nu_r = 3/2$ resonance leading to a radial beam density variation at energies above 450 MeV [10]. To suppress this resonance a delicate machine tune was developed employing two sets of harmonic coils near the extraction radii (HC12 & HC13). This reduced the intensity instability in the unregulated beam lines (BL1A & BL2C) from ±10 to ±2 %.

The ISAC production targets operate at extreme temperatures, very close to the material's destructive damage limit. Therefore they are very vulnerable to abrupt thermal changes and thus sensitive to any beam interruptions causing big changes in power deposition on the target. Instead of the full beam trip previously induced by an over-current or some other abnormal beam condition, a so-called "soft" beam trip was therefore implemented, where the beam intensity is dropped down to 80 % without interruption, allowing the operator to address the anomaly and restore normal production. Also, slow ramping of the beam intensity, (~1 min, up or down), has been introduced to mitigate target thermo-cycling issues.

References

1. Richardson, J.R.: Preliminary considerations on meson cyclotrons. Nucl. Instrum. Methods **18–19**, 457 (1962)
2. Warren, J.B.: The TRIUMF project. In: Proc. 5th Int. Cyclotron Conf., Oxford. ISBN 0 408 70043 2 (Butterworths, 1971), p. 73. http://accelconf.web.cern.ch/accelconf/c69/papers/cyc69a08.pdf (1969)
3. Richardson, J.R., et al.: Production of simultaneous, variable energy beams from the TRIUMF cyclotron. IEEE Trans. **NS-22**, 1402. http://accelconf.web.cern.ch/accelconf/p75/PDF/PAC1975_1402.pdf (1975)
4. Bylinskii, I., et al.: TRIUMF 500 MeV cyclotron refurbishment. In: Proc. 18th Int. Cyclotron Conf., Giardini Naxos, p. 143. http://accelconf.web.cern.ch/AccelConf/c07/PAPERS/143.pdf (2007)
5. Bylinski, Y.: TRIUMF cyclotron rf system refurbishing. In: Proc. 17th Int. Cyclotron Conf., p. 327. Tokyo. http://accelconf.web.cern.ch/AccelConf/c04/data/CYC2004_papers/20C3.pdf (2004)

6. Sekachev, I., et al.: TRIUMF cyclotron vacuum system upgrade and operational experience. In: Proc. PAC09, Vancouver, p. 396. http://accelconf.web.cern.ch/AccelConf/PAC2009/papers/mo6rfp021.pdf (2009)
7. Baartman, R.A., et al.: Commissioning the TRIUMF 300 keV H- vertical injection line, IPAC11, San Sebastian. http://lin12.triumf.ca/text/2011IPAC/2011ISIS/Poster.pdf (2012)
8. Baartman, R.A.: Reliable production of multiple high intensity beams with the 500 MeV TRIUMF cyclotron. In: Proc. 19th Int. Cyclotron Conf., Lanzhou, p. 280. http://accelconf.web.cern.ch/AccelConf/Cyclotrons2010/papers/tum1cco03.pdf (2010)
9. Pon, J.J., et al.: Recent changes in the 500 MeV cyclotron's central control system to reduce beam downtime and beam on/off transitions. In: Proc. ICALEPCS 2009, Kobe, Japan, p. 179. http://accelconf.web.cern.ch/accelconf/icalepcs2009/papers/tup043.pdf (2009)
10. Planche, T., Baartman, R., Rao, Y.-N.: Correction of $\nu_r = 3/2$ resonance in TRIUMF cyclotron. In: Proc. IPAC12, New Orleans, p. 415. http://accelconf.web.cern.ch/AccelConf/IPAC2012/papers/moppd023.pdf (2012)

DOI 10.1007/978-94-007-7963-1_3
Reprinted from *Hyperfine Interactions* Journal, DOI 10.1007/s10751-013-0879-5

ISAC targets

M. Dombsky · P. Kunz

Published online: 3 December 2013

Abstract The TRIUMF-ISAC radioactive ion beam facility was designed and constructed to allow irradiation of thick targets with up to 100 μA proton beam intensities. Since beginning operation in 1998, beam intensities on ISAC targets have progressively increased toward the 100 μA design limit. Routine operation with p^+ intensities up to 75 μA is currently possible for both refractory metal target materials and for composite metal carbide materials; full 99 μA p^+ intensity has been achieved for Nb foil target material. Consideration must be given to the beam power deposition, the power dissipation and the limiting temperature criteria of each target material. Increased beam power dissipation has been achieved by modifying target materials and target containers. Increasing irradiation currents have produced benefits, drawbacks and unexpected results for ISOL operations.

Keywords Radioactive ion beams · On-line isotope separation · High power targets

1 Introduction

The TRIUMF-ISAC facility was designed for radioactive ion beam (RIB) production using protons from the TRIUMF 500 MeV H^- cyclotron as the "driver" beam. ISAC is an ISOL (Isotope Separation On-Line) facility where short lived nuclei are produced in thick targets by nuclear reactions initiated with the p^+ beam and converted to a radioactive ion beam in an ion source. The thick targets are operated at high temperature (up to 2300 °C) to enhance the diffusion and effusion of the

ISAC and ARIEL: The TRIUMF Radioactive Beam Facilities and the Scientific Program.

M. Dombsky (✉) · P. Kunz
TRIUMF, 4004 Wesbrook Mall, Vancouver, BC, Canada, V6T 2A3
e-mail: marik@triumf.ca

P. Kunz
e-mail: pkunz@triumf.ca

product nuclei out of the target. The ion source is closely coupled to the target to decrease delays between production and ion beam generation. Currently, ISAC targets operate with surface ionization sources, resonant laser ionization or FEBIAD plasma sources. Ion beams are extracted with energies up to 60 kV, mass separated by passage through 2 dipole magnets in series and delivered to a variety of low energy experimental stations or post accelerators to achieve higher RIB energies. A general description of the ISAC design is available in Ref. [1].

ISAC first delivered radioactive beams to experiments in 1998 using 1–3 μA p^+ beam intensities on target. In 1999, the beam intensity was increased to 10 μA on a Nb foil target and by 2000, both Nb foil and Ta foil targets were operated with 20 μA p^+ beam intensities. The first compound target material (SiC powder pressed into pellets) was irradiated with 10 μA p^+ during 2000. During 2001 the program to increase beam intensity on target continued with a Nb foil target receiving 30 μA, a Ta foil target receiving 40 μA and a SiC pressed pellet target 15 μA p^+ beam. Development of composite target materials in the form of foils, rather than pressed pellets, allowed p^+ increases to 45 μA for SiC, 40 μA for TiC and 50 μA for ZrC during 2002–2003. Further increases in p^+ intensity were not achieved until 2005 when a modified high power target container was developed. With the increased radiative power dissipation of ISAC high power target containers, p^+ intensities up to 75 μA are currently routine for refractory metal and composite metal carbide target materials. For the case of Nb metal foil, the combination of beam power deposition and dissipation resulting from the target material and container properties, allowed first operation at the full ISAC design criteria of 99 μA p^+ in 2009.

2 Increasing beam power on target

Progress towards increasing proton currents on ISAC targets has dealt with addressing both beam power dissipation through target materials and increasing the passive radiative cooling of the target containers. The standard ISAC target containers design consists of a 19 mm diameter, 19 cm long Ta tube with an integral 3 mm ID Ta transfer line perpendicular to the p^+ beam axis. The transfer line acts both as a surface ionization source and guide channel for resonant laser ionization or can be replaced with a Ta hollow cathode for a FEBIAD plasma ion source. The design has been described in Refs. [2, 3] and is displayed in Fig. 1. Both target and transfer line temperature is maintained by resistive Joule heating using 1000 A/10V and 500 A/10V power supplies, respectively. ISAC targets are fabricated “in house” employing electron beam welding. During initial 1–3 μA operation, ISAC targets were wrapped with 3 layers of thin (0.025 mm) Ta heat shields to increase the resistive heating efficiency.

As a first step to increase p^+ irradiation current, the radiative heat dissipation ability of the target container was increased by simply removing the heat shields and reducing the resistive target heating to balance the power deposited by the p^+ beam. The effective emissivities of a target wrapped with 3, 2, 1 and 0 heat shields were determined by off-line measurements to be approximately 0.18, 0.20, 0.25 and 0.34 respectively [4]. Additionally, the thin (0.025 mm) Nb or Ta metal target foils were cut slightly larger than the inside diameter of the target container. When pressed perpendicularly into the target container, the foils make a tight contact with the

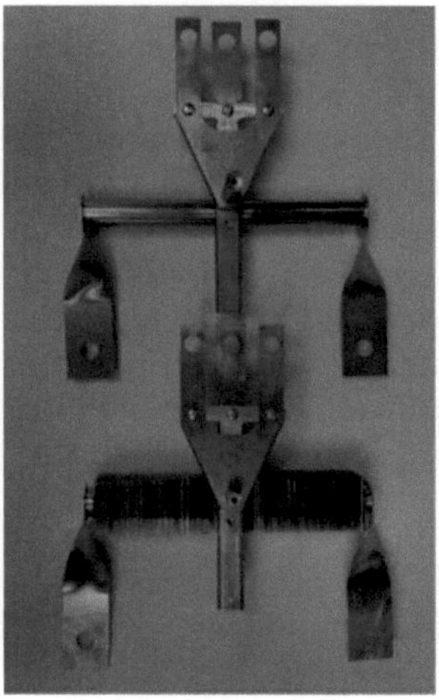

Fig. 1 An ISAC standard target container with a large transfer tube for surface ionization (*top*) and a high power target container with a smaller transfer tube for the attachment of a FEBIAD plasma source (*bottom*). Both targets are 19 cm long

target container wall, insuring good thermal conduction of the deposited p^+ beam power for subsequent radiative cooling. Such simple measures allowed up to 40 μA operation for Ta foil targets and 15 μA operation for pressed pellet SiC targets. The lower SiC p^+ current was dictated by the poor heat transfer between the target pellets and the target container.

To address the low thermal conduction through the pressed powder pellet forms used for compound targets such as SiC, a means of fabricating composite target foils (analogous to the metal foil materials) was developed. Instead of pressing SiC into pellets, the SiC is mixed with small quantities of organic binders and plasticizers in aqueous or organic solvent systems and slip cast onto thin sheets (0.13 mm) of flexible exfoliated graphite foil. After solvent evaporation, the result is a thin flexible composite foil of ceramic powder bonded to graphite sheet which can be cut and treated in a manner analogous to metal foils. Typically, ceramic layers ~0.25 mm thick are achieved, much thinner than can be routinely obtained by pressing powders (~1–2 mm). Slip casting produces ceramic layers with typically 50–60 % theoretical compound density, comparable to that previously achieved for ISAC pressed pellet targets. Details of the fabrication techniques are available in [5]. The still flexible, composite foils are loaded into the target containers and subsequently heated off-line to sinter the ceramic layer and decompose the organic components. The graphite layer serves to increase the effective thermal conductivity and the improved contact to the target container allows composite target foils in bare target containers to operate at p^+ currents up to 35 μA. This technique has been successfully applied to SiC, TiC, ZrC and most recently to TaC materials. The technique has also been applied to ceramics other than carbides. By slip casting Nb_5Si_3 onto thin Nb foil, the composite target material was operated with up to 15 μA p^+. Additionally, oxide ceramics have been cast onto metal foils. For example, UO_2/Nb and NiO/Ni targets have been recently tested on-line.

The final step to achieving full p^+ design intensity was to develop a target container capable of enhanced radiative power dissipation. The Rutherford Ion Source Test (RIST) Project showed that effective emissivities of 0.7–0.8 could be achieved for mechanically tailored Ta surfaces [6]. An old "trick" of calibrating optical pyrometers is to approximate the emissivity of a black body using a cavity with 6 to 1, depth to diameter, aspect ratio. With this in mind, radial fins were diffusion bonded to the standard ISAC target cylinder by heating (in vacuum) at 1500 °C to provide a series of annular radial channels (with 6/1 depth/spacing). The resultant high power target container is shown in Fig. 1. In off-line testing it was measured to have an effective emissivity of ~0.92 and was capable of radiatively dissipating >25 kW of power [7]. Using high power target containers, both metal and composite carbide foil materials routinely operate at 70–75 μA p^+ intensities. In the special case of Nb foils, the favourable combination of beam power deposition and dissipation has allowed target operation at the 99 μA p^+ ISAC design limit.

3 Benefits, detriments and unexpected effects

The most striking beneficial effect observed with increased proton currents on ISAC targets has been the observation of Radiation Enhanced Diffusion (RED) [7, 8]. The kinetics of steady state radiation enhanced diffusion have been reviewed by Dienes and Damask [9] and Sizeman [10]. Under intense proton irradiation, Frenkel pair defects (vacancies and interstitial atoms) are created in the target material lattice. These defects are, themselves mobile, greatly enhancing the diffusion of product nuclides out of the target material. RED is proportional to p^+ flux, resulting in a nonlinear yield proportionality of ${\Phi_{p+}}^{3/2}$ to ${\Phi_{p+}}^2$ rather than merely Φ_{p+} depending on whether the Frenkel pairs recombine or annihilate at lattice boundaries. The effect is most striking for short-lived products, since their survival during transit out of the target material is greatly enhanced. Longer-lived species are unaffected and display the expected linear proportionality to p^+ flux.

However, increased diffusion also produces a detrimental effect by decreasing target longevity. Long term target irradiations with 500 MeV protons produce large quantities of all nuclides lighter than the target elements. Many undesired products can form stable compounds with the target materials or target container, congregating at grain boundary sites and eventually leading to loss of target container integrity. The target foils are also susceptible to such degradation where loss of material integrity leads to reduced thermal conductivity and subsequent overheating. For example, the ISAC Ta #35 HP target of 16.6 g Ta/cm^2 operated for 57723 μA·hrs receiving a total of 1.1×10^{21} protons. Based on the total inelastic cross section for Ta(p,X) at 500 MeV, this would result in an estimated ~41 mg (0.12 %) of the Ta target material being transmuted to impurities by nuclear reactions. ISAC metal foil targets particularly exhibit beam aging with lowered yields towards the end of runs. Composite carbide targets seem less susceptible to aging effects, perhaps because of their already porous nature.

Another both beneficial and detrimental effect of high current operation has been the observation of molecular beams resulting from radiogenic production of both constituents. The formation of MeF^+ ions from injection of CF_4 into targets is a well known technique for enhancing release and mass separating isobars [11–15]. Similarly, release and selectivity of group 14 sulfide ions MeS^+ has been demonstrated with addition of H_2S or CS_2 gases [16, 17]. However, with sufficiently high production under intense irradiation it is possible for such molecular species to form with both radiogenic constituent atoms without any intentional addition of reactant gas. The

$^{18}F^+$ yield from ISAC SiC targets is decreased by formation of AlF^+, MgF^+, SiF^+ molecular species of equal (or greater) intensity than the atomic ion. Similarly, the dilution of both Sr and F ISAC yields across $^{79-88}Sr$ and $^{17-21}F$ constituents of SrF^+ has previously been reported [18]. Observation of $^{27}AlBr^+$ beams at ISOLDE were reported to be at ~5 % of the atomic Br^+ beam intensity, with a steady decrease in molecular beam yield as Al impurities outgassed [19]. However at ISAC, $^{27}AlBr^+$ beams were observed at intensities an order of magnitude higher than the atomic Br^+ beam intensity with no noticeable decrease over the course of a 17 day irradiation. Most surprisingly, the same ZrC target operating with a FEBIAD plasma source under 75 μA p^+ irradiation showed a significant dilution of $^8Li^+$ ($t_{1/2}$ = 844 ms) yield over molecular species. With a measured atomic $^8Li^+$ yield of 1.2×10^5/s, molecular ions of $^8Li^{19}F^+$, $^8Li^{77}Br^+$ and $^8Li^{79}Br^+$ were observed with 9.9×10^3/s, 7.2×10^3/s and 1.9×10^4/s, respectively; accounting for ~32 % of the atomic ion yield. Although ^8Li activity could not be measured at other molecular masses due to intense masking activities, it is possible to speculate that significant amounts of Li are distributed among other radiogenic halogen molecules.

A further effect of intense beam operation is the loss of uniform target temperature. For a nominal 7 mm FWHM Gaussian ISAC p^+ beam profile on an 19 mm diameter Ta target with resistive heating, the thermal gradient from foil center to edge is ~10 °C at 1 μA, but ~800 °C at 75 μA with only p^+ beam heating. The temperature gradient is required to transfer deposited beam power from the centre of the target to the container wall for subsequent radiative cooling. The effects of such a temperature gradient on product release are not fully understood, but are likely detrimental.

4 Recent developments and future directions

At the end of 2011, ISAC received regulatory approval to operate a uranium containing target material at 10 μA. UC_x composite target foils have been prepared and several target irradiations have been performed so far. Operating UC_x targets at such intensity with a 500 MeV p^+ beam is a novel experience. The optimization of the target material properties such as thickness, particle size, porosity and carbon content is an ongoing development. Francium beams are a good indicator for target performance. Its release is not much influenced by fluctuations of the ionization efficiency or by chemical reactions in the target. However, as shown in Fig. 2, there can be still significant differences in the yields of Fr^+ isotopes from two nearly identical targets. A major part of future target developments will be to pinpoint the causes of such yield fluctuations. In order to improve quality control and to identify which particular material properties are most important for the efficient release of certain isotopes, it is important to enhance the diagnostic capabilities for the characterization of UC_x composite target foils. Recently, progress was made with determining the uranium mass distribution within target foils using L-edge densitometry [20].

Additionally, the composite foil fabrication techniques that have proved to be successful for carbide materials need to be extended to other ceramics, particularly oxides. A recent beam development run of a composite NiO/Ni foil target with a FEBIAD plasma source has shown promising first results regarding the release of

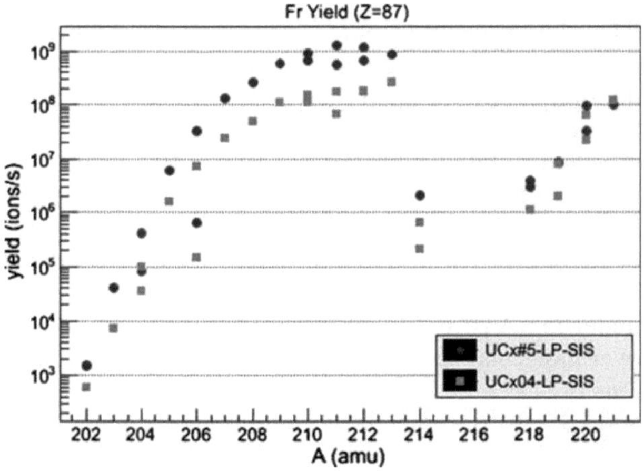

Fig. 2 Yields of Fr^+ isotopes from two successive UC_x targets with surface sources

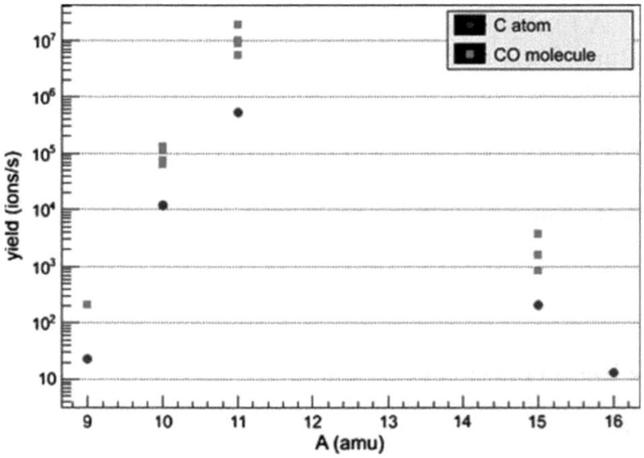

Fig. 3 Yields of CO^+ and C^+ beams from a composite NiO/Ni foil target with a FEBIAD plasma source

radiogenic carbon isotopes. The operating temperature and the p^+ beam current were limited to 1100 °C and 15 μA, respectively. At this temperature the effusion of carbon from the target is only feasible with the formation of molecules such as CO. As shown in Fig. 3, the measured yields of molecular CO^+ beams are about one order of magnitude higher than the yields of atomic C^+ beams, which are most likely the result of CO breakup in the plasma source.

Because of the relatively poor thermal conductivities of oxides, p^+ intensities on such targets are expected to be limited to ~20 μA with the existing passive radiatively cooled target containers. To reach higher irradiation intensities, development of actively cooled targets will be required. An alternative option would be to change the p^+ profile on target. Rather than a relatively broad static p^+ beam, a narrow rotating beam on annular target foils could be used. Such an option was first proposed by the RIST Project [21]. Beam power deposited close to the target container surface would be more efficiently dissipated and a higher beam power density may also enhance product release through increased RED effects. In fact, such a design was tested on an ISAC high power target with a 10 Hz rotating p^+ beam of 3 mm

FWHM. Unfortunately, the target sustained damage when beam positioning was lost. Additionally, the 10 Hz rotation rate has proved to be too slow. The rotating beam option should continue to be developed with higher repetition rates. Finally, it is clear that molecular formation is significant with higher production of radiogenic nuclides. Target chemistry must be evaluated and understood for each new target material brought on line.

Acknowledgement This work is financed by TRIUMF which receives federal funding via a contribution agreement through the National Research Council of Canada.

References

1. Bricault, P., Schmor, P., Stanford, G., Mark, C., Dombsky, M., Gallop, M., Moritz, L., Udy, L.: In: Proc. 15th Int. Conf. On Cyclotrons & Their Applications (Cyclotrons '98), p. 347. Caen, France (Inst. of Physics Publishing) (1999)
2. Dombsky, M., Baartman, R., Bricault, P., Doornbos, J., Jayamanna K., Kuo, T., Mackenzie, G., McDonald, M., Schmor, P., Yuan, D.: Rev. Sci. Instrum. **69**, 1170 (1998)
3. Dombsky, M., Bishop, D., Bricault, P., Dale, D., Hurst, A., Jayamanna, K., Keitel, R., Olivo, M., Schmor, P., Stanford, G.: Rev. Sci. Instrum. **71**, 978 (2000)
4. Dombsky, M., Bricault, P., Schmor, P., Lane, M.: Nucl. Instrum. Methods B **204**, 191 (2003)
5. Hanemaayer, V., Bricault, P., Dombsky, M.: Nucl. Instrum. Methods B **266**, 43341 (2008)
6. Bennett, J.R.J.: Nucl. Instrum. Methods B **126**, 105 (1997)
7. Bricault, P., Dombsky, M., Dowling, A., Lane, M.: Nucl. Instrum. Methods B **204**, 319 (2003)
8. Dombsky, M., Bricault, P., Hanemaayer, V.: Nucl. Phys. A **746**, 32c (2004)
9. Dienes, G.J., Damask, A.C.: J. Appl. Phys. **29**, 1716 (1958)
10. Sizeman, R.: J. Nucl. Mater. **69–70**, 386 (1978)
11. Ravn, H.L., Sundell, S., Westgaard, L.: J. Inorg. Nucl. Chem. **37**, 383 (1975)
12. Putaux, J.C., Obert, J., Kotfila, L., Roussierre, B., Sauvage-Letessier, J., Liang, C.F., Peghaire, A., Paris, P., Giroux, J.: Nucl. Instrum. Methods **186**, 321 (1981)
13. Hellström, M., Fogelberg, B., Jacobsson, L., Spanier, L., Rudstram, G.: Nucl. Instrum. Methods B **70**, 142 (1992)
14. Kawase, Y., Okano, K., Shibata, M., Taniguchi, A.: Nucl. Instrum. Methods B **70**, 146 (1992)
15. Kirchner, R.: Nucl. Instrum. Methods B **126**, 135 (1997)
16. Stracener, D.W.: Nucl. Instrum. Methods B **204**, 42 (2003)
17. Kirchner, R.: Nucl. Instrum. Methods B **204**, 179 (2003)
18. Dombsky, M., Bricault, P.: Nucl. Instrum. Methods B **266**, 4240 (2008)
19. Köster, U., et al.: Nucl. Instrum. Methods B **204**, 303 (2003)
20. Kunz, P., Bricault, P., Dombsky, M., Erdmann, N., Hanemaayer, V., Wong, J., Lützenkirchen, K.: J. Nucl. Mater. **440**, 110–116 (2013)
21. Bennett, J.R.J.: Final Report on RIST. Rutherford-Appleton Laboratory Internal Report RISTMC/P1/97 (1997)

DOI 10.1007/978-94-007-7963-1_4
Reprinted from *Hyperfine Interactions* Journal, DOI 10.1007/s10751-013-0880-z

Rare isotope beams at ISAC—target & ion source systems

Pierre G. Bricault · Friedhelm Ames · Marik Dombsky · Peter Kunz · Jens Lassen

Published online: 10 October 2013

Abstract The present status of the ISAC facility for rare isotopes beams after its first 10 years of operation is presented. Planning for the ISAC facility started in 1985 with the Parksville workshop on radioactive ion beams (Buchmann and D'Auria 1985). It was put on halt by the KAON proposal and planning was only resumed in 1993 after the cancellation of KAON. The ISAC facility was built to satisfy the scientific need for accelerated beams of rare isotopes for use in applications such as nuclear physics, nuclear astrophysics, atomic and condensed matter physics as well as medicine. At the time of the ISAC proposal submission, a number of facilities were either planned or under construction. In order to have an impact in the field, the requirements and specifications for the driver beam intensity on target was set to 100 μA, 500 MeV protons, which for ISAC results in a driver beam power of 50 kW.

Keywords Rare isotope beam · Isotopic separation on-line · Ion sources · Targets

1 Introduction

There are two principal methods of generating rare isotope beams (RIB). One is the so-called in-flight projectile fragmentation where heavy ions are accelerated to high energy and then impinge onto a thin target, which is usually made of a light mass material, such as Be. Due to the conservation of the energy and momentum the fragments produced during the interaction have a velocity similar to the projectile. The fragments are then mass analyzed according to their mass to charge ratio to provide RIB to experiments. The second method is called the isotope separation

ISAC and ARIEL: The TRIUMF Radioactive Beam Facilities and the Scientific Program.

P. G. Bricault (✉) · F. Ames · M. Dombsky · P. Kunz · J. Lassen
Accelerator Technology Division, TRIUMF, 4004 Wesbrook Mall,
Vancouver, BC V6T 2A3, Canada
e-mail: bricault@triumf.ca

on-line (ISOL); in this method the primary beam usually consists of light ions (p, d, ^{3}He, α) impinging on a thick target while at the same time the target is heated to high temperature to speed the release of the rare isotopes created in the interaction between the driver beam and the target nuclei. The products are stopped inside the bulk of the target material. For the atoms to reach the ion source, they have to diffuse to the surface of the target material grains. From there the effusion process bring the atoms to the ion source where ionization takes place and an ion beam is extracted and prepared for mass separation before being sent to the experiments. Experiments either receive low energy ion beams, or require the ions to be injected into accelerators to form beams suitable for experiments with accelerated beams.

The ISAC facility was built with the goal of producing rare isotope beams by using a continuous, up to 100 μA proton beam instead of the usual 2 μA used for more than 40 years at ISOLDE/CERN [2]. The radiation protection measures and radiation resistant target/ion source at the ISAC facility were implemented right from the start in the design of all systems. The intense proton beam used to bombard the production target in an ISOL type facility gives rise to high radiation fields that require substantial shielding. At ISAC a number of well-developed tools for estimating the shielding required around the production target were used. Simulations using FLUKA [3] give good estimates of the surrounding steel activation at saturation. At ISAC it is equivalent to 6 TBq/kW [4]. In addition the target material itself deteriorates through impact of the proton beam and must be removed periodically. The rare isotope production can be estimated using FLUKA or empirical models such as the one by Silverberg and Tsao [5, 6]. The activity per unit target thickness is approximately 1.5×10^9 p Bq/(g/cm^2), where p is the proton beam intensity in μA [4]. For a 100 g/cm^2 target bombarded by a 100 μA proton beam (the ISAC design value) this amounts to 15 TBq. The dose rate from such a target would be approximately 3 Sv/h at a distance of 1 m. Servicing these targets therefore requires remote handling capabilities. Because of the high radiation fields from the target, the ISAC remote handling is quite different than the one in use at ISOLDE [5, 6] or HRIBF [7], for example. Instead of a purpose-built robotic device an overhead crane that was adapted to be remotely controlled is used.

2 ISAC technology for ISOL method

Beams of rare isotopes are a challenge to produce—especially the short-lived isotopes, which do not exist on earth. These isotopes have to be produced artificially in the laboratory. The isotopic separation on-line or ISOL method can be described as a process where the isotope of interest is fabricated artificially by bombarding a target material nucleus with fast projectiles. In a thick target the reaction products are stopped in the bulk of the material. The target container is attached directly or indirectly to an ion source, allowing the reaction products to be quickly ionized and turned into an ion beam that can be mass analyzed and be delivered to experiments. The building blocks for an ISOL facility to produce high intensity RIB are:

- a high energy driver, such as the TRIUMF H^- 500 MeV cyclotron,
- a target material inserted into an oven or target container connected to an ion source,

- an ion source at high voltage to produce an ion beam,
- a high-resolution, high throughput mass separator.

To solve the problem of producing intense rare isotope beams the best suited target material that favors the production of the desired RIB species and that is capable of releasing it must be found. Target materials and the design of the target containers are described in more detail in the contribution by M. Dombsky in this issue. Another important consideration is the contamination of the ion beam by isobars; e.g. isotopes having the same mass number A, but different atomic number Z. Yet another consideration must be the power deposited inside the target material. If the deposited power density is too high, the temperature of the target material will locally increase above safe operation level and the target material will begin to evaporate. This can have unfortunate effects on the ion source efficiency, and is especially noticeable when operating plasma ion sources. At ISAC the target operating temperature is constrained to values at which the vapour pressure does not exceed 10^{-6} mbar. This is mainly due to the fact that we have to operate the target for at least four weeks due to the lengthy target exchange procedures involving remote handling. To avoid excessive power deposition by the incoming beam we do not stop the primary beam in the target. This is accomplished by choosing the target thickness such that the energy degradation of the proton beam is below 300 MeV. A dedicated water-cooled beam dump is located just behind the target to capture the entire proton beam emerging from the target. There are three main nuclear reactions accessible to produce rare isotope beams with a proton beam in this energy range. They are:

1. Spallation, a process where individual nucleons are knocked out of the target material nuclei. The product distribution peaks a few mass units lighter than the target nucleus. Because neutron emission is energetically favored over proton emission (due to the Coulomb barrier that the proton experiences), the production of neutron deficient nuclei is favored. A good example is the high production of Rb isotopes from Nb or Zr targets.
2. Fragmentation is the counterpart of the spallation reaction, where the product is one of the light fragments. The fragmentation method is advantageous for producing light, neutron rich products from heavier target nuclei with high neutron to proton ratios. Because of the neutron excess in the heavy target nuclei, the fragments tend to retain a statistical memory of the neutron to proton ratio and, as a result, have higher neutron to proton ratios themselves. A good example of such a reaction is the production of ^{11}Li from Ta, or the production of ^{32}Na from U targets, respectively.
3. Induced fission occurs when the incoming projectile deposits sufficient energy in the target nucleus to induce a breakup into two roughly equivalent mass products. Unlike neutron-induced fission, this reaction mechanism is open to both fissile target nuclei and to heavy nuclei such as tantalum and lead. Again, because of the higher neutron to proton ratio in heavy nuclei, neutron-rich products in the medium mass region can be effectively produced by this reaction mechanism.

The yield of a specific isotope can be expressed using the following equation:

$$\Phi(A, Z) = \phi_P \sigma(A, Z) N_T \varepsilon_{Dif} \varepsilon_{Eff} \varepsilon_{Ion} \varepsilon_{Trans}, \quad (1)$$

where ϕ_P is the proton beam intensity impinging on target, $\sigma(A, Z)$ is the cross section for the production of the specific isotope, N_T is the number of target nuclei per square cm, ε terms are the efficiencies diffusion, effusion, ionization efficiency and beam transport encountered as the product nuclides are converted into radioactive ion beam.

2.1 ISAC target station description

The ISAC target-handling of the new facility was based on years of experience at operating meson production target stations. The meson production target and beam stop areas of these facilities have power dissipation and radiation levels similar to, or greater than, those expected at ISAC. Meson factory experience shows that the best approach to handle components in high-current and high-radiation areas, is to place the beamline components in tightly shielded canyons. Access to the components is done vertically and repair and service is made in dedicated hot cells. Three important factors not encountered in the meson factory targets have to be addressed. These are:

- containmination from large amounts of mobile radioactivity,
- high voltage required for beam extraction,
- routine replacement of short-lived target systems.

In the present design these issues are solved by placing the target in a sealed self-contained module that can be transferred directly to the hot cell facility for maintenance. The main guidelines for radiation protection considerations are described in a paper by Lutz Moritz [4]. The target stations are located in a sealed building serviced by an overhead crane. The target maintenance facility includes a hot cell, warm cell, decontamination facilities and a radioactive storage area. The target area is sufficiently shielded so that the building is accessible during operation at the maximum proton beam current. Beam-line elements near the target are installed inside a large T-shaped vacuum chamber surrounded by close-packed iron shielding blocks. This general design eliminates the air activation problem associated with classical high current target areas by removing all the air from the proton beam path. The design breaks naturally into modules: an entrance module containing the primary beam diagnostics; an entrance collimator and a pumping port; a beam dump module containing a water cooled copper beam dump; a target module containing the target/ion source, extraction electrodes and first steering component and heavy ion diagnostics; and two exit modules containing the optics and the associated diagnostics for the transport of heavy ion beams. Figure 1 shows a plan view of the target stations area. The vacuum design seeks to eliminate the need for radiation hard vacuum connections at beam level by using a single vessel approach. The front-end components, with their integral shields, are inserted vertically into the T shaped single large vacuum vessel. Most vacuum connections are situated where elastomer seals may be used. Only two beam-level connections exist: one at the proton beam

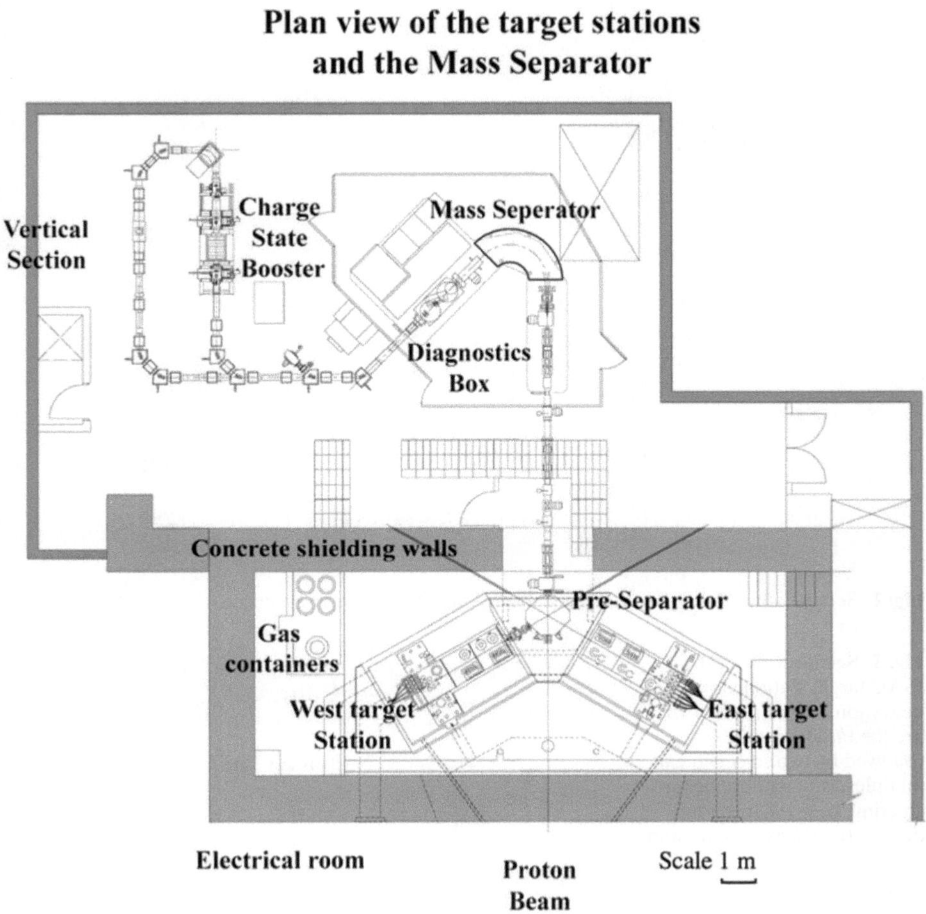

Fig. 1 Plan view of the ISAC target stations with the pre-separator and switch yard in a high radiation area and the main mass separator area, which includes a charge state breeder. Indicated are also the penetrations through the concrete shielding to the pre-separator for laser beam transport

entrance and one at the heavy ion beam exit. At these locations only metal seals can be used. The target stations' vacuum envelope is separated from the proton beam line by a water- cooled 0.75 mm thin Al window. The target stations are shielded by approximately 2 m of steel placed close to the targets. Outside this steel shielding the operating radiation field is sufficiently low so that radiation damage to equipment is not a concern. The target station shielding is a mixture of steel and concrete. The steel shielding is approximately 1.4 m thick and is surrounded by an additional 2–4 m of concrete, which provides the required personnel protection during operation.

Figure 2 shows a section view along the heavy ion beam optics and Fig. 3 shows a section view along the proton beam axis. To service the targets, shielding above the target station is removed, giving access to the services at the top of the steel shielding plugs. Residual radiation fields at this level are low enough to allow hands-on servicing. The ISAC target module is the heart of the production facility. It has to

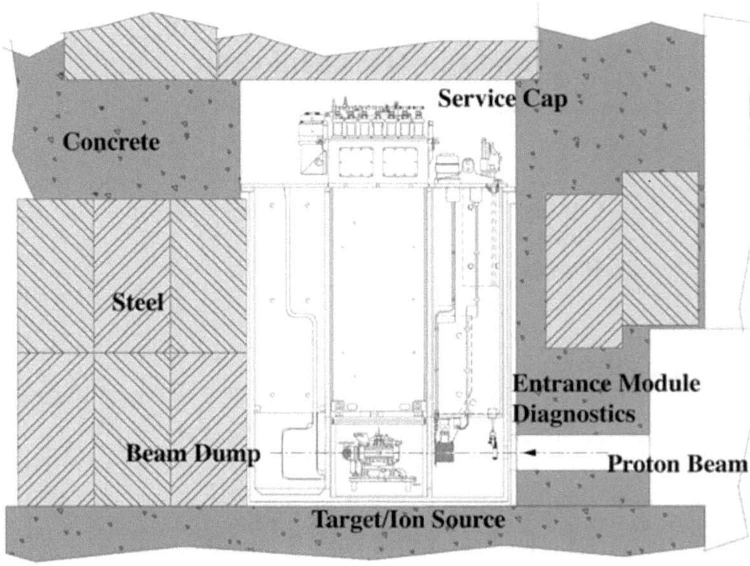

Fig. 2 Section view of the ISAC target station along the proton beam line

Fig. 3 Section view of the ISAC target station along the heavy ion beam line. Shown are the target module, followed by two so called exit modules, which house beam steering, beam shaping and diagnostics (Faraday cups and harp monitors)

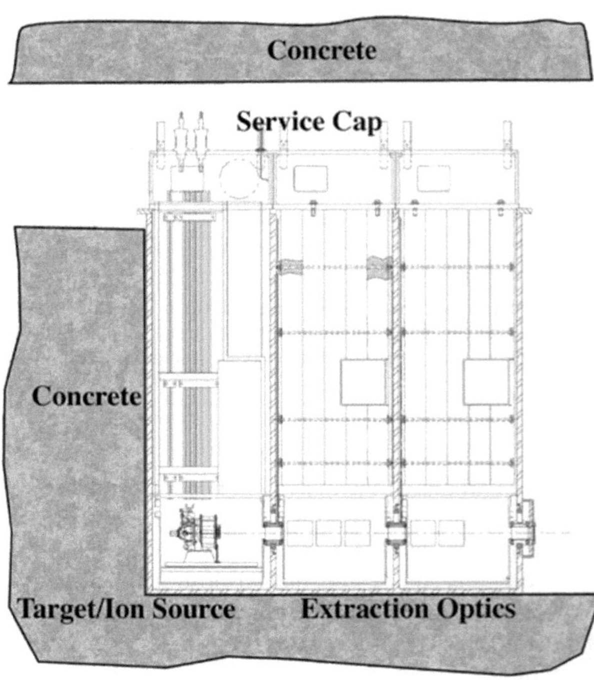

operate in a hostile radiation environment and provide all the necessary services to operate the target and ion source, namely, power for the target oven and ion source, high voltage for the ion source, cooling for the beam window and target heat shield, ion extraction optics and vacuum. Figure 4 provides a 3-D view of the target module.

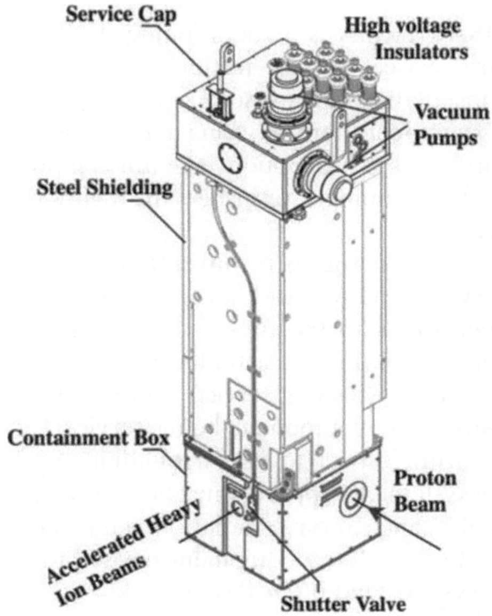

Fig. 4 3-D view of the ISAC target module consisting of the target ion source, enclosed by the containment box and connected to the service cap by steel shielding that reduces the radiation to levels that are suitable for standard vacuum seals and instrumentation

2.2 Remote handling

An effective remote handling and servicing system is required for quick and frequent target exchanges. All modules in the target area will have high levels of residual activity and will be potentially contaminated with mobile activity. Both aspects are considered in the handling design. Target component maintenance involves disconnecting services and craning the module to the hot cell. Removing the concrete blocks covering the target station gives the overhead crane access to the modules. While the target module is pulled out of the canyon personnel are excluded from the area. Target module transfers to the hot cell must therefore be done completely remotely. The operators follow the progress of the transfer using video cameras and the crane position along the axis of the target hall is constantly monitored using a laser positioning system. The connection and disconnection of the target module services can be done manually since the shielding of the module is thick enough to allow hands-on operation. The mobile contamination produced in the target area is normally contained within the target module. The target module and the two exit modules are equipped with a so-called, containment box, which is only mechanically sealed with pillow-seals to avoid migration of the mobile activity. Nevertheless, contamination of the target building is considered possible. This building must therefore be considered as an extension of the hot cell complex and all entrances must be controlled and provided with appropriate contamination control. The air within the building must be maintained at reduced pressure and filtered using a combination of HEPA and charcoal filters. Exhaust from the vacuum pumps of the target stations goes into a storage tank during the run and the gas is then transferred to a decay storage tank prior to controlled release. The interior surfaces of the target hall are painted with a special expoxy paint to allow for decontamination. All fluid drains go to a so-called active sump tanks for monitoring before disposal. A module storage area is located between the hot cell and the target station. A recent addition to the ISAC facility is a conditioning station, some sort of silo—providing all the necessary services for the testing and preconditioning of targets and ion sources before installation for an isotope production on-line. This area is fully accessible during beam operation; servicing and testing of new target module is therefore possible during beam production. The hot cell provides facilities to remotely maintain, replace, decontaminate, or inspect the highly-radioactive components removed from the target area.

The hot cell design is a conventional design with concrete shielding walls, lead glass viewing windows, and sealable roof ports to allow crane access to the hot cell. Personnel access to the top of the cell is possible, if required. The hot cell bay is equipped with directly actuated master-slave manipulators. The mechanical bay includes remote viewing, service equipment, and an elevating turntable to support and position the component being serviced. The hot cell is kept under negative pressure with respect to its surroundings by its own HEPA-charcoal-filtered air handling system. A support annex houses the remote handling control room, offices, personnel change rooms, radiation safety monitoring equipment, and target hall entry air-locks. All the equipment needed to control the remotely-operated crane, viewing systems and other devices is located in a dedicated control room, which is accessible only by authorized personnel to avoid distraction of the operators while performing handling tasks. Cameras are mounted in strategic locations throughout the building and on the cranes. An air-lock to the target hall is provided for transfer of equipment.

3 On-line ion sources

Experience at operational ISOL facilities clearly shows that there is not a universal target/ion source combination for the production of all required isotopes for their physics programs. Several types of ion sources foreseen at the ISAC facility have been implemented [8, 9]. The target module design was done with this idea in mind and flexibility has been provided in the system to allow their successful implementation. To maximize the yield of a particular species of short-lived isotopes all beam losses must be minimized. This means that the ion source has to be closely coupled to the target oven. This fact has enormous implications on ion sources. The hostile environment dictates that the ion source be both simple and small to minimize surface area and radioactive waste. The simplest and most efficient of the ISOL ion sources is the hot cavity surface ion source. It works well for elements having an ionization potential below 6 eV, such as alkali and some rare earth elements. All other elements can in principle be ionized using a plasma ion source. The most common of these for RIB production is the arc discharge ion source such as the successful FEBIAD (Forced Electron Bombardment Arc Discharge) ion source developed by Kirchner and Roeckl at GSI [10] and modified at ISOLDE-CERN [11]. Another type of plasma ion source is the ECR (Electron Cyclotron Resonance) ion source [12], which is particularly efficient for light gaseous elements and noble gases. Element selective ionization can be achieved by resonant laser ionization using a laser ion source as presented at the Parksville radioactive beams workshop [13]. This technique uses a simple on-line surface ion source, where the surface ionizer is used as a hot cavity in which laser ionization of neutral atoms is performed.

3.1 Hot surface ion source

The simplest ion source we can imagine is the hot cavity surface ion source. An atom coming from the target enters the transfer tube, which connects the target volume to the ion source. It is designed from a re-entrant electrical conductor fabricated from electron beam welded concentric Ta tubes. The re-entrant geometry is used to minimize the magnetic field in the exit region that is generated by the DC heating current as it passed in the opposite direction through concentric conductors [14]. In addition, the re-entrant geometry serves as a heat shield, providing excellent heating to the inner conductor—the surface ionizer tube. A rhenium foil inside this tube acts as a high work function surface ionizer. The surface ion source is typically operated at a temperature of around 2200 to 2400 °C. The multi electrode extraction column consists of the ionizer tube a fixed Ta plate with a 3 mm aperture that is the reference for the extraction system. Next to it is a fixed, water cooled copper electrode with an identical aperture that acts as the puller electrode followed by the ground electrode. Both the ionizer and the reference electrode are biased to the same high voltage. The extraction electrodes system is suspended at the bottom of the shielding plug and is pre-aligned to the ion source during installation. The high radiation fields at the target station preclude the use of electrical components, such as positioning motors, on the extraction electrode. Mechanical positioning of the extraction electrode through 2 m of shielding is excluded as impractical and imprecise. This gives rise to, perhaps, the most significant departure from traditional ISOL designs, namely the use of fixed geometry, multi-electrode extraction systems for all ISAC ion sources. The optimum beam extraction is accomplished by tuning the voltages of the component electrodes rather than the relative positions of the ion source and extraction electrode. While this approach solves the problem represented by the high radiation field environment, it presents new problems in the need for precise and reproducible alignment and voltage regulation. An ion test stand allows to evaluate various ion source extraction geometries and optics for our ongoing program of ion source research and development. One problem not foreseen at the time of the extraction system design is that at high proton beam intensity the amount of radiogenically produced atoms creates problems over time of continuous operation. The material thus produced condensates onto the extraction electrode and forms asymmetric electric fields. This has caused enormous tuning problems because of the asymmetry of the electric field directly at the extraction. Attempts to clean the extraction electron on board of the target module in the hot-cell gave mixed results. At one point the hot-cell operator used a rotating brush to remove a big blob of material on the electrode. It created an enormous problem due to the contamination spread all over the hot-cell and resulting external contamination of the target module. The operation of the hot-cell was stopped to allow a complete cleaning of the entire hot-cell. To mitigate the problem the design of the target module target support and extraction was modified. In the new design the extraction electrode is mounted onto the target/ion source assembly and consequently it is changed as often as the target itself. The other advantage of this design is the fact that the most sensitive part of the extraction system can be aligned in the laboratory instead of relying on jigs.

The other problem that was discovered later was the fact that the alignment of the concentric Ta tube forming the hot cavity of the ion source is not stable if the temperature of the target changes. It has been observed that the vertical steering of the extracted beam had to be adjusted as a function of the proton beam intensity. The target temperature affects the position of the tip of the surface ion source. To limit that motion the external Ta tube has been restrained radially, resulting in fewer pointing stability related beam intensity fluctuations.

3.2 Resonance ionization laser ion source

A resonant ionization laser ion source was foreseen from the beginning of the ISAC radioactive ion beam facility [13]. As such the implementation of the TRIUMF Resonant Ionization Laser Ion Source (TRILIS) was smooth with first radioactive ion beam in 2004 from the west target station. The particular appeal for resonant laser ionization of atoms inside a cavity comes from the high ionization efficiencies that can be obtained, element selective ionization, and the fact that all critical ion source components are far removed from the high radiation environment of the target-ion source region [13, 15].

The laser beams are injected into the transfer tube that connects the target volume to the ion source extraction electrodes, e.g. a standard surface ion source. The lasers, the most complex part of the resonant ionization laser ion source, are located far away from the prompt radiation area. This allows for easy servicing and control during operation. The ISAC laser system is located 20 m away from the west target and 27 m from the east target. The combined laser beams are brought to the ionization region within the target ion source's hot transfer tube (e.g. surface ionizer) via a window port on the pre-separator magnet chamber [15]. Because of operational considerations, cost of ownership, operational stability and ease of operation, an all solid-state laser system was chosen, based on high repetition rate, tunable titanium sapphire lasers, pumped from a single 50 W frequency doubled NdYag laser. All lasers and beam shaping optics are located on a 3.6 m $\times$ 1.5 m optical table in an air-conditioned (regulation $\pm$0.5 C) clean room laser laboratory, which ensures stable operating conditions and low system maintenance requirements.

The price for the approach to use all solids state lasers, at the time novel, was that a significant number of laser excitation schemes, which were established at ISOLDE [16] could not be used, and off-line laser spectroscopy had to be performed to develop suitable laser ionization schemes. By 2012 isotopes from 20 different elements have been delivered by TRILIS on-line (Fig. 5) and more than 50 % of beam time in a given schedule typically are now from TRILIS. Current developments are addressing 24/7 operational requirements and improved beam purity. Examples for such developments are the successful implementation of surface ion rejecting - laser ion source (Sir-LIS) in 2013 and a fast kicker for gating out isobaric background between the laser ionized bunches ions. In the Sir-LIS the hot cavity of the surface ion source is replaced by a set of repeller electrodes and a RF-ion guide. The repeller electrodes repell ions generated on the hot target surfaces, only allowing neutral atoms to enter the laser ionization region. Ions created here are radially confined by the ion guide and extracted through the standard ion extraction optics. This results in laser ion source beams of unprecedented purity.

Off-line development of laser ionization schemes is done at a dedicated test stand [17], which allows for systematic studies [18], laser development [19] and student

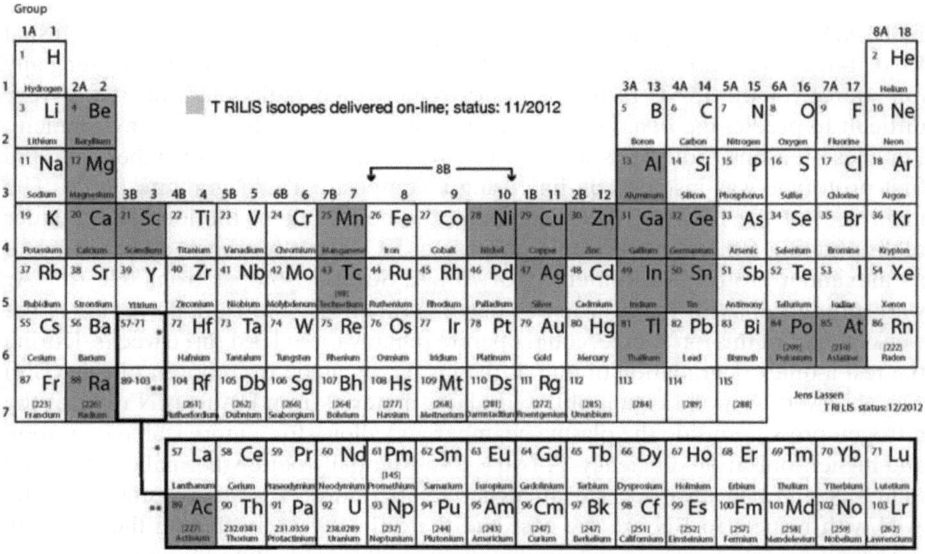

Fig. 5 Elements from which isotopes have been delivered using TiSa laser based resonant ionization (status 12/2012). The current development goal is set at two new beams per year and to continually improve operational reliability. Beams of lanthanides will become accessible through the surface ion rejecting - laser ion source (Sir-LIS) which substantially reduces isobaric background as demonstrated by recent Penning trap mass measurements of Mg and Al isotopes

training. The test stand laser laboratory also serves as a pool for emergency backup equipment.

The TRILIS also serves as the basis for a science program in "in-source laser resonance ionization spectroscopy". Its most noteworthy result so far has been the successful ionization and laser spectroscopy of astatine [20, 21], as well as the spectroscopy of atomic states in Actinium and Polonium.

3.3 Plasma ion sources

For the elements which have a high ionization potential, e.g. He, B, C, N, O, F, Ne and for molecules formed to extract refractory elements, a hot surface ion source or a laser ion source cannot be used. For those elements a plasma ion source is used. In such an ion source, the plasma is generated using fast electrons that can be either produced by a hot cathode as in the Forced Electron Beam Induced Arc Discharge (FEBIAD), or by a radio frequency induced discharge as in the Electron Cyclotron Resonant (ECR) ion source. During the past several years a FEBIAD ion source and a new generation of ECR ion sources for on-line application have been developed

3.3.1 FEBIAD ion source

Originally the FEBIAD source was mainly developed to produce the F and Ne isotopes necessary for the study of the proton capture reactions, $^{18}F(p,\gamma)$ and $^{18}Ne(p,\gamma)$, for nuclear astrophysics, but the program was soon extended to other isotopes and elements. The design is a combination of the Kirchner [9] and Sundell [10] type FEBIAD ion sources that were developed at GSI and CERN. The ISAC

FEBIAD is a very compact version which is housed in the same volume as the usual ISAC target/ion source assembly heat shield. The biggest challenge in designing any ion source for the ISAC facility is the extremely high radiation field. It was very difficult to predict the behavior of insulator materials in such a harsh environment. For these reasons prototypes were built and tested on-line. In November 2006 the first prototype was installed on-line using a composite TiC/C graphite target. The main goal of the first test was to examine the behaviour of the ion source under real operating conditions. For this purpose a TiC target has been operated up to its maximum proton beam intensity of 70 μA. Several isotopes from ^{6}He to ^{45}Ar were produced. With this test it was demonstrated that the cathode, which is the weakest part of this ion source, could operate reliably for at least three weeks. During the test a quick degradation of one of the insulators made from boron nitride (BN) was observed. A modification of the insulator material from BN to AlN reduces the nitrogen pressure inside the plasma chamber and allows for a more stable operation and higher ionization efficiency. AlN has a substantially better thermal conductivity than BN and can also operate at higher temperature. The disadvantages of the AlN with respect to BN are the cost and the difficulty in machining the material. A second prototype was tested on-line in June 2007. This time the ion source was combined with a high power composite SiC/C graphite target that was operated up to 70 μA proton beam intensity. During this development run $^{18}F^{+}$ for an $^{18}F(p,\gamma)$ measurement has been produced. During this run the out gassing of the insulator was reduced; however the ionization efficiency decreased with time. This problem was caused by insufficient cooling of the ion source permanent magnets causing the axial magnetic field to degrade over time.

The present version incorporates several upgrades: a radiation resistant electromagnet, a solid Ta grid replacing the previous W wire anode grid and a water-cooled heat shield protecting the insulators. The electromagnet is made of 6 × 6 mm copper conductor insulated with the same fiber material used for the MISTIC ion source (see next section). Calculations shows that with 100 A and 16 turns coil we can produce a 400 Gauss magnetic field in the centre of the plasma chamber. The modified grid is made from 1 mm thick Ta foil and the slots are machined using a computer-controlled machine. The grid transparency is 75 %. A thick steel plate directly attached to a water-cooled copper plate is placed between the high power target and the FEBIAD ion source. Figure 6 shows a section view of the ISAC target/ion source assembly equipped with a FEBIAD ion source.

On-line tests during November 2007 were conducted using another high power SiC/C graphite target. The FEBIAD ion source produced 5.9×10^{7} $^{18}F^{+}$ per second, and the beam was used by the TUDA group for the $^{18}F(p,\gamma)$ radiative proton capture experiment. During that development run also ^{8}He was produced for a high accuracy Penning trap mass measurement with the TRIUMF ion trap for atomic and nuclear physics (TITAN) facility. Since then several target materials have been used with a FEBIAD source and a number of beams could be delivered to experiments.

Lately an optimization of the plasma volume was done [22] by measuring the Ne ionization efficiency as a function of the distance between the grid and the plasma electrode before the extraction. The same FEBIAD ion source was used for all measurements to avoid renormalization issues. Each position was done by adding shims at the mounting connections between the cathode legs and the target/ion source assembly mounting plate. Results show that there is an optimum position of 10.5 mm between the grid and the plasma electrode.

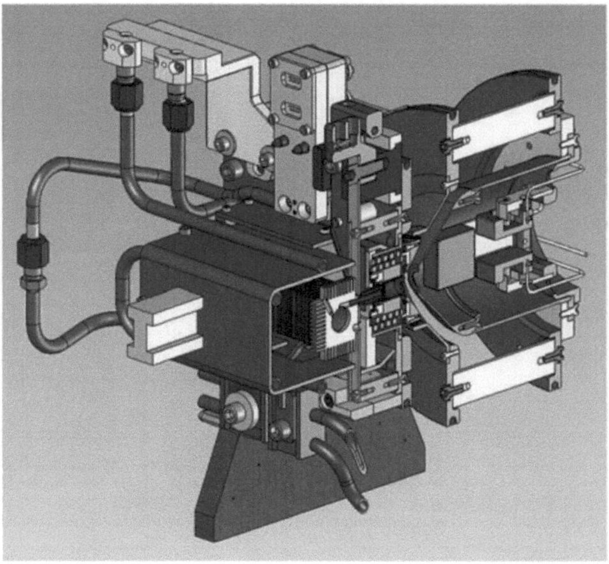

Fig. 6 3-D section view of the ISAC target/ion source assembly equipped with the FEBIAD ion source

3.3.2 ECRIS development

In order to have access to elements which have high first ionization potentials like the noble gas elements a powerful ion source is needed. The best one for gaseous elements is the Electron Cyclotron Resonance Ion Source (ECRIS). A 2.45 GHz ECRIS [23] has been installed on-line (ECR-1), and tested for Ne and Ar production. It was built using a single mode TE111 cavity operating at 2.45 GHz. Due to radiation hardness constraints we ended up with a source where the axial magnetic profile was very shallow and there was no radial confinement because the required sextupole permanent magnet would not have resisted long in the high neutron flux. ECR-1 performances were not satisfactory; it was very sensitive to the pressure increase when the proton beam impinges on the target [24]. The other drawback was the difficulty in igniting the source. A very high pressure was needed to ignite the plasma. This may be a consequence of the low magnetic field confinement. Therefore a new ECRIS for on-line applications is being developed.

The main goal is to design an ion source capable of withstanding the high pressure coming from the target material and be resistant to the high level of prompt radiation produced by the interaction of intense proton beam on the target. Since permanent magnets cannot be used close to the target, we are developing a 6 GHz ECRIS similar to the 2.45 GHz constructed at GANIL [25] where the radial confinement is accomplished using two sets of coils. They reported ionization efficiencies of 72 % for Ar and 35 % for Ne. The new ECRIS called MISTIC (Monocharged Ion Source for TRIUMF and ISAC) [26] operates at 5.6 GHz and has been tested successfully off-line [27] at TRIUMF. The main results are high ionization efficiencies for gaseous elements and ease of ignition. Since MISTIC uses symmetric coils to produce the radial confinement it does not have the usual ECR plasma configuration. The advantage of the symmetric field is the relatively small emittance of the extracted beam from this source. The next step is to build a new ECRIS on board of a target module. Figure 7 shows a schematic view of the MISTIC ion source implementation

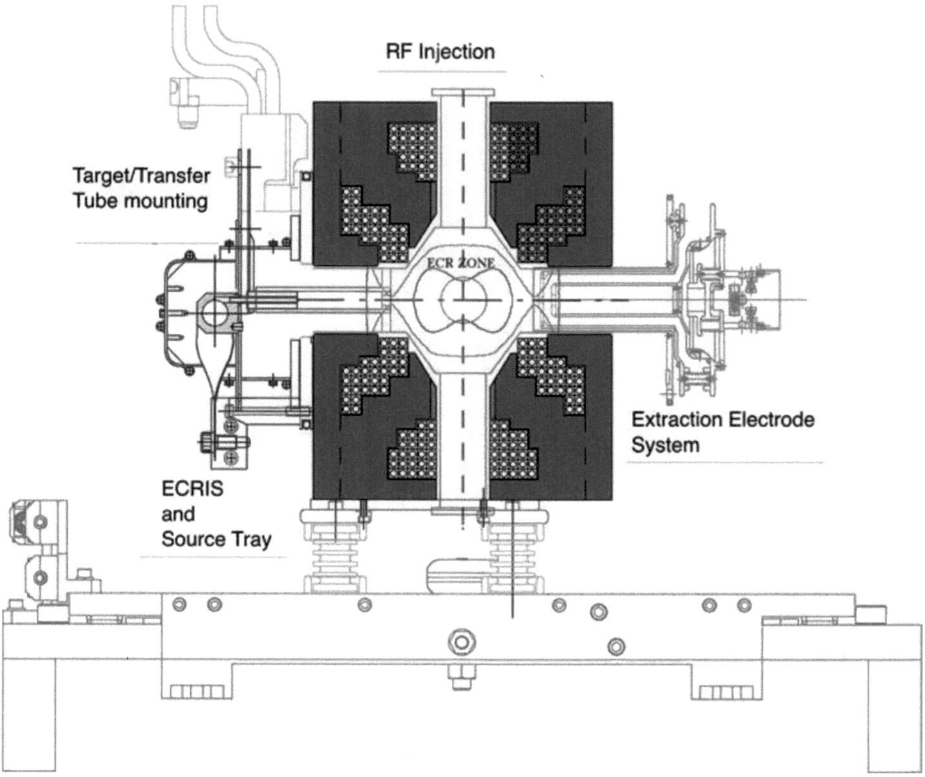

Fig. 7 Schematic view of the MISTIC implementation in an ISAC target module. This type of ion source is so complex that it requires permanent installation on a target module and as such requires a degree of radiation hardness that surpasses that of the surface- or the FEBIAD ion sources

into the target module. In contrast to all other ISAC target-ion source combinations that are all single use, disposable units, the on-line MISTIC ion source must be serviceable and operate reliably in a mode where only the target and serviceable parts are replaced between runs.

4 Mass separator

The ISAC mass separator system has been designed to handle beams in the mass range up to 250 amu and source extraction voltages between 10 and 60 kV. Preliminary mass selection is achieved using a 60° pre-separator magnet. The pre-separator is followed by three matching-sections that allow enough flexibility to adapt the beam in order to obtain the same mass dispersion from either one of two target stations. The front end of the main mass separator includes an electrostatic triplet followed by a doublet. The ion optics calculations were performed up to the third order. The mass separator magnet is the former Chalk River magnet [3], which became available after the physics branch of the Chalk River Laboratory was closed in 1997. The mass separator magnet including the entrance and exit matching sections are on a high voltage platform in order to allow for a reduction of cross contamination and to ease the magnet tuning. However, so far this feature has not been used in on-line beam delivery.

4.1 Pre-separator section

The optics devices following the ion source are suspended at the bottom of the exit modules shielding plugs. The limited space allowed by the size of the shielding module complicated greatly the optics design. An electrostatic triplet just after the ion source matches the beam from the different ion sources. Downstream an electrostatic doublet prepares a parallel beam for the pre-separator magnet. The role of the pre-separator is to act as a cleaning stage in order to limit the contamination in the rest of the mass separator. The mass dispersion of this stage is about 0.3 m. The magnet is designed in such a way that the yoke steel is used as a shield against the contamination deposited inside the vacuum chamber. The pre-separator is also equipped with vacuum viewports that allow for a line of sight into the target-ion sources for the laser beams from the resonant ionization laser ion source. Similarly, the concrete shielding surrounding the target and pre-separator section has penetrations in line of sight of the target for ready laser beam access.

4.2 High resolution mass separator section

The mass separator magnet is the former Chalk River mass separator. Only the entrance and exit arms were modified in order to adapt the magnet to our needs. The magnet is equipped with so called alpha and beta coils. The alpha coil allows correction of the magnet index and the beta coil allows adjustment of the second order correction provided by the pole face curvature. The magnet is on a high voltage platform that will allow rejecting the cross contamination. Such cross contamination comes from ions of different mass which have the same momentum and different energy, for example, due to collision with the residual gas. After acceleration these ions will have different energies and momenta. Consequently, a mass separation becomes possible. The mass dispersion at the exit slit is about 2 m. With typical beams from the ISAC ion sources a mass resolution $M/\Delta M = 2000$ can be achieved. This is sufficient to separate some isobars further away from stability in the low mass region below 30 amu. For higher masses the relative mass difference becomes smaller and this is not any longer possible.

After the mass separator selection slit, the beam is injected into the low-energy-beam-transport line (LEBT). The ion beam from the mass separator is to be switchable between the Low Energy (LE) experimental area and the accelerator. At the same time, there is an off-line source, which is switchable between the same two areas, although its primary purpose is for commissioning the accelerators. A switch-yard has been designed which meets all these goals. At the heart is a cross-over switch which allows the off-line source to supply beams to either the RFQ or the LE, while simultaneously, the mass separator can supply beams to the LE or the RFQ, respectively.

5 Measured RIB yield

Intensities and composition of radioactive nuclear beams (RIB) are determined at a dedicated facility, the ISAC Yield Station. Here, the RIB can be implanted in

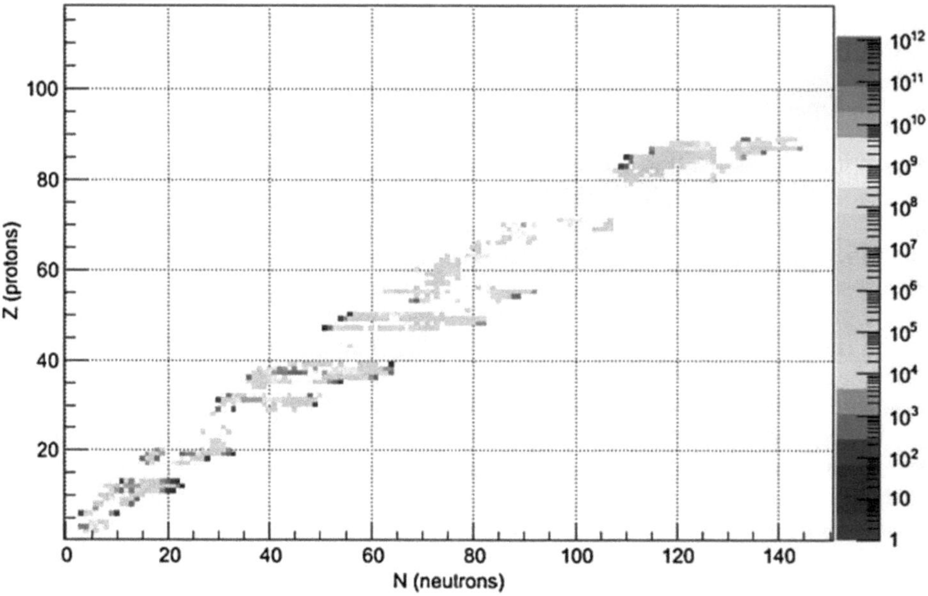

Fig. 8 Radioactive ion beam yields at ISAC from different target and ion source combinations (status 2012) presented as a chart of nuclides. The yields from actinide targets—e.g. all isotopes with $Z > 80$ are recent additions to our production chart

a pulsed mode on an aluminized Mylar tape using a fast kicker system located just after the mass separator. An event-by-event data acquisition system enables the simultaneous measurement of α, β and γ-decays, utilizing 2 windowless PIN diodes, 4 plastic scintillators and a 40 % HPGe detector. The detection efficiency for β–decays can be controlled by the selective employment of individual plastic scintillators with varying efficiency between 0.1 % and 50 %. The geometrical efficiency of the HPGe detector can be continuously adjusted by changing its distance from the source by moving it in or out with a motorized linear guide. Together with customizing the pulsed implantation sequence, these measures allow the monitoring and characterization of beams from a few ions per second up to intensities of more than 10^{11}/s without significant saturation effects or dead-time losses. Figure 8 shows the maximum yields of all RIB measured at ISAC on a nuclide chart with all the known isotopes displayed as light yellow background. The gaps between the yield measurements indicate mostly refractory or chemical reactive elements. Due to their thermo-dynamical or chemical properties these elements are effectively retained in the target material matrix and don't effuse quantitatively to the ion source.

With the FEBIAD and resonant ionization laser ion source in place the number of isotopes from different elements that can be delivered has drastically increased. By 2012 (2013) there are 20 (24) elements for which TRILIS has delivered isotopes on-line. With the addition of actinide targets (the current licensing limit for operation is 10 μA p+ on target with an integrated of 5,000 μAh) the complete spectrum of ISOL produced isotopes should become accessible for future experiments.

6 ARIEL project

So far the ISAC facility can only deliver one RIB species at a time, making it a single user facility. TRIUMF has been looking at expanding the production of RIB using for example another proton extraction port to make a new beam line in parallel to the ISAC beam line and built another target station with the goal to simultaneously deliver two independent radioactive ion beams to different users. Such capability would allow doubling the science output and use of the sophisticated experimental apparatus. The Advanced Rare IsotopE Laboratory (ARIEL) project will not only provide a new 500 MeV, 100 μA proton beam line but will also utilize a 500 kW, 50 MeV electron beam (contribution by J. Dilling in this issue). With ARIEL, TRIUMF will be able to deliver three RIB species simultaneously to ISAC users, one beam from the present facility, one beam from the ARIEL proton beam line and finally one beam from the electron LINAC. One of the directions in nuclear physics at present is the production of more intense neutron rich rare isotope beams for various research fields. The goal is to be capable to reach the r-process path. Currently, only the CARIBU [31] facility at Argonne National Laboratory can reach the r-process path using 252-Cf fission products. This is a unique case because of the doubly magic region the Cf fission products can cover. Presently at ISAC, with 10 μA of 500 MeV proton beam on a UCx target, isotopes relevant for the the r-process cannot be reached.

There are two possible approaches: photo-fission using high intensity electron beam or using intense proton beams on UCx. Because of the high power density involved, both methods will require using the latest development in high power target technology at ISAC, e.g. composite targets where a carbide disc is bonded onto an exfoliated graphite foil which carries away the heat to the so-called high power target container capable of dissipating up to 20 kW.

6.1 ARIEL target station technology

The ARIEL project will be based on some of the technologies that were developed over the past 14 years at ISAC and lessons learned from operations. As with ISAC, the target stations are located in a sealed building serviced by an overhead crane. The target maintenance facility includes a hot-cell for target exchange, hot-cell for target diagnostics and storage preparation, a hot-cell for maintenance, decontamination facilities and radioactive storage vault. As in ISAC, the target will be placed at the bottom of a shielding steel plug and the target station will be surrounded by steel and concrete blocks. With the ARIEL project we have an opportunity to improve upon the current design and upgrade to a second-generation target station for high power RIB production which incorporates the experience of ISAC target station operation and RIB production. During ISAC construction we had to invent or improve most of the technology from the meson production target. Before designing the ARIEL target station we must re-evaluate the existing ISAC technologies and practices. To perform the analysis we apply a method used in product manufacturing called Design Failure Mode and Effect Analysis, DFMEA. Potential or experienced failures, potential or experienced effects of the failure mode and the causes of the failure are described. Then the severity (S) of the failure is given a number from 1 to 10, 1 being benign and 10 severe. The occurrence (O) of the failure is also given a number from 1 to 10, 1 being 1 in 10^3 cycle and 10 is a failure that arises once every running period. The ease of detection (D) to prevent the failure is given a number from 1 to 10, 1 being easy to detect the failure and 10 being very difficult. The product of S*O*D represents the risk priority given for that failure mode. From there we can decide on the criticality of that particular failure mode and recommend action or proceed to a design modification. All changes are recorded using the Engineering Change Order (ECO) associated with the action. The person responsible for the action and the date of completion is also recorded in the DFMEA documents. This document follows the products in its life cycle. The following findings were made during a design failure mode effect analysis, DFMEA, of the target station and target module.

On the positive side:

1. The modular approach allows us to operate at the design proton beam intensity, meaning 100 μA. The non radiation resistant components, such as o-rings, turbo pumps, actuators, cable, are well protected by the module and target station shielding.
2. The two stage mass separator is composed of a low- and a high-resolution mass separator in cascade. The first set of selection slits located at the focal plane of the first separator inside the heavily shielded target hall allows the elimination of most of the unwanted radioactive beam isotopes in a well defined manner, limiting the contamination accumulated on the slits of the high resolution mass separator.
3. We never have to change the optics in the two modules for the heavy ion beam transport from the ion source to the pre-separator and the entrance and beam dump modules.

On the negative side:

1. The vacuum system is complex. There are two different pressure volumes, the primary and the secondary vacuum envelopes. Pillow seals are used to seal the target box volume to the exit module vacuum for the heavy ion beam line.
2. The target box housing the target/ion source assembly is not hermetically sealed. This requires that the target module transfer of the spent target from the target station to the hot cell is done at atmospheric pressure. There is a risk of spreading contamination of volatile species during the transfer. In addition, target materials that are sensitive to moisture in air may react and change their composition potentially resulting in some contamination of the target hall.
   - In the new design the target box will be a hermetically sealable vessel preventing any contamination during the transfer of the irradiated target from the target station to the hot-cell.
   - The fact that the target box is hermetically sealable will allow the elimination of the two vacuum zones. Only one single zone is necessary, thus reducing the complexity of the vacuum system.
   - This will allow to fully condition the high voltage of the target module equipped with the new target and keep the target module under vacuum during the transfer, thereby limiting the risk of spreading contamination and shortening the pump down time of a new target.

3. The mechanical and electrical service connections necessary to operate the target/ion source assembly have to be disconnected manually. In order to permit a person to disconnect the services we have to allow for a cool down period. A complete cycle for the target exchange takes about 3 to 4 weeks.
   - A built in remote service connection and disconnection has to be incorporated into the initial system design.
4. Since the target exchange is time consuming we are forced to operate the target/ion source much longer than desired. We have observed that after two weeks of continuous operation at 70 μA and above, the isotope production yield drops significantly. Radiation damage is clearly visible on the target container and it reduces the overall target performance, e.g. cracks on the target container, which can lead to radioactive atoms to leave the target container before reaching the ion source. One of the main reason for the target container radiation damage observed is the fact that in order to increase the proton beam intensity on target we must defocus the proton beam. Nominally, to operate above 55 μA the proton beam size has to increase to 7–8 mm FWHM. This means that a large fraction of the proton beam goes directly through the Ta target container.
   - For ARIEL we shall install right from the beginning a proper proton beam rastering system to avoid radiation damage to the target container.
5. The servicing of the pre-separator slits is done hands-on in the present design. This limits greatly the maintenance capabilities.
   - In the new design the selection slits as well as all the target station beam diagnostics will be accessible with the overhead crane, allowing maintenance and repair in the hot-cell.

The new target station design addresses the mentioned findings by implementing new vacuum joint technology and remote services connectivity.

The remote vacuum seal requirement specifications are:

- Provide vacuum seal with a leak rate lower than $5*10^{-9}$ mbar liter/s
- Remote actuation, connection and disconnection of all the services to operate the target/ion source assembly.
- The vacuum seal shall last for at least 5 years in the high radiation field generated at the target.

Figure 9 shows a 3-D view of the next generation target station proposed for the ARIEL project. The new target station will be made of a target module, heavy ion beam line, pre-separator and beam dump. Contrary to the current ISAC target station there will be only one module, which is the target module. The entrance diagnostics and beam dump will be stand-alone devices which will not share the same vacuum envelope as the target station.

In the ARIEL project ion optics modules will be replaced by beam pipe sections containing the optics components with a vertical pumping duct and surrounded by shielding blocks composed of half steel pellet and half concrete. As in ISAC, the pumping duct will also serve to insert diagnostics at the beam level, Faraday cup, slit and beam profiler and be the route for signal and high voltage cables. The rationale is to keep the turbo pumps and sensitive equipment as far as possible from the high-

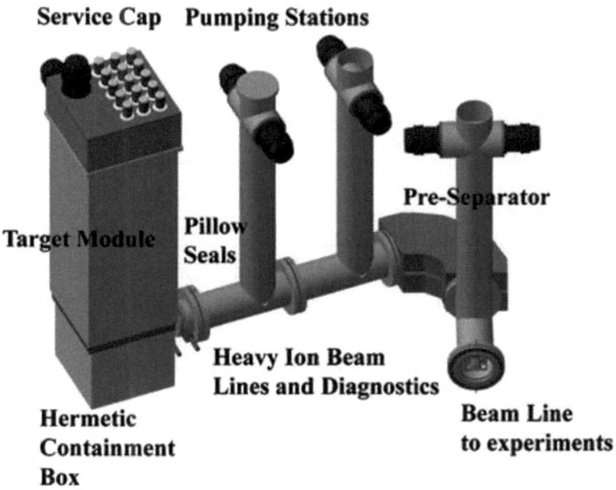

Fig. 9 Schematic 3-D view of the ARIEL target station concept (Nov. 2012). The radiation shielding around the target station is not shown for reasons of clarity

prompt-radiation field. The entrance and beam dump modules will be replaced by a stand-alone diagnostics box, and a water-cooled copper plug, respectively. Again, the rationale is that these devices have never had to be replaced since ISAC went into operation 15 years ago. These decisions simplify the vacuum envelope of the target station compared to ISAC and will reduce the cost of the target station since 4 module steel plugs are replaced by less expensive shielding, e.g. cast steel plugs and concrete blocks.

The vacuum envelope is defined by the target module containment box and the heavy ion beam line up to the experimental hall. The front end of the heavy ion beam line will be built in sections that can fit into the hot-cell for maintenance and repair. The vacuum envelope will be sealed using all metal vacuum joints.

6.1.1 Vacuum joint options

There are two vacuum joint technologies envisaged for the ARIEL target stations. The pillow seal, which was originally developed at the Paul Scherrer Institute (PSI) [33] in Switzerland, uses a single thin metal foil electron beam welded on a flange. The pillow is inserted between two flanges. High pressure He is injected to inflate the thin foil, also called a pillow. The seal joint is made of two flanges equipped with pillows and two concentric bellows welded to each side of the flanges. The helium gas used to inflate the pillow is also injected into the concentric bellows, pushing the flanges outward toward the fixed flanges on each side. A newer design developed for the T2K beam window uses a double pillow set-up, allowing a pump-down in between the pillows. They reported [34] a leak rate of 10^{-8} Pa m^3/s. Unfortunately, this is larger than desirable for constructing the entire heavy ion beam line with this technology. There will be eight of these vacuum joints in the heavy ion beam line, rendering the total leak rate much larger than acceptable. One can imagine having only one pillow seal for the target module vacuum connection and all the other using the actual ISAC technology used for the proton beam line connection to the target station. This will make this solution possible. Another option is to use HELICOFLEX, C-FLEX seals. This seal technology relies on elastic deformation of the C shape metal joint. During compression a contact point to each sealing surface is obtained. Although, this technology allows much lower leak rates, the key to the success of this joint is to provide the proper compression force and to maintain a very good surface finish, which may be difficult to maintain in the target station environment.

6.2 ARIEL science phases

6.2.1 First ARIEL beam, phase-I

The ARIEL project has to be built in phases, the first phase being a 100 kW, 25 MeV electron beam on a target. The electron beam will impinge onto a converter made of water cooled Ta discs; an Al disc placed after the converter will stop the remaining electrons before they can reach the target. The target stations, shielding, hot-cells, storage vault, front ends and mass separator are not part of the initial funding agreement Since most of the services required to operate actinide target will not be present in the first phase it is advantageous to consider using a non actinide target to do the ARIEL target and front end commissioning. It is advantageous to consider the production of ^{8}Li using ^{9}Be target via the following reaction:

$$^{9}\mathrm{Be}\left(\gamma,\mathrm{p}\right)^{8}\mathrm{Li}. \tag{2}$$

Advantages of using ^{8}Li on a ^{9}BeO target are:

- The level of radiation is not extremely large and the ^{8}Li has a very short half-life of 840 ms only.
- The longest half-life nuclei produced is ^{7}Be ($T_{1/2} = 53$d).
- With ^{8}Li the experimental program can be started while preparing for operation with an actinide target.
- This will allow a complete commissioning of the whole system from target station to experimental facility.

The ^{8}Li beam is used by the β-NMR material science community. The ^{8}Li beam is (nuclear spin) polarized using a collinear optical pumping system in which the polarized light from a laser beam is directed along the heavy ion beam axis. This method is well established at ISAC [28]. The first step is to neutralize the ^{8}Li ion beam by passing it through a Na vapour jet. The neutral atom then drifts nearly 2 m in the optical pumping region in presence of a small longitudinal magnetic holding field of 1 mT. Then the beam goes through a He cell where a large fraction of the now polarized ^{8}Li is ionized and then sent to the β-NMR or β-NQR station [28].

Results from FLUKA [29] simulations [30] show that we can produce close to 10^{9} ^{8}Li /s in target during phase-I.

6.2.2 Phase-II

The second phase will upgrade the energy to 50 MeV and the power to 500 kW. For this power, the simple water cooling of Ta discs approach will no longer be sufficient. Due to the large power density it is preferable to use a liquid metal converter, such as Hg or Pb because of their high Z. Lead is preferable to mercury because of health concerns, radioactive waste disposal, environmental issues and reactivity of the mercury metal with stainless steel pipe. Furthermore, mercury produces more

Table 1 FLUKA results of in-target production of key nuclei using 5 kW proton beam and 500 kW electron beam on UC_x target

Nucleus	5 kW p^+	500 kW e^-
Ni-72	3.8E+08	2.0E+08
Zn-78	1.4E+09	3.4E+09
Kr-91	5.3E+10	2.3E+11
Kr-94	1.3E+10	1.3E+11
Rb-97	7.4E+09	1.1E+11
Sn-132	1.1E+10	2.5E+10
Sn-134	1.0E+09	2.4E+09
Xe-142	1.1E+10	5.2E+10
Xe-144	1.0E+09	7.9E+09
Cs-144	6.8E+09	6.0E+10
Cs-146	5.0E+07	9.2E+08

The target thickness was set to 70 g/cm^2 UC_x for the proton beam target and 9.5 g/cm^2 for the electron beam the target, respectively

long lived isotopes by (γ, n) reactions than lead. For this phase, the second target station dedicated uniquely to high electron beam power will be built to the east of the phase-I target station. The proton beam line will be installed up to the first target station and will be dedicated to proton beam.

7 ARIEL expected yields for neutron-rich isotopes

At TRIUMF we are fortunate to have a high energy and high intensity proton driver with the H- 500 MeV Cyclotron and a high intensity, 50 MeV, 10 mA, electron driver under construction. We can produce neutron-rich isotopes using three different techniques: 5 kW direct proton onto the UCx target, 500 kW two-stage photo-fission using the electron driver and a 50 kW two-stage neutron-induced fission using the proton beam on a converter and UCx target.

7.1 5 kW proton on UCx

This is the current situation, where the high energetic proton beam impinges onto the UCx target discs. There are several reaction processes as mentioned earlier. This setup is less suitable since the spallation reaction produces isobars that contaminate the desired neutron-rich species. Furthermore, the high energetic proton beam induces fission that yields to less neutron-rich isotopes than pure low energy neutron induced fission. The limitation to 5 kW is only due to regulatory considerations, not to the capability of power dissipation of the UCx target, which can reach 20 kW.

7.2 500 kW electron, two-stage photo-fission of 238-U

ARIEL, as mentioned earlier, aims to accelerate an electron beam up to 50 MeV and 500 kW total beam power. The electron photon converter for such power can only be made using liquid lead. A first simulation shows that 375 kW is deposited into the converter and 75 kW in the target itself [32]. This is 7 times larger than the power we can handle in one single target. To handle the power deposition in the UCx we will use the composite target technique as described earlier; the uranium carbide will be deposited onto an exfoliated graphite foil. Furthermore, the target may be divided into several smaller target containers and helium gas may be used to cool the whole target assembly.

The expected yield of some key nuclei on the neutron-rich side are listed in Table 1.

8 Concluding remarks

To satisfy the demand for higher RIB intensity the incident proton beam intensity on target at ISAC has been raised compared to other facilities. This was only possible by pushing the technologies for material target and target container to sustain the high power deposition in the ISOL RIB production. We have developed a technique to produce a composite target material capable of dissipating larger power deposition. This technique increases the overall thermal conductivity of the target material. The target material is deposited onto a highly conductive substrate, exfoliated graphite foil for the carbide target material and metal foil for oxide target material. At present we routinely operate the proton beam at 35 to 50 kW, level [33, 34]. The target container has to extract heat from the target material and dissipate that heat to the target cooling system. We use radiative cooling to get the heat out of the target. A finned target container has been developed for this purpose and it is capable of dissipating up to 20 kW of beam power deposited in the target. With the ARIEL project we have the opportunity to build the next generation of high power target station for producing high intensity ISOL RIB. An analysis of the present ISAC target station has been performed and the findings will be applied to the design of the next generation of target station. The ARIEL first beam will be generated using a BeO target to produce ^{8}Li for the material science community. We used FLUKA to obtain the production rate of ^{8} Li in target; the estimated intensity is well above the need for performing beta-NMR studies. There are several new projects around the world with the aim of producing intense RIB using fission reaction. The power density inside the U target in all of these projects exceeds the present capability of carbide material thermal conductivity. A high thermal conductivity target material is mandatory for the success of these projects. The development of composite target material at TRIUMF/ISAC is an effective way to achieve such high thermal conductivity allowing high power RIB production at the ISOL facility. Furthermore, for the dissipation of the power to the cooling system it is necessary to use a target oven capable of dissipating the power release from the target material to the cooling environment. The ISAC high power target has proven its capability to dissipate up to 20 kW. To go beyond that we need to use other means such as helium gas cooling as anticipated at ARIEL photo-fission target. Finally, one of the frontiers in physics is the capability of producing intense and pure neutron-rich RIB. These beams can be produced using fission reactions. At TRIUMF we will have access to the fission products by using photo-fission and a two-stage target system using 50 kW proton beam on a neutron converter target.

Acknowledgements The authors wish to acknowledge the excellent work of the target/ ion sources and remote handling staff for their excellent work for the ISAC operation and target development. TRIUMF's core operations are supported via a contribution from the federal government through the National Research Council Canada; the Government of British Columbia provides building capital funds. Several of the co-authors receive NSERC grants for target and ion source related research.

References

1. Buchmann, L., D'Auria, J.M. (eds.): Accelerated Radioactive Beams Workshop. Parksville, BC, Canada, TRI-85-1, 5–7 Sept 1985
2. Kugler, E.: The ISOLDE facility at the CERN PS booster. Nucl. Inst. Methods **79**(1–4), 322–325 (1993)
3. Ferrari, A., Sala, P.R., Fasso, A., Ranft, J.: FLUKA: a multi-particle transport code. CERN-2005-10, INFN TC05 11, SLAC-R-773 (2005)
4. Moritz, L.: ISAC shielding requirements specification. ISAC Safety Analysis Report, TRIUMF (1998)
5. Silverberg, R., Tsao, C.H.: Ap. J. Suppl. **313**(25), (1973)
6. Silverberg, R., Tsao, C.H.: Ap. J. Suppl. **58**, 873 (1985)
7. Kugler, E.: Hyperfine Interact. **129**(1–4), 23–42 (2000)
8. Beene, J.R., Bardayan, D.W., Galindo Uribarri, A., Nazarewicz, W., et al.: J. Phys. G: Nucl. Part. Phys. **38**, 024002 (2011)
9. Bricault, P., Ames, F., Labrecque, F., Lassen, J., Li, R., Dombsky, M., Mjos, A., Minor, G., Teigelhoefer, A.: Recent development of the on—line ion sources ar TRIUMF-ISAC. Rev. Sci. Instrum. **83**(2), (2012)
10. Kirchner, R., Roeckl, E.: Nucl. Inst. Methods **133**, 187–204 (1976)
11. Sundell, S., Ravn, H.: Nucl. Inst. Methods Phys. Res. B **70**, 160–164 (1992)
12. Geller, R.: In: Proc. 1st Int. Conf. Ion Sources, Saclay, p. 537 (1969)
13. Kluge, H.-J., Ames, F., Ruster, W., Wallmeroth, K.: In: Buchmann, L., D'Auria, J. (eds.) Accelerated Radioactive Beams Workshop, 119. Parksville, BC, Canada, TRI-85-1, 5–7 Sept 1985
14. Dombsky, M., Baartman, R., Bricault, P., Doornbos, J., Jayamanna, K., Kuo, T., Mackenzie, G., et al.: Rev. Sci. Instrum. **69**, 1170 (1998)
15. Lassen, J., Bricault, P., Dombsky, M., Lavoie, J.P., Geppert, C., Wendt, K.: Hyperfine Interact. **162**, 69–75 (2005)
16. Koester, U., Fedosseev, V.N., Mishin, V.I.: Spectrochim. Acta B **58**(6), 1047–1068 (2003)
17. Lavoie, J.P., Li, R., Bricault, P., Lassen, J., Chachkova, O., Teigelhoefer, A.: Rev. Sci. Instrum. **84**(1), 013306 (2013)
18. Raeder, S., et al.: Hyperfine Interact. **216**(1–3), 33–39 (2013)
19. Teigelhöfer, A., Lassen J., Abboud Z., Bricault, P., Heggen, H., Kunz, P., Li, R., Quenzel, T., Raeder, S.: Yttrium ionization scheme development for Ti:Sa laser based RILIS. Hyperfine Interact. **216**(1–3), 65–70 (2013)
20. Rothe, S., et al. Measurement of the first ionization potential of astatine by laser ionization spectroscopy. Nat. Commun. **4**, 1835 (2013)
21. Teigelhoefer, A.: Laser resonance ionization spectroscopy of yttrium and astatine. M.Sc. thesis, Physics, University of Manitoba (2012)
22. Labrecque, F.: Development of radiation hard plasma sources for ISAC laboratory. PhD Thesis, Les Presses de l'Université Laval, Québec
23. Jayamanna, K., et al.: Rev. Sci. Instrum. **73**, 792 (2002)
24. Bricault, P., Jayamanna, K., Yuan, D.H.L., et al.: In: Proc. 16th Int. Workshop on ECR Ion Sources. Berkeley, California, USA, 16–20 Sept 2004
25. Leroy, R., et al.: In: Proceedings of the 17th International Conference on Cyclotrons and their Applications. Tokyo, Japan 2004, 261 (2004)
26. Labrecque, F., Lecesne, N., Bricault, P.: Nucl. Instrum. Meth. B **266**(1920), 4407–4410 (2008)
27. Labrecque, F.: Development of radiation resistant plasma sources for rare isotope production. PhD thesis, Les Presses de l'Université Laval, Québec, Québec, Canada (2012)
28. Levy, C.D.P., Pearson, M.R., Morris, G.D., Chow, K.H., Hossain, M.D., Kiefl, R.F., Labbé, R., et al.: Hyperfine Interact. **196**, 287–294 (2010)
29. Ferrari, A., Sala, P.R., Fasso, A., Ranft, J.: FLUKA: a multi-particle transport code. CERN-2005-10, INFN/TC-05/11, SLAC-R-773 (2005)
30. Bernier, N.: Study of the photoproduction of ^{8}Li with the reaction ^{9}Be(γ, p) ^{8}Li. Master's thesis, U. Laval (2013)
31. Van Schelt, J., Lascar, D., Savard, G., Clark, J.A., Bertone, P.F., Caldwell, S., Chaudhuri, A., Levand, A.F., Li, G., Morgan, G.E., Orford, R., Segel, R.E., Sharma, K.S., Sternberg, M.G.: First results from the CARIBU facility: mass measurements on the r-Process path. Phys. Rev. Lett. **111**, 061102 (2013)

32. Lebois, M., Bricault, P.: Simulation for the future converter of the e-linac for the TRIUMF ARIEL facility. In: Proceding INPC, Vancouver (2010). J. Phys. Conf. Ser. (2011)
33. SIN report, Jahresberitch. JB11-18 (1984)
34. Rooney, M., Densham, C., Filton, M., Francis, V., Yamada, Y., Mark, C., Gallop, M.: 3rd High-Power Targetry Workshop, Paul Scherrer Institute, Sept. 10–14. Bad Zurzach, Switzerland (2007)

DOI 10.1007/978-94-007-7963-1_5
Reprinted from *Hyperfine Interactions* Journal, DOI 10.1007/s10751-013-0881-y

Off line ion source terminal

K. Jayamanna

Published online: 16 October 2013

Abstract The off-line ion source (OLIS) terminal provides beams from stable isotopes to ISAC (see Fig. 1) experiments as well as for accelerator commissioning and for pilot beams for radioactive beam experiments. The OLIS terminal (see Fig. 2) is equipped with a microwave driven cusp source for single and double charge ions, a surface ion source for low energy spread alkali beams, and a multi-charge ion source.

Keywords Ion sources · Ion beams · Highly stripped ion beams · Multi-charge ion beams · Radioactive ion beams · RIB · Stable ion beams · SIB · Plasma sources

1 Introduction

The Off-Line Ion Source [1] (OLIS) facility consists of a high voltage terminal containing (see Figs. 1 and 2b) a microwave cusp ion source [2], a surface ion source [3] or a hybrid surface-arc discharge ion source [3] (which can operate in a hybrid arc-discharge mode) and a multi-charge ion source [4, 5]. The system also includes an electrostatic switch that allows the selection of any one of the sources without mechanical intervention. Figure 1 shows the schematic drawing of the OLIS terminal and the electrostatic switching box. These sources provide a variety of +1 or +n beams up to A/Q = 32 (A: atomic mass and q: charge state) for ISAC experiments, commissioning the accelerators, setting up the radioactive experiments and for tuning the beam lines. The primary accelerator RFQ (Radio Frequency Quadrupole) is designed to accept beams at a fixed injection velocity corresponding

ISAC and ARIEL: The TRIUMF Radioactive Beam Facilities and the Scientific Program.

K. Jayamanna (✉)
TRIUMF, Canada's National Laboratory for Particle and Nuclear Physics,
4004, Wesbrook Mall, Vancouver, BC, Canada V6T2A3
e-mail: Keerthi@triumf.ca
URL: olis.triumf.ca

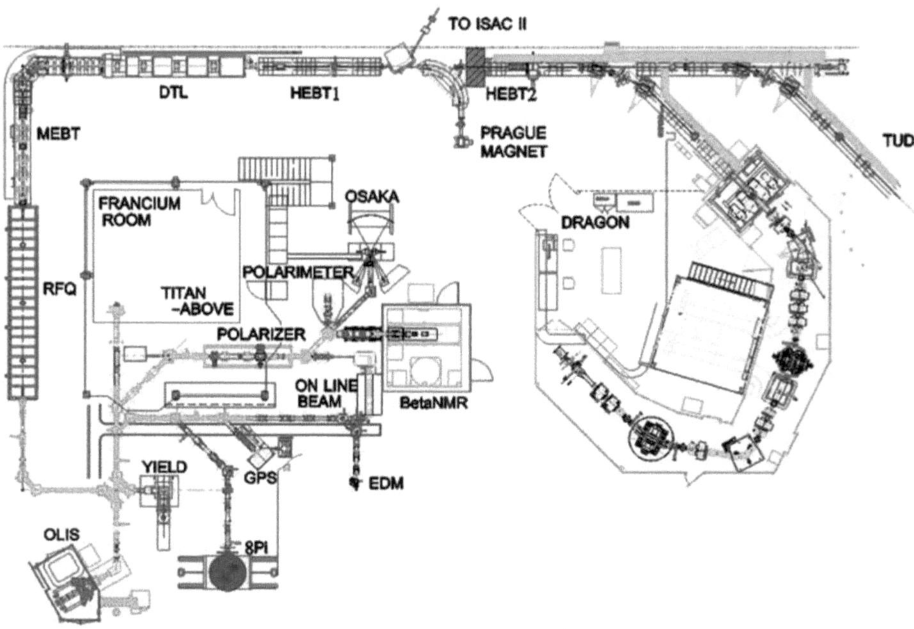

Fig. 1 Technical drawing of ISAC 1 where OLIS supplies beams from stable isotopes to various experiments. The ISAC II experimental hall is not shown in the picture. The stripping foil mechanism is located at the MEBT section before the bending/analyzing magnet

to 2.04 keV/u. Moreover, the secondary accelerator Drift Tube Linac [6] (DTL) requires a mass-over-charge ratio between 3 and 6 ($3<A/Q<6$). However, the source extraction voltage is limited to 65 kV. The multi-charge ion source was installed to deliver beams above mass 32 and with a A/Q value up to 6, without the need of a stripper foil between the RFQ and DTL. With this addition, OLIS can provide ion beams from all stable elements and satisfy all ISAC demands.

ISAC has two separate high and low energy areas. The online ion source terminal provides beams through a junction cross in order to provide beams simultaneously to high-energy and low-energy beam lines. While radioactive ion beams are delivered from the online terminal to a low-energy area, the stable beams from the offline terminal can be diverted to a high-energy area and vice versa.

2 Ion sources

2.1 Microwave ion source

The microwave ion source (see Fig. 3) has been operational since 1995 and provides only singly and some doubly charge ion beams from various stable isotopes. Originally, its prime goal was to provide beams from gaseous elements, but later two ovens and a sputtering system were added in order to provide beams from liquids and from solids. This source terminal is now automated for start up and for mass selection. It is capable of providing positive ions from stable isotopes for months without

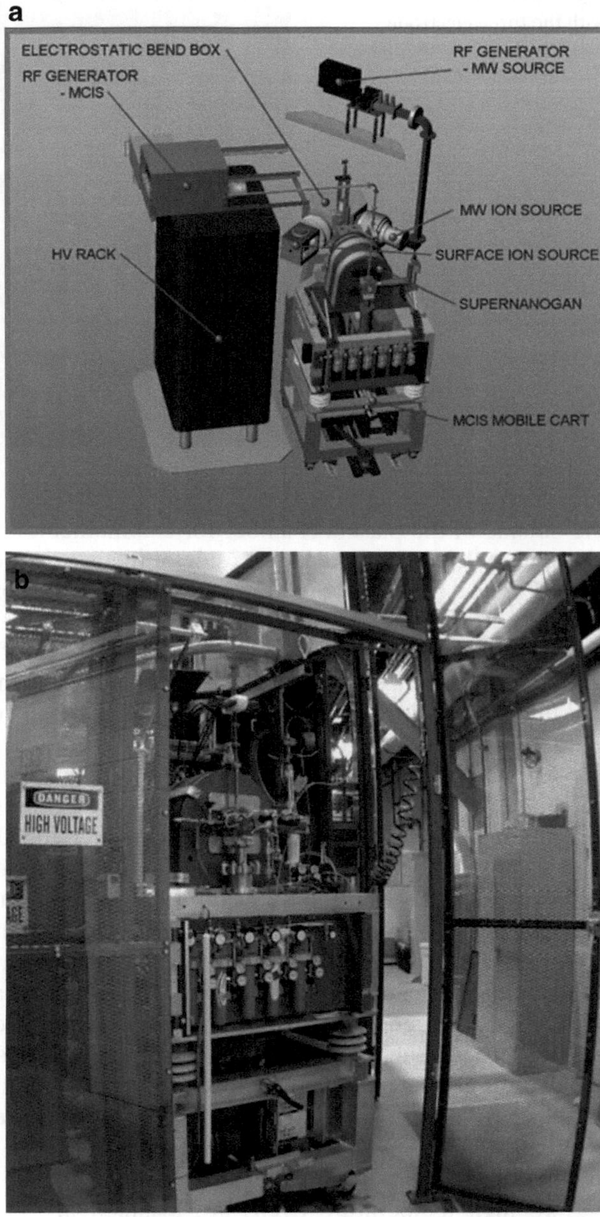

Fig. 2 **a** Engineering model of the OLIS terminal and the electrostatic switching box. **b** Photograph of the OLIS terminal with the high voltage cage

maintenance and even negative ion beams if required. To date, over 40 different isotopes including many rare isotopes have been delivered to various experiments.

The microwave ion source is the most used ion source in the OLIS terminal. It consists of two vacuum chambers. One is a cylindrical cavity with a quartz liner, which serves as a plasma chamber. This chamber is 15 cm in length and 10 cm in diameter. The second chamber consists of two 90° bend wave guide systems as a windowless RF coupling chamber. Ten water-cooled $SmCo_5$ magnetic bars are installed around

Fig. 3 Microwave ion source with the three electrode extraction system and the RF injection system

the outside of the plasma chamber for confinement. Four more magnetic bars with alternate poles are installed in the side of the back plate to achieve continuous cusp lines. For the positive ion extraction, the magnetic configuration has a cylindrical symmetry. For the negative ion extraction, the magnetic configuration is changed by flipping the polarity of the last magnets in a pair of diametrically opposed rows in order to create a strong virtual filter. This divides the plasma chamber into high and low energy electron regions. Two additional pairs of magnets are placed in the extraction region so that the total integral field is equal to zero.

Microwave power at 2.45GHz is introduced axially to the source in between two back plate magnetic bars through a 10 mm by 72 mm rectangular shape matching diaphragm placed in between two chambers mentioned above. A Teflon™ window with a choke flange in between the microwave generator and the source serves as both a vacuum break and 65 kV electrical isolation. A one meter long waveguide and two 90° bends stop any plasma or particle migration to the Teflon window. An axially symmetric three electrode system was developed to extract beams with an extraction voltage range from 6 kV to 60 kV without significant change in the emittance values.

A gas system equipped with 12 different gases is connected to the source via remotely controllable valves via two high precision gas flow controllers. When the source is used to provide pilot beams or for commissioning the accelerators, a mix of various gases can be injected into the source to provide any mass from the gaseous elements required by the accelerator at any given time. If the source is used only for gaseous elements, it can run with minimal maintenance for years.

Two high temperature ovens are installed in the source, so that high vapor pressure elements could be injected into the source via the back plate. Quartz liners are placed in the source chamber so that, contaminations can be easily removed from the source after each run and before the next run begins.

The source can also be easily converted into a sputter source simply by attaching a standard sputter disk made of required material to the back plate of the ion source and applying 300 V to it with respect to the plasma. DRAGON [6] and TUDA [7] have been the main users of the OLIS microwave ion source for years and H_3^{+1}, $^{12}C^{+}$, $^{20,21,22}Ne^{+}$, $^{16,17,18}O^{+}$, $^{36,40}Ar^{+}$, He^{+}, $^{40}Ca^{+2}$, $^{24,25,26}Mg^{+}$, and $^{27}Al^{+}$ were the common beams delivered to the experiments. Some difficult beams (considering purity), such

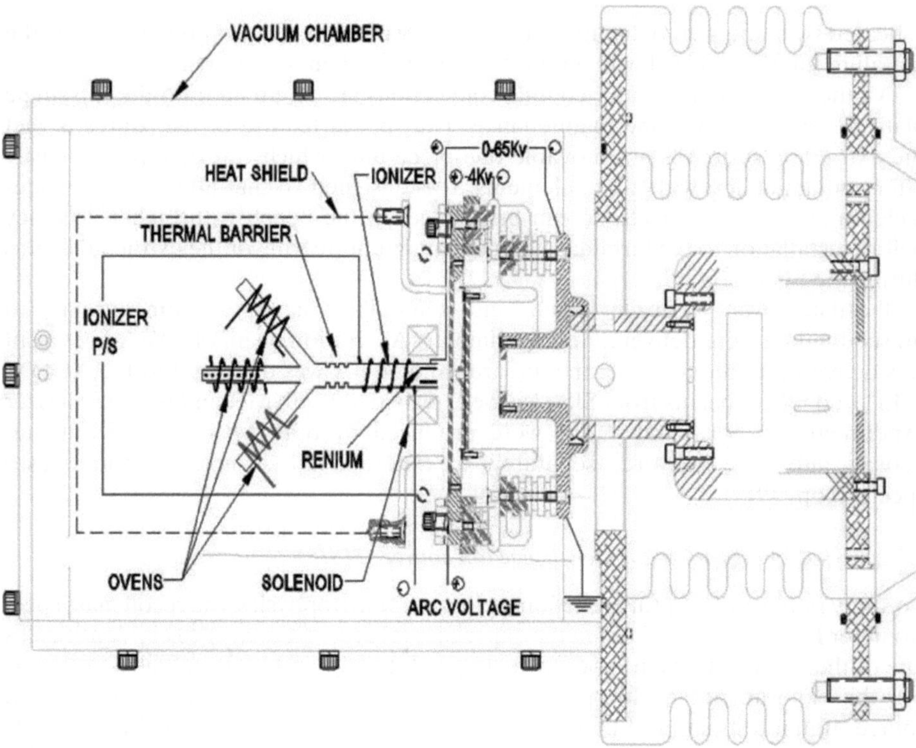

Fig. 4 Surface ion source with three heaters and an ionizer. It is capable of converting to a hybrid surface arc discharge ion source by switching on the arc voltage and axial magnetic field

as $^{10}B^{+}$, $^{28}Si^{+}$ and $^{18}O^{+}$ were also provided for the experiments in the high energy area. During the $^{10}B^{+}$ beam, boron trifluoride was used as the trace gas. The source needed to produce enough boron beam current so that suppression of the impurities ($^{20}Ne^{+2}$ and $^{40}Ar^{+4}$) was possible. During the $^{28}Si^{+}$ beam, it was very difficult to suppress $^{28}N_2^{+1}$ completely, therefore $^{28}SiH_3^{+1}$ was extracted from the source and delivered up to the stripping foil as mass 31 for subsequent down stream selection of the pure multi-charge Si^{+7} beam. The SiH_3^{+} beam was then delivered to the RFQ for DRAGON at 63.24 kV, which is above the designed upper limit of the OLIS high voltage system. $^{139}La^{+}$, $^{141}Pr^{+}$, $^{23}Na^{+}$, $^{39}K^{+}$, $^{75,77}Rb^{+}$ and $^{19}F^{+}$ from the microwave ion source were also among the other beams delivered to the low energy areas [8].

2.2 Surface ion source (SIS) and arc-discharge mode

2.2.1 Surface ion source

The surface ion source (see Fig. 4) mounted onto the second port of the electrostatic switching box is equipped with an ionizing chamber and three ovens. The three ovens give it the flexibility to run three different temperature regions simultaneously (25–600°C, 600–1200°C, 1200–2000°C). The ionizing chamber is a 5 mm diameter, 5 cm long tantalum tube narrowed down to 3 mm diameter at the extraction side, which is heated by a tungsten filament. For elements with higher ionization potentials, the tantalum tube is replaced with a tungsten tube with rhenium inserts.

All the electrical and water-cooling connections are mounted onto the top single plate for easy removal and maintenance. The special feature of the source is that the extraction system is a removable one-piece unit, which can be replaced easily after contamination with alkali elements. It is designed so that all three electrodes including the ground electrode can be aligned on the bench with higher accuracy before installation to the terminal allowing minimum steering of the extracted beams during the operation.

To date over a dozen alkali and semi-alkali isotopes from lithium to praseodymium were delivered to experiments. A ^{23}Na ion beam of 10 μA in current was delivered to the 8π experimental area and 10 nA was delivered to DRAGON. $^{6,7}Li^+$ and $^{23}Na^+$ beams from SIS also were provided to βNMR and to the polarizer experiments[7]. $^{139}La^+$ and $^{141}Pr^+$ were produced by both MIS and SIS simultaneously and were delivered separately to experimental set-ups for atomic laser spectroscopy [9].

2.2.2 Arc-discharge mode of the surface ion source (HSIS)

To study radioactive capture reactions relevant for astrophysics, the recoil mass spectrometer DRAGON was built in the experimental area. $^{40}Ca(\alpha,\gamma)^{44}Ti$ is identified as one of the key reactions in supernovae to produce ^{44}Ti and is given highest priority. For this experiment, an ultra pure Ca^{+2} beam was requested from the off-line ion source. Initial tests showed that when using conventional ion sources, ^{40}Ar and ^{40}K were the impurities that were most difficult to eliminate. In order to overcome this problem, a new concept was needed and the hybrid surface arc discharge ion source (see Fig. 4) was born. This ion source consists of a small surface ionizer and an arc discharge placed in a magnetic field produced by a small solenoid.

During surface ionization $^{40}Ar^+$ can be filtered out since it has a high ionization potential of 15.76 eV. Only $^{40}Ca^+$ and the $^{40}K^+$ is present in the arc discharge chamber since ionization potentials are 6.11eV and 4.34 eV respectively. By carefully selecting the arc voltage and the arc current, $^{40}K^{+2}$ with an ionization potential of 31.63 eV can be reduced while optimizing $^{40}Ca^{+2}$, which has lower ionization potential equal to 11.87 eV.

Beam contamination can only be measured at high energy levels in an ionization chamber at the end of the recoil mass separator since the mass difference between ^{40}Ar and ^{40}Ca is less than 10^{-5} amu. A very low $^{40}Ar/^{40}Ca$ ratio of 8×10^{-5} for the hybrid surface arc discharge ion source was measured using an the DRAGON ionization chamber whereas the microwave ion source gave much higher values ($>5 \times 10^{-3}$). No indication of ^{40}K contamination was found from either ion sources. The current for the $^{40}Ca^{+2}$ beam was about 50 nA, enough to successfully perform the $^{40}Ca(\alpha,\gamma)^{44}Ti$ experiment [10].

2.3 Multi-charge ion source (MCIS)—supernanogan

The design objective of the Supernanogan (a commercially available multi-charge ECR ion source from PANTECHNIK) addition to the OLIS ion source terminal is to provide multi-charge ion beams from the Supernanogan ion source while minimizing the impact on the microwave and surface ion source operation. The concept adopted

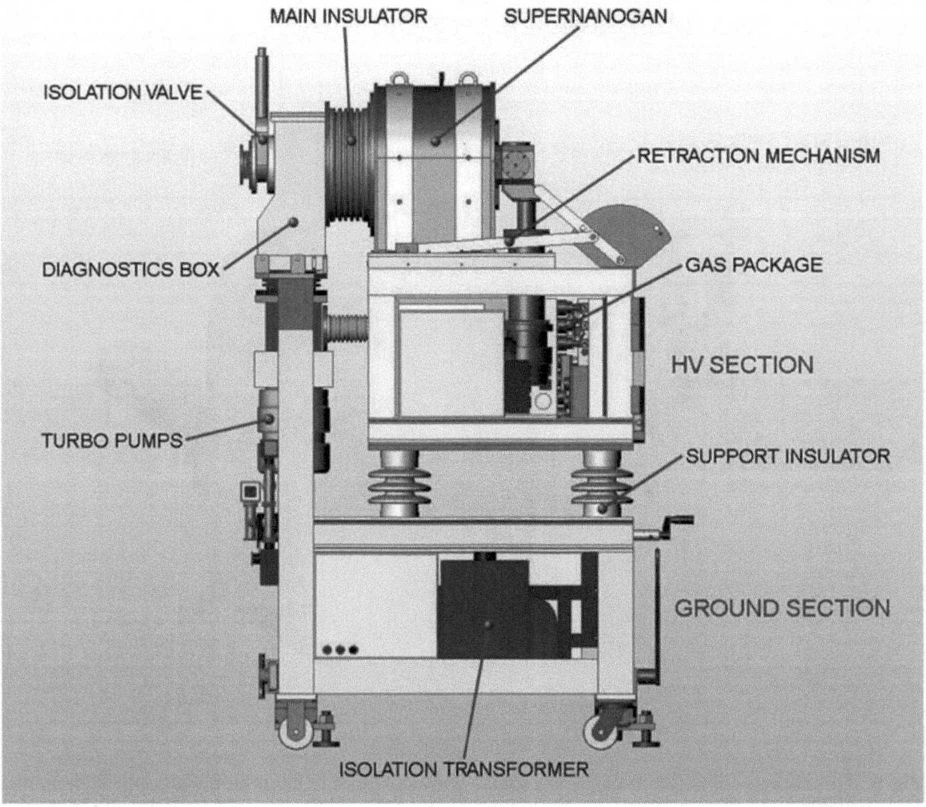

Fig. 5 The mobile station, referred to as the "cart", is a self-contained ion source system that has all the necessary power supplies, vacuum components, diagnostic devices and control systems

for this functionality is a mobile, virtually self-contained ion source station (see Fig. 5). This station consists of two main sections, one at ground potential and the other at a high voltage bias of up to 20 kV. The ground section contains a high voltage isolation transformer, two turbo pumps and their controllers, an ion gauge controller, a vacuum box for optics and services, power distribution, computer controls and monitoring. The HV section contains the Supernanogan ion source and shielding, dry scroll vacuum pump, gas supply system, DC power supplies for Supernanogan operation, power distribution and computer control and monitoring. The HV section communicates with the ground section controls via an optical link, and then the ground section controls connect to the OLIS controls system.

The cart rolls into the OLIS HV enclosure and obtains a vacuum tight connection to the OLIS electrostatic switching box at the central port. The cart is then connected to services such as power, RF, water, air, gas, vacuum roughing system and controls. The operation of the Supernanogan is then much like the other OLIS ion sources, and utilizes the same interlock and safety systems already in use. When the Supernanogan is in need of servicing, it can be disconnected (from the OLIS system) and removed without affecting the integrity of either the OLIS or Supernanogan vacuum. When outside the HV enclosure, the Supernanogan vacuum system can continue to

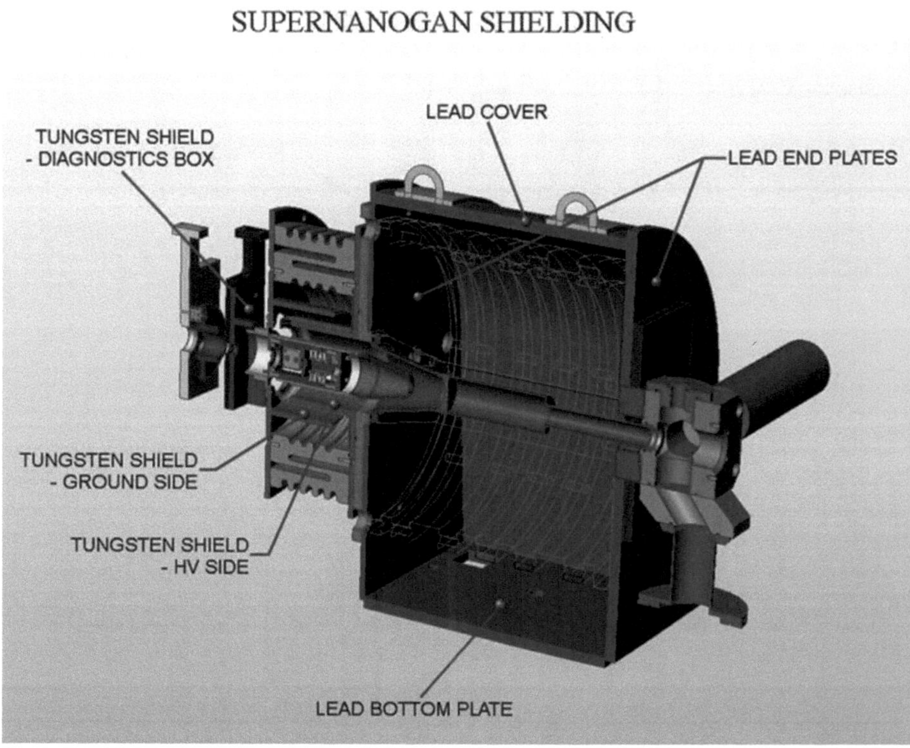

Fig. 6 This cutaway illustration of the ion source shows the extraction region with optical elements, HV insulator and X-ray shielding. Local shielding of the Supernanogan is achieved by a lead shroud with end and bottom plates. Shielding the extraction region is with Elkonite alloy since the X-ray shielding penetrates the vacuum space

operate, and the long conditioning times required for high charge states can be realized. The Supernanogan vacuum conditioning and maintenance time spent outside of the OLIS facility has no impact on the operation of the other ion sources. When the Supernanogan is required to provide beam to ISAC, disruption of OLIS beam operation by introducing the cart into the facility is minimal. Figure 6 is showing the extraction optics of the Supernanogan ECR ion source as well as its x-ray shielding.

2.4 A magnetic stabilization system

A magnetic field stability system was developed for the mass separator magnet at the OLIS. The magnet is connected to a main power supply capable of providing up to 500A. This power supply has a resolution of about 50 mA. The lack of resolution of the main magnet power supply, temperature effects to the magnet material and power supply and various other factors affect the magnetic field produced. This creates an unstable beam over short and long periods of time. Through the use of a PID (proportional-integral-derivative) controller, a small power supply with a better resolution is used to correct for changes in the magnetic field over time. An additional power supply is connected in parallel with the main power supply to provide

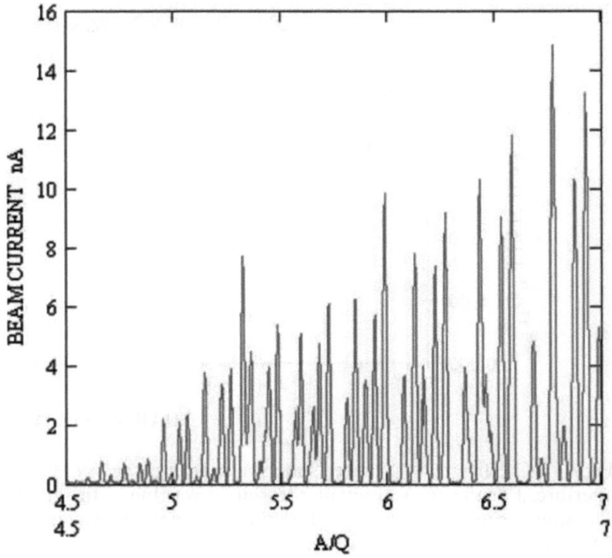

Fig. 7 Xenon multi-charge spectrum with 1 mm object and image slits. Most of the xenon peaks consist of isobaric and different charge xenon peaks. For example the peak at A/Q = 5.6 consist of a $^{128}Xe^{+23}$, $^{134}Xe^{+24}$, $^{129}Xe^{+23}$, $^{130}Xe^{+23}$, $^{136}Xe^{+24}$ and a little bit of $^{132}Xe^{+24}$, $^{131}Xe^{+23}$

a current resolution in the magnet of about 1 mA. The magnetic field was measured with a Group 3 MPT-141 Hall probe with a Group 3 DTM-151 digital Gauss meter. The second power supply is capable of correcting for as much as 40 Gauss in changes to the magnetic field with a minimum of 0.01G with 1 mA resolution in this particular set up. The PID controller is run continuously to correct for changes in the magnetic field as read by the magnetic field probe. It was found that, over a period of a week of operation, the magnetic field was kept steady to within +/− 0.2G using the PID controller to stabilize the magnetic field. This stabilizing system is paramount hold the exact A/Q for a prolong periods. For example holding at the $^{18}O^{+1}$ within the tail of the H_2O^{+1} peak and away from the maximum H_2O^{+1} peak is only possible with this stability method.

3 Results and discussion

3.1 Multi-charge ions and their efficiency studies

Multi-charge ionization can be achieved by either step-by-step ionization or by Auger transition. ECR plasma at higher frequencies is able to deliver enough energetic electrons (T_e) and electron densities (n_e) high enough to produce highly charge ions. Most importantly, the Golovanivsky's boundary condition [11] ($n_0/n_e \leq 7 \times 10^3 \xi T_e^{-3/2} A^{1/2}/Z$) must be fulfilled in order to increase the population of the given charge state, where Z: desired charge state, A = atom mass number and ξ = total number of electrons in the outer shell. Since ISAC needs A/Q ~ 6, the background neutral density must reach 10^7 cm^{-3} or below, if the electron density is to be in the range of 10^{11} cm^{-3}. A very delicate balance of electron temperature versus electron density must be maintained in order to produce significant amount of higher charge ions in the range over mass 100 and A/Q<6.

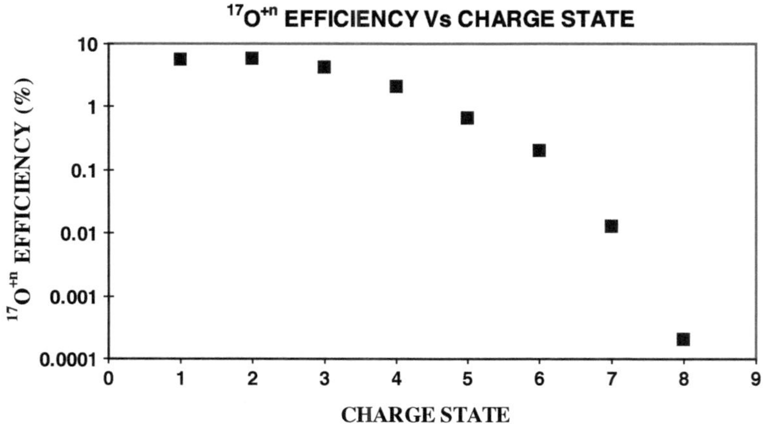

Fig. 8 ^{17}O efficiency vs. charge state was measured at 18 kV extraction with 2 mm diameter plasma aperture. For this measurement, the source was tuned for minimum emittance of charge state +3. The source pressure with the ^{17}O flow and when the high voltage is on and off are 4×10^{-8} Torr and 5.5×10^{-8} Torr respectively. Beam current is significant enough to affect the pressure

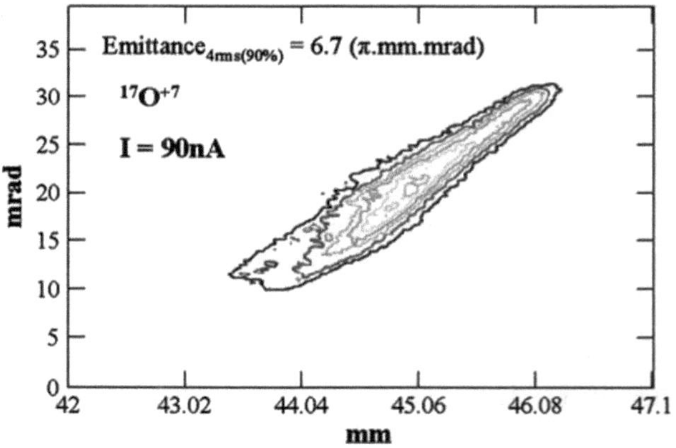

Fig. 9 Emittance of the $^{17}O^{+7}$ beam measured with a 2 mm diameter plasma aperture

Efficiency of the source was measured using calibrated leaks from oxygen and neon isotopes. In order to achieve maximum efficiency, the gases were sent directly into the source chamber through the hollow coaxial antenna of the source. The efficiency of each charge vs. mass is shown in Fig. 7. In order to match the acceptance of the ISAC RFQ, the plasma aperture was reduced to a 2 mm diameter.

3.2 Emittance vs. charge state studies

For emittance measurements, an improved version of the electric sweep scanner, originally proposed by Allison [12], was used. The emittance figure for $^{17}O^{+7}$ is shown in Fig. 8. It was found that the emittance of the extracted multi-charge ions varies with different charge states of the same element. Since the variation was very little, a new and more accurate data processing method had to be developed in order to minimize the noise reduction and emittance calculation error to ≤ 1 %. Figures 9

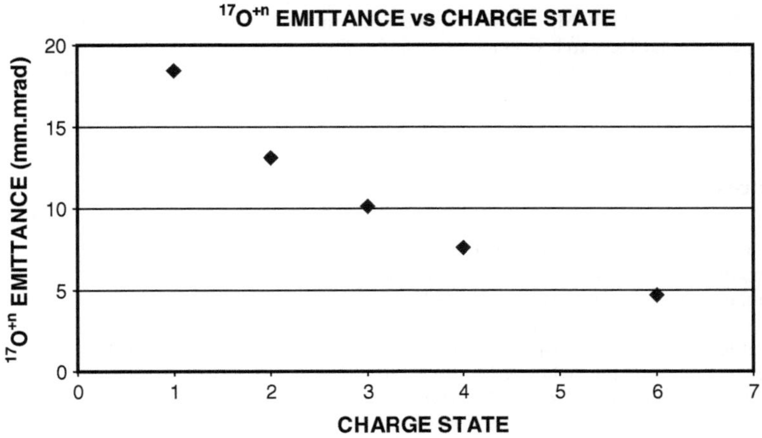

Fig. 10 ^{17}O emittance vs. charge state was measured at 18 kV extraction. The concentration of higher charge states near the centre of the plasma may explain the observed lower emittance of higher charge states

and 10 show the variation of the emittance values for different charges of ^{17}O isotope. Detailed analysis of the data is beyond the scope of this paper and will be published elsewhere.

4 Discussion and summary

The OLIS Terminal (with three ion sources, a microwave ion source, a surface ion source and multi-charge ion source) is capable of delivering from virtually any stable isotope close to A/Q = 6 or higher. Each ion source can be serviced outside the OLIS terminal while other sources provide the beam to the ISAC. To date a large number of isotopes have been delivered (see OLIS web page for details) to ISAC experiments including exotic isotopic beams such as $^{13}C^{+1}$, $^{17}O^{+3}$, $^{18}O^{+4}$, $^{33}S^{+6}$, $^{15}N^{+2}$, $^{21}Ne^{+5}$, $^{36}Ar^{+7}$, $^{74}Se^{+14}$, $^{80}Kr^{+15}$ and $^{130}Xe^{+24}$ from the multi-charge ion source, $^{17}Be^{+1}$, $^{18}O^{+2}$, $^{19}F^{+1}$, $^{15}N^{+1}$, $^{21}Ne^{+1}$, $^{24,25,26}Mg^{+1}$, $^{28,29}Si^{+1}$, $^{36}Ar^{+1}$, $^{35,37}Cl^{+1}$, $^{48}Ti^{+1}$, $^{58,60}Ni^{+1}$, Kr^{+1}, $^{139}La^{+1}$, $^{141}Pr^{+1}$ from the microwave ion source and most of the alkali beams from the surface ion source.

Regarding the multi-charge ion source, the ability to vary the frequency with very fine intervals has led to a significant enhancement of source performance. A few watts of RF power can be sufficient to provide the necessary multi-charge beams to ISAC. The emittance of multi-charge ions was measured. Observed lower emittance for higher charges can be explained by the concentration of higher charges towards the center of the plasma. With the installation of the Supernanogan a stripping foil is no longer necessary and the ISAC beam current capability for nonradioactive beams has increased by more than ten folds. Recently ISAC achieved one of it's milestones 1 GeV using $^{131}Xe^{+24}$ beam from OLIS. There are some discussions about using OLIS terminal for radioactive isotopes since it can produce large amounts from pre-prepared samples.

Acknowledgements TRIUMF/ISAC receives federal funding via a contribution agreement through the National Research Council of Canada. The author would like to express their sincere gratitude to the ISAC Controls group for their unwavering support during commissioning and operation.

References

1. Jayamanna, K., et al.: Rev. Sci. Instrum. **79**, 02C711 (2008)
2. Jayamanna, K., et al.: Rev. Sci. Instrum. **67**, 1061 (1996)
3. Jayamanna, K., Rev. Sci. Instrum. **79**, 02711 (2008)
4. Jayamanna, K., et al.: Rev. Sci. Instrum. **81**(2), 02A331 (2010)
5. Laxdal R.E., et al.: Nucl. Instrum. Methods Phys. Res. B **204**, 400–409 (2003)
6. Engel, S., et al.: Nucl. Instrum. & Methods Phys. Res. A **553**, 491 (2005)
7. Parpottas, Y., et al.: Phys. Rev. C **71**, 025802 (2005)
8. Levy, C.D.P., et al.: Nucl. Instrum. & Methods A. **580**, 1571 (2007)
9. Cocolios, T.E., et al.: MAI. **45**(04), 88 (2006)
10. Vockenhuber, C., et al.: Nucl. Instrum. & Methods. B **259**, 688 (2007)
11. Golovanivsky, K.S.: Instrum. Exp. Tech. **28**, No 5, part 1, 989 (1986)
12. Allison, P.W.: IEEE Trans. Nucl. Sci. NS **30**, 2204 (1983)

DOI 10.1007/978-94-007-7963-1_6
Reprinted from *Hyperfine Interactions* Journal, DOI 10.1007/s10751-013-0882-x

Charge state breeding of radioactive isotopes for ISAC

F. Ames · R. Baartman · P. Bricault · K. Jayamanna

Published online: 10 October 2013

Abstract For the acceleration of radioactive isotopes with a mass greater than 30 amu charge breeding with an electron cyclotron resonance ion source (ECRIS) is being used at ISAC. Singly charged ions from the target ion source combination are injected into a 14.5 GHz ECRIS from PANTECHNIK and charge bred to highly charged ions with a mass to charge ratio around 6. Efficiencies from 1 to 8 % could be reached for different isotopes. The article describes the set- up of the system and reports on results obtained for efficiency as well as purity of the beam. Methods to improve the purity are discussed.

Keywords Radioactive isotopes · Charge state breeding · Electron cyclotron resonance ion source

1 Introduction

At the ISAC facility at TRIUMF radioactive nuclides are produced by bombarding solid target materials with high energy protons from the TRIUMF cyclotron. The targets are operated at high temperature to allow the products to be released into an on line ion source (see contribution by P. Bricault et al.). Mainly singly charged ions are produced. They can be extracted and accelerated up to an energy of 60 keV, separated by their mass and injected into a post accelerator. The accelerator consist of a room temperature radio frequency quadrupole accelerator (RFQ), a room temperature drift tube section and a superconducting linac, which allow final

ISAC and ARIEL: The TRIUMF Radioactive Beam Facilities and the Scientific Program.

F. Ames (✉) · R. Baartman · P. Bricault · K. Jayamanna
TRIUMF, 4004 Wesbrook Mall, Vancouver, BC, V6T 2A3, Canada
e-mail: ames@triumf.ca

energies up to about 10 MeV/u. The RFQ section can accept ions with a mass to charge ratio up to 30 amu/e. The following drift tube and superconducting cavities require mass to charge ratios less than 7 amu/e. Therefore the ions have to pass a stripper foil after the first acceleration at an energy of 150 keV/u to increase the charge state. If ions with a mass greater than 30 amu are to be accelerated their charge state has to be increased already before entering the accelerator. In order to avoid further losses from the stripping process, which becomes less efficient for high masses, it is preferable to start with mass to charge ratios less than 7 amu/e.

The charge state of low energy ions can be increased by injecting them into an ion source for highly charged ions. Two types of ion sources are used for this purpose at different facilities: an electron beam ion source (EBIS) or an electron cyclotron resonance ion source (ECRIS). In both sources the ions are stopped and confined while more electrons are removed in collisions with high energy electrons. An overview describing the different techniques and compiling results from different facilities is given for example in [1]. At TRIUMF an ECRIS (PHOENIX booster from PANTECHNIK) has been chosen as it allows a continuous mode of operation and can handle high current throughput. Thus, it is well adapted to the accelerators and the target ion sources. An rf-field at a frequency of 14.5 GHz heats up the electrons in the source to several 100 keV, sufficient to reach high charge states even for heavy isotopes. Charge state breeding with an ECRIS has been described in more detail for example in [2].

The velocity acceptance of the RFQ requires an energy of 2.04 keV/u for the injected ions. That means with a desired charge to mass ratio below 7 amu/e both the ion source for the singly charged ions and the charge state breeder have to operate at a voltage of less than 14 kV.

2 Set-up

The charge state breeder (CSB) source is installed in a shielded room just following the mass separator for the singly charged ions. If ions with a mass higher than 30 amu are to be accelerated, an electrostatic bender can be inserted into the beam path to direct the ions into the charge breeder. The plasma chamber of the source is floated at an electrical potential close to the one of the target ion source. This allows the ions to be decelerated before they are finally stopped in the plasma. The highly charged ions are extracted at the opposite side of the source and accelerated again to ground potential. Both the injection and extraction electrodes have been changed from the original PHOENIX design to a two step configuration. This allows an easier adaptation to different operating voltages. The following combination of a magnetic dipole and two electrostatic benders acts as a Nier type spectrometer and allows the selection of the desired mass to charge ratio. The highly charged ions are then injected into the transport beam line to the accelerator. Details of the ion optics design can be found in the contribution by R. Baartman. Figure 1 shows the layout of the beam lines after the ISAC mass separator before the ions are send to the experimental hall and the accelerator.

A small surface ion source for Cs ions is located in front of the charge breeder to allow set up and commissioning independent from the operation of the ISAC target ion sources.

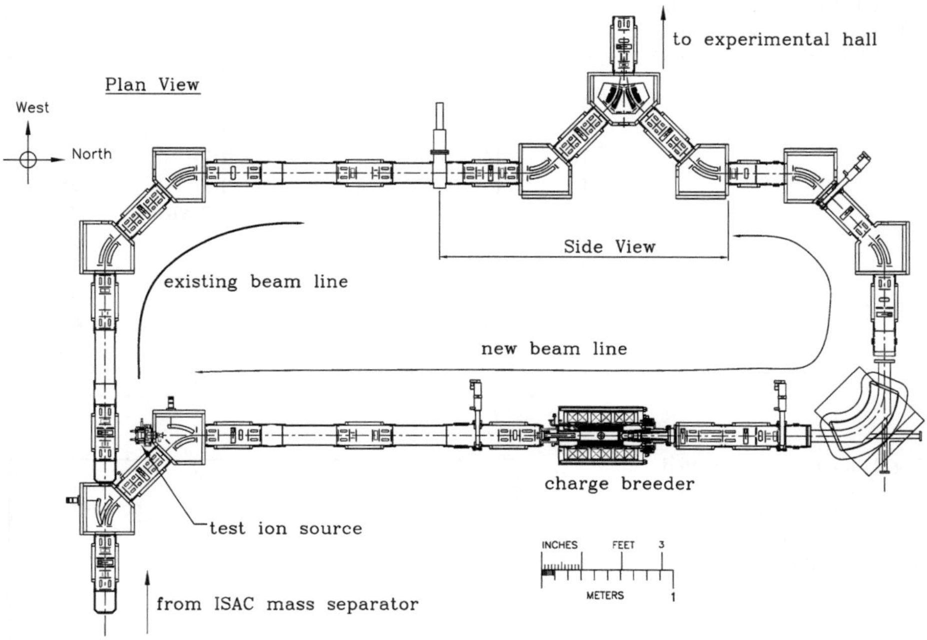

Fig. 1 Layout of the CSB beam lines in the ISAC mass separator room

3 Results

3.1 Charge breeding efficiency

During the commissioning phase, several stable and radioactive ions have been injected into the charge breeder and the charge state distribution of the highly charged ions and the breeding efficiency has been determined. Table 1 shows a list of the radioactive ions, which have been charge bred so far together with the maximum efficiency obtained. In general the charge state breeding efficiency is higher for gaseous elements than for condensible ones. If an ion hits the wall of the plasma chamber in case of a gaseous element it can be released and ionized again, whereas for condensible elements it will stick and is lost. In some cases it is beneficial to inject ions not in the atomic state but in a molecular form. This may reduce background from isobars, which do not form this specific molecule or increase the extraction efficiency out of the on line target. The method has been used for the charge breeding of ^{78}Br which has been injected as AlBr. The main isobaric contaminant ^{78}Rb does not form this molecular ion. Most of those results have been published before [3].

3.2 Beam purity

The major problem encountered while operating the charge breeder source is the background of ions from the residual gas or from sputtered material from the plasma chamber wall or the surrounding electrodes. The intensity of those ions can be several orders of magnitude higher than the intensity of the charge bred

Table 1 Charge state breeding efficiency and background levels for radioactive isotopes

Isotope	q	A/q	Efficiency [%]	I (in) [1/s]	Background [pA]
^{46}K	9	5.11	0.5	4.0E4	340
^{64}Ga	13	4.92	0.7	8.4E4	150
^{64}Ga	14	4.57	0.75	8.4E4	210
^{74}Br	14	5.28	3.1	3.2E7	10,000
^{74}Br	15	4.93	2.1	3.2E7	25
^{78}Br	14	5.57	4.5	2.8E7 AlBr	20
^{74}Kr	15	4.93	6.2	2.1E6	25
^{76}Rb	15	5.07	1.68	3.8E6	15
^{80}Rb	13	6.15	1.17	5.7E7	35
^{80}Rb	14	5.71	1.1	5.7E7	70,000
^{122}Cs	19	6.42	1.1	3.1E5	6
^{124}Cs	20	6.2	1.37	2.75E7	50

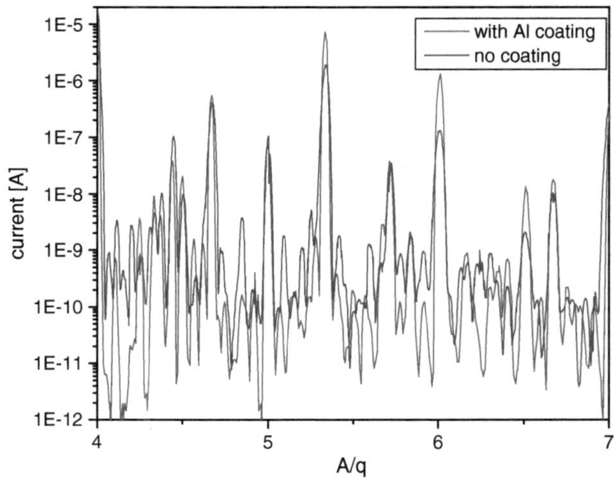

Fig. 2 Ion current extracted from the CSB as function of mass to charge ratio for an aluminum plasma chamber and with additional pure aluminum coating

radioactive ions. This presents a big challenge for the experiments performed after the post acceleration. In most cases additional filtration and cleaning techniques in the acceleration chain have to be applied. In general this is accompanied with a loss in the efficiency for the desired ions as well. In order to reduce this background from the source the original stainless steel plasma chamber of the PHOENIX source and all electrode surrounding the plasma chamber have been replaced by aluminum. Additionally, the plasma chamber has been coated with ultra clean aluminum. In Table 1 background levels from the original stainless steel chamber are given as well. In some cases charge states with a smaller breeding efficiency beside the optimal one have to be chosen to keep the total intensity below acceptable limits.

First results with the aluminum coated plasma chamber have been obtained. Figure 2 shows a comparison of mass spectra with and without the aluminum coating in the relevant mass to carge ratio range between 4 and 7. Although, the intensities of ion beams from multicharge components of the residual gas like C,O and N are still the same or higher, those from other components in between are reduced. More detailed measurements will be performed in the next few months.

References

1. Wenander, F.: Charge state breeders: on-line results. Nucl. Instrum. Meth. **B266**, 4346 (2008)
2. Lamy, T., et al.: Status of charge breeding with electron cyclotron resonance ion sources. Rev. Sci. Instrum. **77**, 03B101 (2006)
3. Ames, F., et al.: Commissioning of the ECRIS charge state breeder at TRIUMF. In: Proceedings of ECRIS2010. JACOW Conference Proceedings, Grenoble, France (2010)

DOI 10.1007/978-94-007-7963-1_7
Reprinted from *Hyperfine Interactions* Journal, DOI 10.1007/s10751-013-0883-9

ISAC LEBT

Richard Baartman

Published online: 10 October 2013

Abstract The ISAC electrostatic Low Energy Beam Transport (LEBT) system is described.

Keywords Charged particle optics · electrostatic

1 Introduction

In the isotope separation on-line (ISOL) technique, radioactive ions are created at rest and accelerated with a static potential to an energy of only a few tens of keV. At this energy, the most efficient way to transport the particles to the experimenter is with electrostatic bending and focusing elements. The reason for this can be understood from the Lorentz force $\mathbf{F} = q(\mathbf{E} + \mathbf{v} \times \mathbf{B})$ (E is the electric field and B the magnetic; v is the velocity). For $v \ll c$, and typically attainable fields, the first term is much larger than the second. Moreover, up to the sparking limit, electric fields ($\sim$10 kV/cm) are far more economical than magnetic fields (1 T). Lastly, electrostatic fields have advantages when it is required to transport ions of widely different mass and charge. Given only electric fields, all particles that have been accelerated from rest by the same potential will follow the same trajectory irrespective of mass or charge.

One is free to choose any length scale for the optics that transports and focuses the radioactive ion beam particles: widely separated quadrupoles with large apertures can in principle transport as well, with same acceptance, as small apertures and many quads, and the overall cost is little different. However, the more focusing there is

ISAC and ARIEL: The TRIUMF Radioactive Beam Facilities and the Scientific Program.

R. Baartman (✉)
TRIUMF, 4004 Wesbrook Mall, Vancouver, BC V6T 2A3, Canada
e-mail: baartman@triumf.ca

per unit length, the less effected is the beam by perturbations such as stray magnetic fields and misalignments. This favours many small quadrupoles over few large ones.

2 Standard modules

In ISAC, we chose a length scale of 1 metre as a good compromise between economics and sensitivity. By this is meant that a focusing unit, consisting of two quadrupoles, has a length of 1 metre. The Twiss β-function is thus also roughly 1 metre, and beam radius for a typical emittance of $\pi\epsilon = 10$ πmm-mrad is 3 mm. ISAC quadrupoles have length of 50 mm and aperture of 50 mm, with generally 25 mm dia. aperture ground planes at entrance and exit. This requires the electrode voltages to be ± 2 kV for a 60 keV particle energy, and yields acceptance in the 100 to 200 $\pi\,\mu$m range. Bends are also electrostatic, and their electrodes are spherical. This gives equal focusing in the bend and non-bend planes. Bend radius is 25.4 cm in all cases. There are 10 identical 90° bend sections, consisting of two 45° benders sandwiching a triplet tuned to make the section achromatic.

The periodic transport sections have cells consisting of two quads; each cell has length 1.0 m, but the quads, rather than equally separated, are arranged as doublets (see Fig. 1). This allows that a 45° spherical bender switch can be inserted into any long drift to create a spur line, for example to an experiment area. Insertion of the typical 90° bend section displayed in Fig. 1 into the periodic section requires the periodic section to reverse parity (from FODO to DOFO), since the bend section is mirror-symmetric. But this same quality allows to use two 90° bend sections in sequence if they are in opposite planes. This trick is used in bringing the ISAC beam from the separator to the main floor: from the mass separator, the beam is matched to the periodic section and needs no other matching until it arrives on the main floor, in spite of four 90° bend sections.

In every case, the beam must be matched out of the separator, through the transport system, and matched out of the transport into the experiment. Originally, the transport system consisted of only 4 types of modules: the periodic section, the 90° achromatic bend section, the low-β insertion, and the order-reversing section. The low-β section is used for the buncher of the radiofrequency accelerator (RFQ) and for the switch cross that switches between the low energy experiment area and the RFQ. The order-reversing section matches from FODO periodic section to DOFO.

The 45° benders of the bend sections can function as switches, but this requires that the outer electrode move out of the way when the bend is turned off. See Fig. 2; this is achieved with an air-actuated piston that pivots the electrode. This feature is used in a number of locations in ISAC: switching between an experiment such as Yield or GPS2, Francium, or carrying on straight ahead to another location.

Mechanical switching is adequate for hourly or daily switching, but too slow for time sharing on the scale of seconds or less. Moreover, mechanical switches do not permit co-axial overlapping of the radioactive ion beam with for example a laser beam as is needed in the optically-pumped polarizing section. For these cases, an electrostatic switch was developed by modifying a single 45° bender, splitting it into a 9° electrostatic deflector and 36° spherical bender. By combining such a 36° bender with its mirror image, a 3-way switch was built. A photograph of the 3-way bender is shown in Fig. 3, and an example application in Fig. 4.

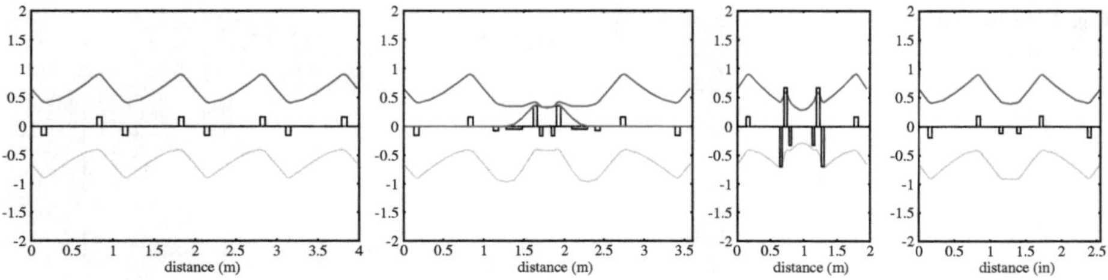

Fig. 1 *Left to right*, x and $-y$ envelopes, in cm (*red and green*), for (resp.) the periodic, the 90° bend section (*dispersion in magenta*), the low-β insertion, and the order-reversing section. Emittance is $\pi\epsilon = 50\,\pi\,\mu$m. The *stepped function in blue* is the focal strength in arbitrary units

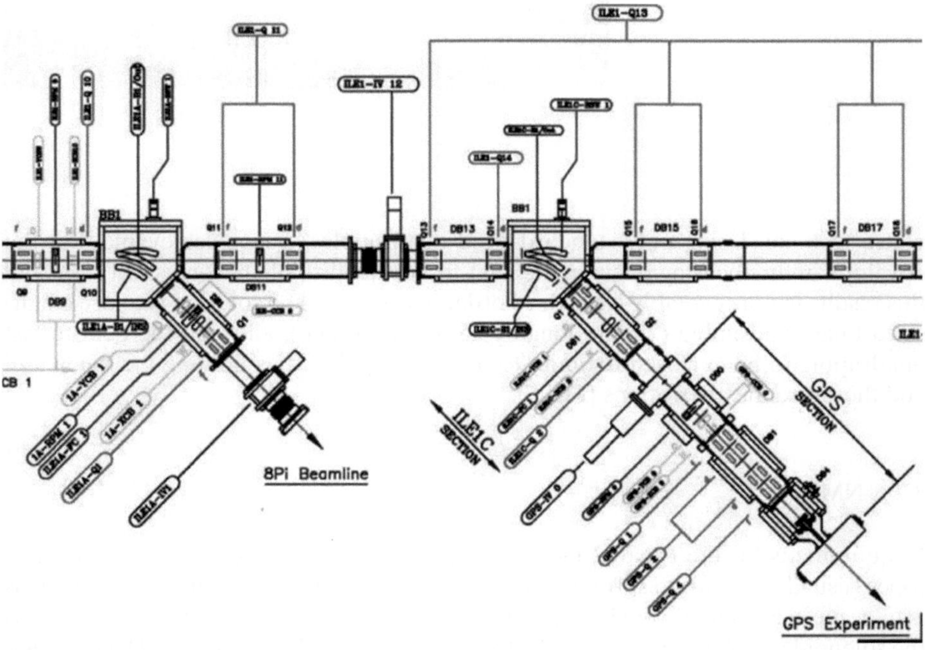

Fig. 2 ISAC periodic section, showing 5 periods, and 2 mechanical beam switches. *Quadrupoles* are *red*, *steering correctors green*, and *diagnostics blue*

3 Special sections

Each experiment has its own requirements on target. Usually, these are met with a single 4-quadrupole matching section. However, there are at least four sections with special properties.

3.1 RFQ match

Matching from the periodic section to the 150 keV/u RFQ requires an optics scale change as the RFQ is very strongly focusing: the Twiss β-function changes from 1

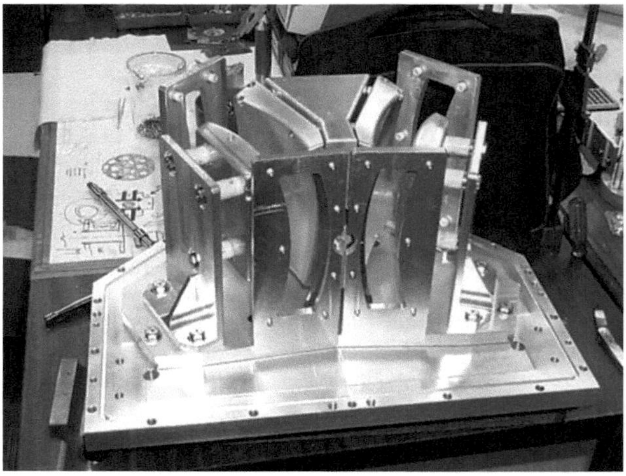

Fig. 3 Spherical 3-way bend unit, on the bench

metre to 6 cm. Using only standard-sized quadrupoles, this is achievable, but the envelope in the focusing plane in the next-to-last quadrupole becomes too large and this results in unacceptably large third order aberration. The solution is to use a half-sized final quadrupole (25 mm long by 25 mm bore dia.), allowing the next-to-last quadrupole to be in close proximity to the match point, thus limiting the beam size and therefore the aberrations (Fig. 5).

3.2 β-NMR

Requirements for the β-NMR facility are particularly challenging. Radioactive isotopes (usually ^8Li) are to soft-land at energies down to 100 eV, or up to 90 keV. This is to occur in a magnetic field of up to 9 Tesla. Einzel lenses are used for matching, as the solenoid and the deceleration electrodes are also axially symmetric. Deceleration to the lowest energies results in very strong radial electric forces which can over-focus the beam. This can be compensated with the focusing effect of the solenoid (see Fig. 6), but for example, it is not possible to focus on sample while decelerating below about 2 keV with the magnetic field off (see Fig. 7, left). However, one can make use of cycloid-shaped modulations of the beam envelope in the case where the solenoid is powered. For example, the lowest magnetic field that can be used when soft-landing at 500 eV is 1.70 Tesla (see. Fig. 7, middle).

3.3 Polarizer

The polarizer requires the radioactive ions to be neutralized and so they cannot be focused for a 2.5 m drift length of beamline. This requires a matching section that expands the beam and decreases its divergence (Fig. 8).

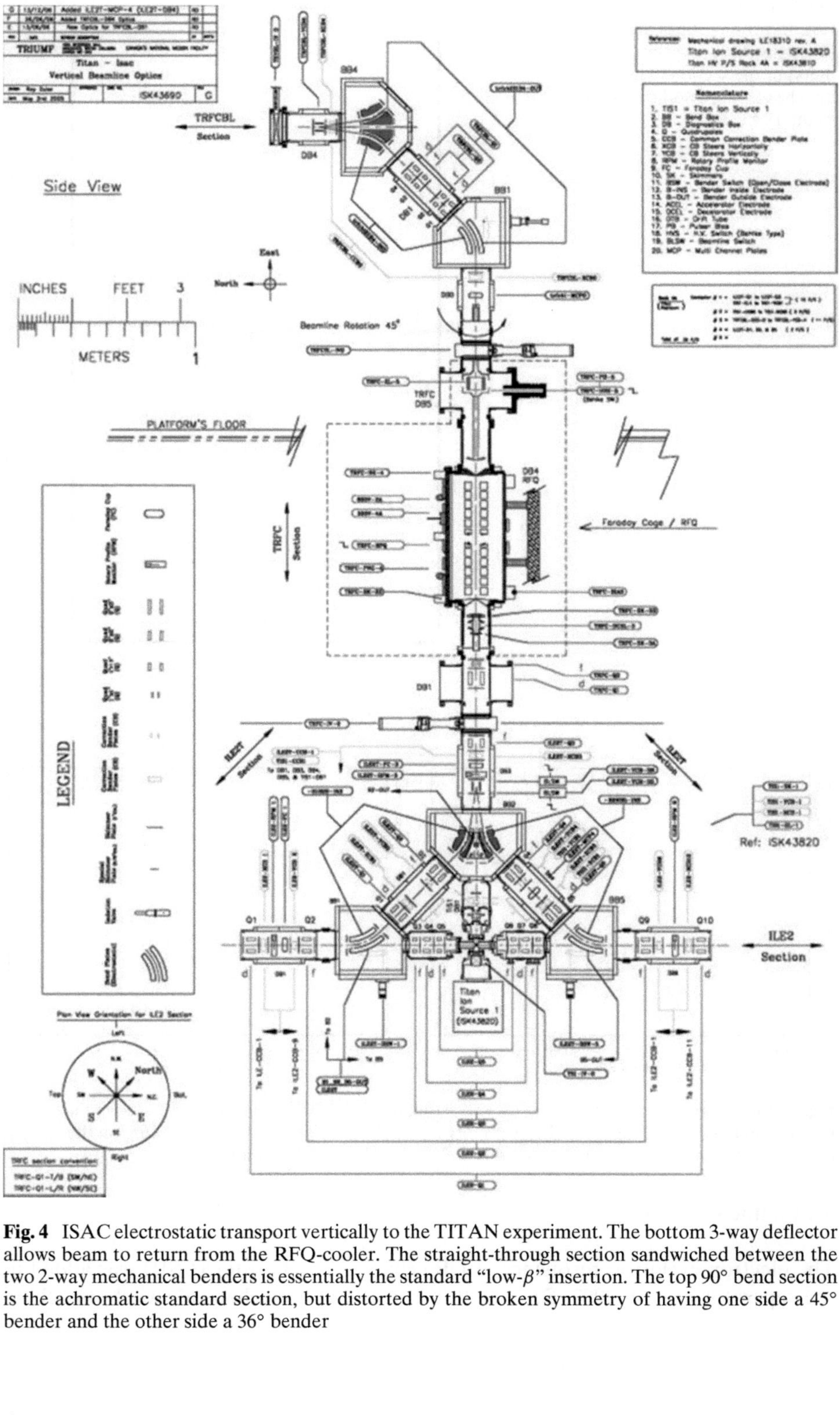

Fig. 4 ISAC electrostatic transport vertically to the TITAN experiment. The bottom 3-way deflector allows beam to return from the RFQ-cooler. The straight-through section sandwiched between the two 2-way mechanical benders is essentially the standard "low-β" insertion. The top 90° bend section is the achromatic standard section, but distorted by the broken symmetry of having one side a 45° bender and the other side a 36° bender

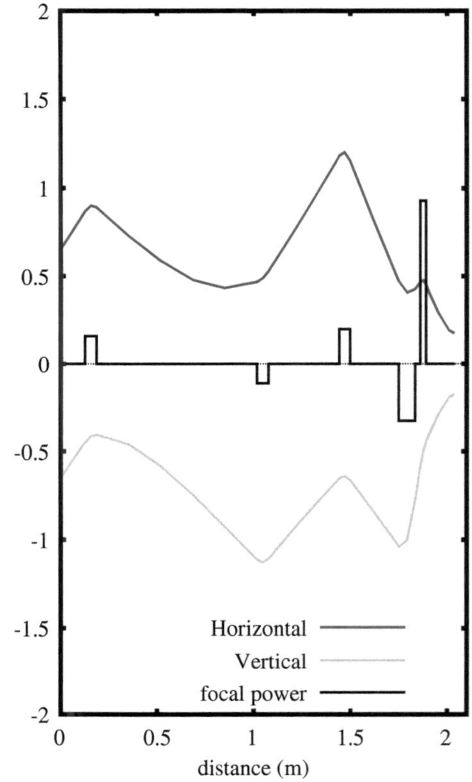

Fig. 5 Beam envelopes (in cm) for $\epsilon = 50\mu$m and focal strengths (blue, in arbitrary units) of the section that matches from periodic section to the ISAC RFQ

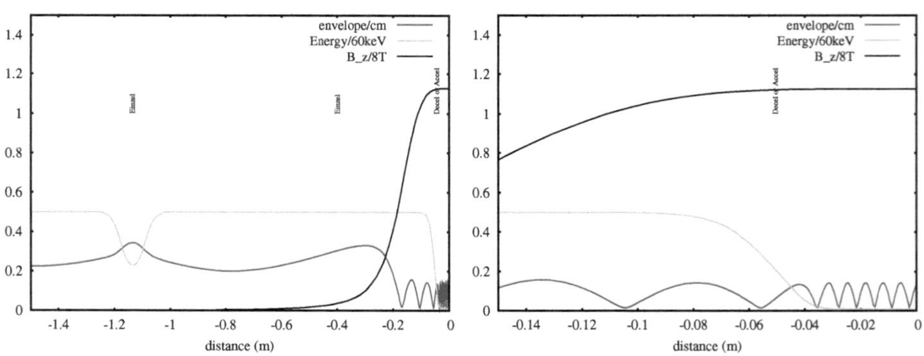

Fig. 6 Envelopes for emittance of $\pi\epsilon = 12.5\ \pi\mu$m, for the β-NMR platform; the *plot on the right* is with expanded z scale. The initial beam energy is 30 keV, decelerating for soft landing on the sample to 500 eV. The superconducting solenoid is at 9 Tesla. The sample is mounted near $z = 0$

3.4 Charge state booster

The electron cyclotron resonance (ECR) charge state booster (CSB) was added in the separator area in 2007. See Fig. 9; it consists of a mechanical switch, matching

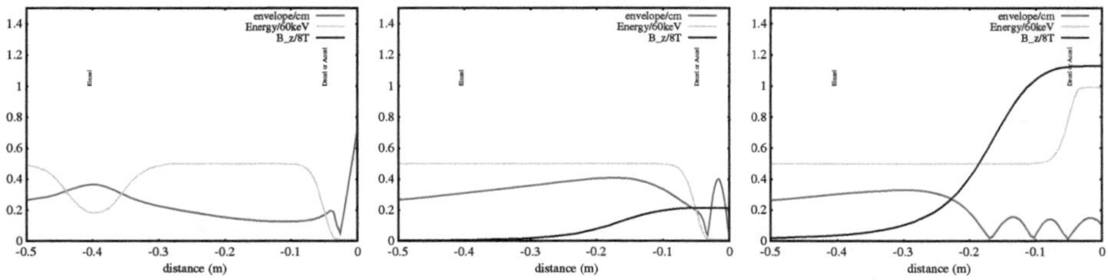

Fig. 7 Three other options for implanting onto the β-NMR sample. *Left to right*: Deceleration to 500 eV with solenoid off, deceleration to 500 eV with solenoid at 1.7 Tesla, acceleration from 30 keV to 60 keV with solenoid at 9 Tesla. *Note*: the final einzel lens at $z = -0.4$ m cannot be used when the solenoid is powered, because it initiates a Penning discharge

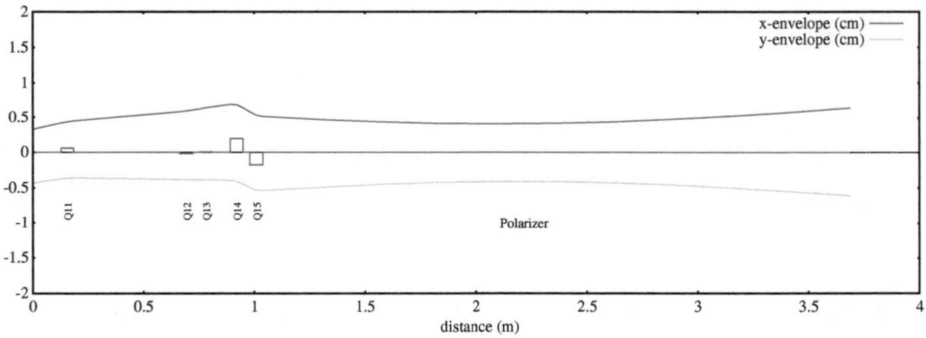

Fig. 8 Envelopes for section of beamline to match into a large, low divergence beam. $\epsilon = 12.5\ \mu$m

to decelerate the beam to soft-land into the ECR, matching out, and a Nier-type separator and finally, the bender to the vertical section was modified to allow switch between the (old) bypass, or to send beam from the CSB.

Of note is the section of beamline from the ECR to the vertical section, which contains the Nier separator. From the ECR to the magnetic dipole, the ion beam is dominated by ions of the support gas, up to 500 μA, so space charge becomes an important effect. As this is the match into the Nier separator, a technique was needed to make the match independent of space charge. This was achieved by allowing the beam at the match point just upstream of the dipole to be large for high current cases, but small for low current cases. See Fig. 10.

The Nier separator has a total bend of 180°. The principle is that the energy dispersion of the magnetic bend is precisely compensated by the energy dispersion of the electrostatic bends. Since mass dispersions are different for the two types of bends, they do not cancel, resulting in a good mass separation in spite of a comparable fractional energy spread. Ours uses two standard ISAC LEBT electrostatic spherical bends and a magnetic dipole of largest possible size in the confined space. As mass resolution scales with size, the limited space allows a resolution of only roughly 200, depending upon ECR source emittance. See Fig. 11.

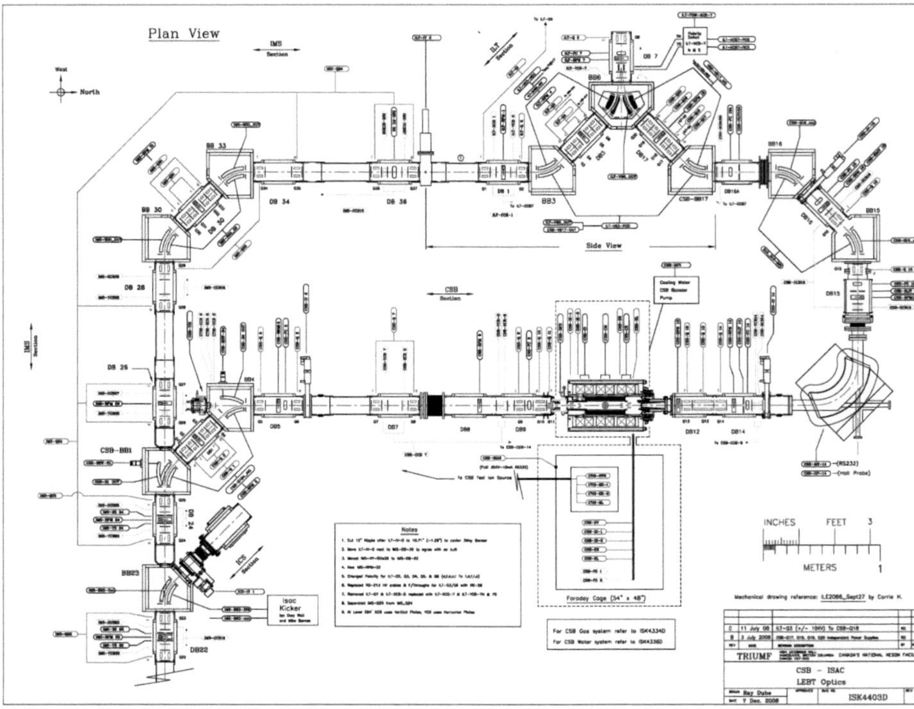

Fig. 9 Layout of the CSB. Note that most of the drawing is a plan view, but the bend system in the *upper*, *right-of-centre portion* is in a different plane, bending vertically

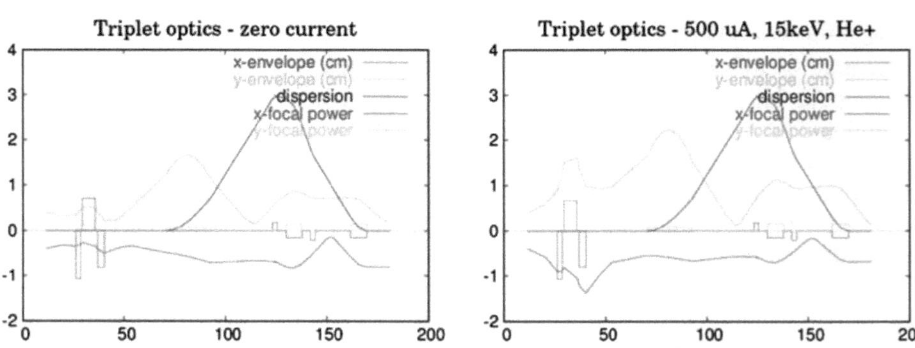

Fig. 10 Beam envelopes for two different levels of space charge, with equal settings of the quadrupoles. *Left* is for zero current, and *right* is at 500 μA

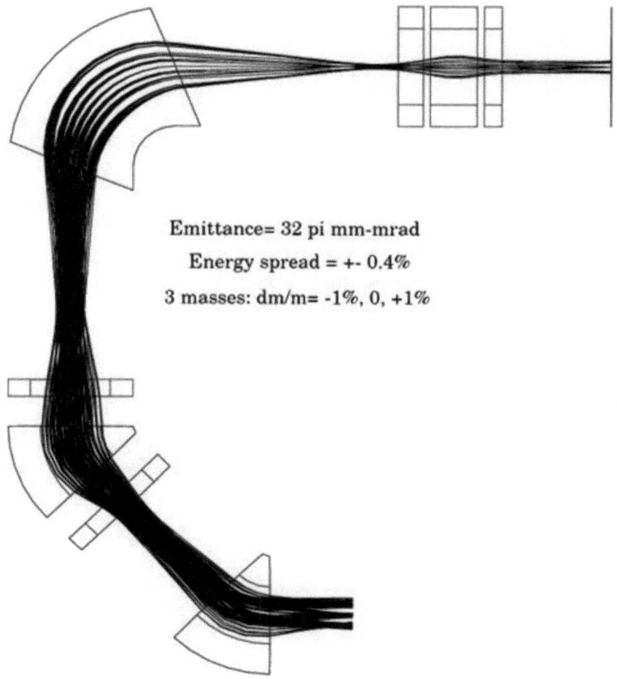

Fig. 11 Trajectories for three masses through Nier separator, as calculated to third order using GIOS, showing a clear separation at resolution of 100 for a pessimistically large emittance and energy spread. (Note that the transverse size is not to scale, but enlarged for clarity. Beam proceeds from *upper right, counterclockwise*)

DOI 10.1007/978-94-007-7963-1_8
Reprinted from *Hyperfine Interactions* Journal, DOI 10.1007/s10751-013-0884-8

The ISAC post-accelerator

R. E. Laxdal · M. Marchetto

Published online: 13 November 2013

Abstract The acceleration chain of the ISAC facility boosts the energy of both radioactive and stable light and heavy ions for beam delivery to both a medium energy area in ISAC-I and a high energy area in ISAC-II. The post-accelerator comprises a 35.4 MHz RFQ to accelerate beams of A/q $\leq$ 30 from 2 keV/u to 150 keV/u and a post stripper, 106.1 MHz variable energy drift tube linac (DTL) to accelerate ions of A/q $\leq$ 6 to a final energy between 0.15 MeV/u to 1.5 MeV/u. A 40 MV superconducting linac further accelerates beam from 1.5 MeV/u to energies above the Coulomb barrier. All linacs operate cw to preserve beam intensity.

Keywords ISAC · ISAC-I · ISAC-II · pre-buncher · RFQ · Chopper · Stripping foil · DTL · Superconducting linac · Post accelerator

1 Introduction

In the ISAC-I facility 500 MeV protons from the cyclotron at up to 100 μA impinge on one of two production targets to produce radioactive isotopes [1, 2]. The isotopes migrate in a short heated tube to an adjoining source where they are ionized and accelerated off the biased platform. Typical source geometries include a surface source and a FEBIAD source while laser ionization is used to enhance ionization of certain species. The resulting beam is mass-separated and transported in the low energy beam transport (LEBT) electrostatic beamline to either the low energy experimental area or through a series of room temperature accelerating structures (RFQ, DTL) to the ISAC-I medium energy experimental area [3]. The 35.4 MHz RFQ accelerates ions with 3 $\leq$ A/q $\leq$ 30 from 2 keV/u to 150 keV/u and the 106.1 MHz

ISAC and ARIEL: The TRIUMF Radioactive Beam Facilities and the Scientific Program.

R. E. Laxdal (✉) · M. Marchetto
Accelerator Division, TRIUMF, 4004 Wesbrook Mall, Vancouver, BC V6T 2A3, Canada
e-mail: lax@triumf.ca

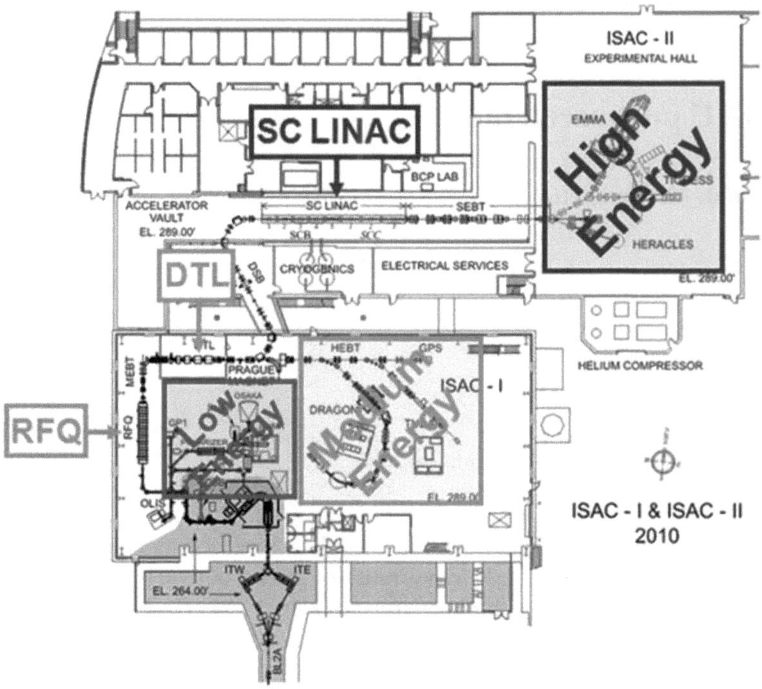

Fig. 1 Overview of the ISAC facility at TRIUMF. The ISAC II linac is superconducting while in ISAC I the RFQ and the DTL are room temperature accelerators

DTL accelerates ions with A/q $\leq$ 6 to energies fully variable from 117 keV/u to 1.8 MeV/u. To reduce the required acceleration voltage the charge state of the ions is increased by passing the beam through a thin carbon foil in the medium energy beam transport (MEBT) between the RFQ and the DTL. The accelerated beam can also be transported to the ISAC-II 40 MV Superconducting Linear Accelerator (SC-linac) for acceleration above the Coulomb barrier and delivered to the ISAC-II high energy area. The ISAC electron cyclotron resonance (ECR) charge breeder [4], CSB1, installed downstream of the ISAC mass-separator, is used to boost the charge state of masses with A > 30 to allow acceleration in the RFQ. The accelerator chain is shown in Fig. 1.

The eight meter long RFQ structure [5] (Fig. 2) is composed of nineteen split ring resonators supporting the four electrodes.The RFQ does not have a bunching section; the beam is pre-bunched at the entrance with a three harmonics RF buncher, the fundamental being 11.78 MHz. This configuration produces a high quality longitudinal emittance after the RFQ (0.5 πkeV/u·ns) and a beam time structure convenient for experiments with a period of 85 ns. The RFQ accepts 80 % of the beam with the rest of the beam unaccelerated and lost in collimators in the MEBT.

After the RFQ the ion charge state is increased by means of stripping through a thin carbon foil (4 μg/cm^2) with charge selection done in the MEBT dipoles. In order to reduce the emittance growth due to multiple scattering and energy straggling in the stripping foil, the beam is focused transversely and bunched in time. The dipoles accept ions with A/q $\leq$ 7. The efficiency of the stripping foil depends on the mass of

Fig. 2 ISAC-I radio frequency quadrupole

Fig. 3 ISAC-I drift tube linac

the stripped ions; in most of the cases it ranges between 30 % to 50 % for A/q $\leq$ 30 with most probable charge state given by $\overline{q} \simeq 0.12A + 1.5$.

The DTL [6] is a variable energy machine (Fig. 3) covering the entire range of energies between 150 keV/u $\leq$ E $\leq$ 1.5 MeV/u for A/q = 6. The DTL maximal final energy is somewhat A/q dependent where the relation $E[\mathrm{MeV/u}] \simeq -0.074\mathrm{A/q} + 1.95$ gives a reasonable fit with limits from $2 \leq \mathrm{A/q} \leq 7$. As well the lower limit can be decelerated to 117 keV/u. The DTL is a separated function machine composed of five interdigital H-mode accelerating cavities and three split ring bunchers located between the first four cavities. This layout produces good beam quality at every energy in the range. Transverse focus through the linac is provided by quadrupole triplets between each cavity. The transmission through the linac is greater than 95 %.

Fig. 4 ISAC-II superconducting linac vault final installation

The SC linac [7, 8] (Fig. 4) boosts the beam energy with an accelerating potential of up to 40 MV. The SC linac is composed of eight cryomodules. The first five cryomodules house four superconducting cavities and one superconducting solenoid. The last three cryomodules house respectively six, six and eight supercoducting cavities and each of them one superconducting solenoid. The superconducting cavities are bulk niobium quarter wave resonators at 106.08 MHz and 141.44 MHz operating at 4.4 K. The cavities are independently phased and have a broad velocity acceptance and thus acceleration can be optimized for each ion with lighter ions able to reach energies of up to 16 MeV/u. The maximum final energy capability is roughly given by $E[\mathrm{MeV/u}] \simeq 1.5 + 35\ \mathrm{q/A}$ so that all ISAC ions can be accelerated above the Coulomb barrier.

2 RFQ

The design of the RFQ is dominated by three considerations. Firstly, the low charge-to-mass ratio of the ions dictates a low operating frequency to achieve both an adequate transverse focusing and a reasonable cell length to bore ratio at injection. Secondly, continuous wave (CW) operation is required to preserve beam intensity. Thirdly, an external buncher and novel RFQ design produce a low longitudinal emittance requested by experimenters.

In a cw linac voltage considerations are dominated by the maximum rf power dissipated per unit length. In the ISAC RFQ the maximum voltage of 75 kV corresponds to a conservative Kilpatrick factor of 1.15. This corresponds to an rf power of 85 kW required or about 11 kW per meter with over 90 % of the power dissipated in the split ring resonators and electrodes.

In order to reduce the structure length, the buncher and shaper sections are completely eliminated in favour of a discrete three-harmonic saw-tooth pre-buncher located 5 m upstream. The fundamental frequency of the prebuncher is chosen at 11.8 MHz, the third sub-harmonic of the RFQ. The choice gives a reasonable spacing of 85 nsec between bunches for time of flight discrimination—useful for certain experiments. The elimination of the bunching section reduces the longitudinal emittance at the expense of the capture. The RFQ synchronous phase is fixed at

−25° through the acceleration so that the front end of the RFQ 'chops' the time modulated beam density leading to a brighter core but with an acceptance of only 80 %. The removal of the bunching section reduces the overall length by about 4 m from 12 m to 8 m.

To maintain a reasonable acceptance, the vane design has characteristic radius to pole tip $r_0 = 0.741$ cm. An exit taper was substituted by a much shorter transition cell, and a transition cell was introduced between the radial matching section (RMS) and the booster. To minimize machining costs, vanes were fabricated with a constant transverse radius of curvature of $\rho = r_0$. This leads to a significant departure (up to 35 % for ISAC) from the two term potential. The cell parameters a and m were systematically corrected to compensate for this effect using an electrostatic solver.

2.1 Structure

The low frequency of the RFQ dictates that a semi-lumped resonant structure be used to generate the required rf voltage between the electrodes. In the ISAC RFQ a post and split ring structure is used with nineteen split ring resonators distributed uniformly along the length. The structure is a variant of the 4-rod structure developed at the University of Frankfurt. In the ISAC case each end of the split ring terminates in a claw with each claw holding opposing electrodes. The electrodes are vane shaped and to symmetrize the structure the vanes are held at 45° as opposed to the traditional 0°. Magnetic quadrupoles before and after the RFQ are used to rotate the lab frame to 45° and back again to avoid emittance growth. The structure has a relatively high specific shunt impedance, good mechanical stability and eliminates unwanted even-type transmission line modes.

The vacuum tank is square in cross-section, made from mild steel with a Cu plating on the inside surface. It is split diagonally by an o-ring flange into two parts, the tank base and the tank lid, that gives full unobstructed access to the RFQ modules for ease of installation and alignment. An alignment tolerance of ±85 μm was confirmed with a bead pull down the outside of each of the electrode pairs.

2.2 Commissioning tests and performance

The RFQ was commissioned in two stages in 1998 through 2000. In 1998 an interim beam test was completed with the first 7 ring section (2.8 m) accelerating beams to 55 keV/u [9]. In 1999 the final 12 rings were added [10]. The RFQ was operated in cw mode for all beam tests. The RFQ has been operated with high reliability since 2000.

Beam capture has been measured as a function of RFQ vane voltage for several ions and for both unbunched and bunched input beams. The MEBT quadrupoles are used as a velocity filter to remove the unaccelerated beam. The results for atomic and molecular Nitrogen are given in Fig. 5 along with predicted efficiencies based on PARMTEQ [11] calculations. The RFQ capture efficiency at the nominal voltage is 80 % in the bunched case (three harmonics) and 25 % for the unbunched case in reasonable agreement with predictions. A separate measure of the timing pulse train obtained from scattering the beam in a gold foil shows that 5 % of the accelerated beam is distributed in the two 35.4 MHz side-bands. These bunches are routinely removed with the 11.8 MHz chopper in the MEBT. This means that after chopping the overall capture efficiency in the 11.8 MHz bunches will be 75 %.

Fig. 5 RFQ beam test results showing capture efficiency for beams of $^{14}N^{1+}$ (*squares*) and $^{14}N_2^{1+}$ (*circles*) as a function of relative vane voltage. The beam capture for both bunched and unbunched initial beams are recorded and are compared with PARMTEQ calculations (*dashed lines*)

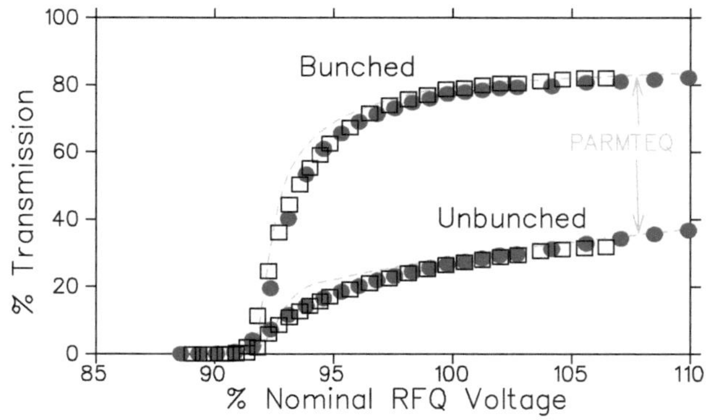

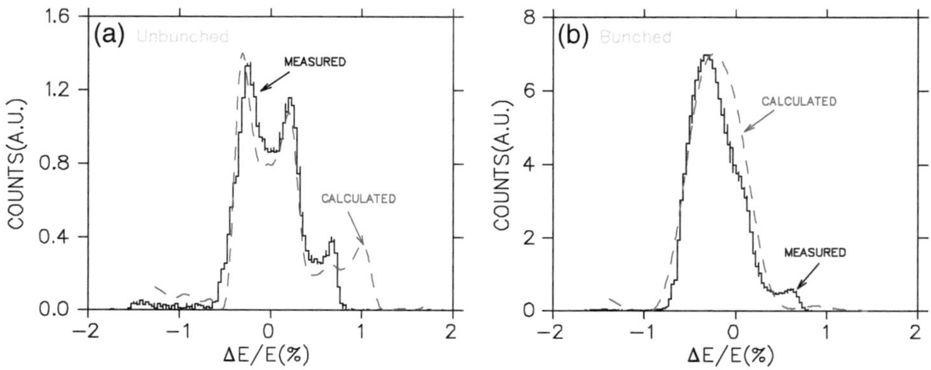

Fig. 6 Results of energy spread measurements of accelerated $^{14}N^{1+}$ beams for both **a** unbunched and **b** bunched cases. PARMTEQ simulation results are plotted for comparison

By varying the MEBT rebuncher while measuring the product of energy spread and time width at an energy or time focus gives an estimate of the longitudinal emittance.The results indicate an emittance of 0.5 πkeV/u·ns in agreement with simulations. The measured energy of 153 keV/u also is in agreement with design and energy spread values for both bunched and unbunched cases compare well with PARMTEQ simulations (Fig. 6) Transverse emittances were measured before and after the RFQ. The results show that, when the matching is optimized, the emittance growth in both planes is consistent with zero for the 7-ring configuration for an initial beam of 15 π µm. In the 19-ring test the emittance scanner was moved after the 90° bend in MEBT. In this case the emittance growth was non-zero but less than a factor of two.

The transverse and longitudinal acceptances were explored with a so-called pencil beam defined by two circular apertures of 2 mm each separated by 0.7 m placed in the RFQ injection line. One set of steering plates ('x' and 'y') was available downstream of the collimators to steer the pencil beam around the RFQ aperture. In the case of the longitudinal acceptance the energy and phase of the incident beam was varied while recording the beam transmission.

Based on the steering/transmission data the transverse acceptance was estimated to be equal to 140 π µm. The longitudinal acceptance was measured for both a

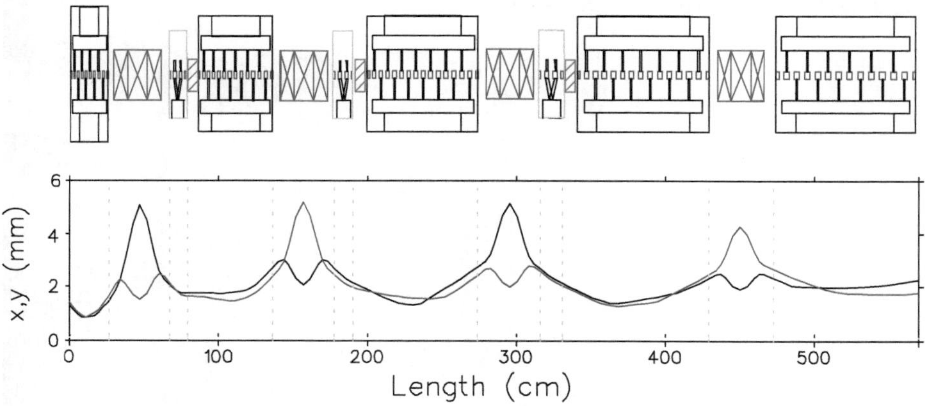

Fig. 7 Schematic drawing of the ISAC variable energy 106 MHz DTL (*upper figure*) and corresponding beam envelopes (*lower figure*). The beam envelopes define the x (*black*) and y (*red*) maximum half sizes of the beam for matched elliptical emittances of 0.25 π μm (*normalized*)

centered and an off-centered beam ($A_c = 2.7$ mm) at the nominal RFQ voltage using the pencil beam. The energy and phase settings where the acceptance dropped to 50 % of the peak value were used to define the longitudinal acceptance contour. The acceptance of the centered beam was estimated to be 180 π%·deg at 35 MHz or 0.3 πkeV/u·ns. The acceptance opens up for off-centered beams with values of 400 π%·deg at 35 MHz or 0.7 πkeV/u·ns. The expected longitudinal acceptance based on PARMTEQ simulations is 0.5 πkeV/u·ns. In general the beam test results demonstrate a strong confirmation of both the beam dynamics design and the engineering concept and realization.

3 DTL

The separated function DTL (Fig. 7) has sufficient flexibility to provide good beam quality at the design voltage settings but also at detuned settings to achieve variable energy [12].

At full voltage matched beams enter each accelerating section converging in all dimensions and after acceleration the now diverging beams are refocussed before the next accelerating section. The final energies after each tank in this mode are 0.237, 0.439, 0.781, 1.149 and 1.53 MeV/u. The bunchers are positioned downstream of Tanks 1, 2 and 3. They are tuned at a synchronous phase of $\sim -60°$ and so provide both bunching and acceleration with design energies of 0.254, 0.461 and 0.803 MeV/u after Bunchers 1, 2 and 3 respectively. In order to maximize longitudinal acceptance the intertank length is minimized. Each triplet consists of quadrupoles with effective lengths of 5.8, 8.7 and 5.8 cm with maximum gradients of 6.7, 7.5, 6.7 T/m and a total length of 37.5 cm. Horizontal and vertical steering coils are wound in the central quadrupole.

Variable energy is achieved by turning off downstream tanks to be within the energy range of the last operating tank, then reducing the voltage and adjusting the phase negatively with respect to the synchronous phase so that as the particles lose

Fig. 8 DTL Tank 1 structure

step with the synchronous particle and drift to more positive phases they gain the required energy. The upstream buncher is used to match the beam to the detuned tank. The buncher following the tank is then used to capture the diverging beam.

3.1 Structure

The accelerating tanks utilize the Interdigital H-Mode. The five IH tanks are cylindrical with lengths of 260, 500, 770, 900 and 980 mm. Two ridges, a top ridge and a bottom ridge, support the drift tubes with neighbouring drift tubes supported from the opposite ridge (Fig. 8). The tanks and ridges are fabricated from mild steel. Cooling channels are drilled in the ridge and in the tank end plates in the vicinity of the beam entrance/exit and ridge ends. Stems are fabricated from copper and cooled through two water channels drilled from the stem base to near the drift tube and fed from the ridge. The interior of the tank is copper plated with a bright acid finish to a thickness of 0.25 mm. Tuning is done through a capacitive plate with a servo-drive controlled via an rf pick-up loop.

Maximum accelerating gradients are determined by restricting the total power per unit length to less than 20 kW/m based on shunt impedance calculations with an electro-magnetic solver. Accelerating tank power levels in the five IH tanks at full voltage are 3.3, 7.8, 14, 17, 19 kW for a gradient of ~2.4 MV/m in cw mode. The drift tube aperture is significantly larger in Tank 2–5 (14, 16, 16, 16 mm) than in Tank 1 (12 mm) with a wall thickness of 5 mm in each case. The increased capacity of the larger tubes requires a lower tank radius in Tanks 2–5 (380 mm) than that of Tank 1 (460 mm) to obtain the desired resonant frequency. Ridge parameters, in particular the ridge base width and ridge length, were varied in EM-modeling to alter the inductance for frequency correction.

The bunchers for the DTL are three independent triple gap split ring resonators operating at 106.08 MHz. These were designed, developed and constructed at

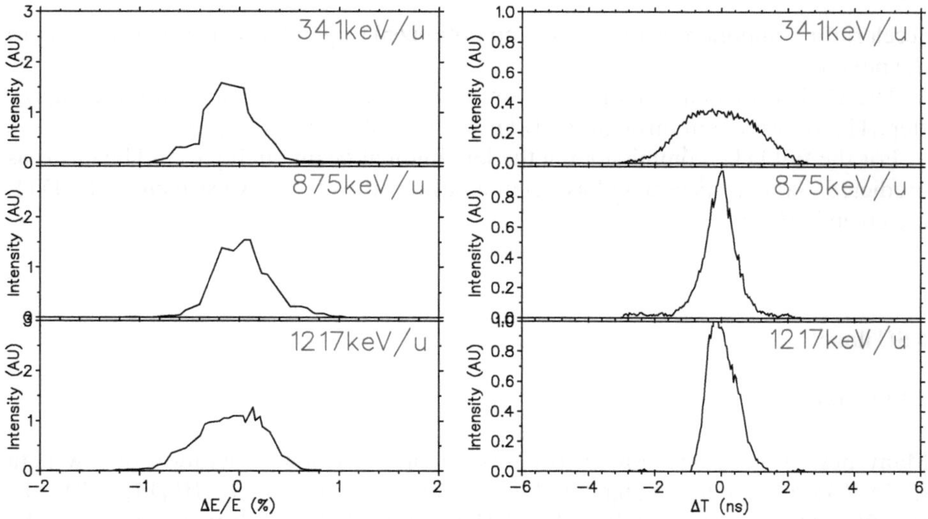

Fig. 9 A sample of energy spectra and associated pulse width measured 5 m from the DTL

INR-Troitsk and fully tested at TRIUMF. The three bunchers operate in cw mode with gap to gap distance optimized for ion velocities of $\beta = 2.3$, 2.7 and 3.3 % respectively.

3.2 Commissioning

Beam tests were completed in 2000 both for Tank 1 alone and for the full DTL [13]. The diagnostic station downstream of the DTL included Faraday cups (FC) for beam transmission measurements, a slit and harp transverse emittance rig (ε_x,ε_y), a 50Ω co-axial fast Faraday cup (FFC) for pulse width and time of flight (TOF) measurements and a 90° bending magnet with object and image slits and dispersion of 3 cm/%(Δp/p) to analyze the energy and energy spread in the beam.

The transverse emittances injected into the DTL are typically 0.06 π µm (normalized). Emittance growth is measured at ≤20 %. The longitudinal emittance is measured to be (95 % included) 0.6 πkeV/u·ns for the unstripped beam and about two to three times this for the stripped beam.

Longitudinal emittance measurements of the accelerated beam are done by varying the Buncher 1 voltage and measuring the energy and time spectra at an energy spread minimum. These spectra give emittance values of 0.5–0.6 πkeV/u·ns consistent with little or no longitudinal emittance growth over the Tank 1 energy range. Transverse emittance measurements were completed for Tank 1 operation with $^{14}N^{1+}$. The emittance increases ∼ threefold in the stripping foil with only a small increase (∼ 15 %) during Tank 1 acceleration.

Beam commissioning in Jan. 2001 with the complete DTL using $^{4}He^{1+}$ beams established the DTL rf parameters and beam optics settings for over twenty different energy set-points covering the full specified operation range. Initial phase and amplitude set points are determined empirically by beam energy measurement; each accelerator component is turned on and optimized sequentially before advancing to the next device.

The DTL triplets are adjusted to optimize transmission with each large energy step. The transmission through the DTL was over 95 % for all cases.

For the full DTL commissioning the longitudinal beam quality for all beams is as predicted. Sample spectra and associated pulse widths 5 m downstream of the DTL are given in Fig. 9.

4 SC linac

4.1 Overview

Many new projects are now in progress involving a new generation of low beta (5–15 %) superconducting light and heavy ion linacs including FRIB [14], SPIRAL-II [15], SARAF [16], IFMIF [17] and HIE-ISOLDE [18]. All these facilities take advantage of the early developments, production and operation of quarter wave resonator (QWR) niobium cavities at ATLAS [19] and later at INFN-LNL [20] and JAERI [21].

Prior to ISAC-II, cw operation was limited to peak surface fields of 20–25 MV/m. The TRIUMF ISAC-II superconducting linac is the first realization of a new generation facility with a design goal to operate at a peak surface field of 30 MV/m [22].

The TRIUMF heavy ion superconducting linac is an extension to the ISAC facility and adds ~40 MV of accelerating voltage to the existing room temperature linac capability of 1.5 MeV/u for ions with $A/q \leq 6$. The superconducting linac is composed of bulk niobium, quarter wave, rf cavities, for acceleration, and superconducting solenoids, for periodic transverse focussing, housed in several cryomodules.

The ISAC-II Phase I linac [7] was commissioned in 2006 and the Phase II upgrade [8] was commissioned in 2010. The Phase-I linac consists of twenty quarter wave cavities housed in five cryomodules (SCB1-5) with four cavities per cryomodule. The first eight cavities have a geometric beta of 5.7 % and the remainder a geometric beta of 7.1 %. The cavities operate at 106 MHz and produce an effective acceleration of 1.1 MV for a cavity power of 7 W at 4.2 K. The cavities were fabricated by Zanon in Italy. The Phase-II upgrade also consists of twenty quarter wave resonators housed this time in three cryomodules (SCC1-3), six cavities in each module for SCC1-SCC2 and eight cavities in SCC3. These bulk niobium cavities have a geometric beta of 11 % and are resonating at 141.44 MHz. One 9 T superconducting solenoid is installed in the middle of each of the eight cryomodules in close proximity to the cavities. The cavities were fabricated by PAVAC Industries of Canada.

A schematic representation of the expansion is given in Fig. 10.

4.2 Cavities

The ISAC-II superconducting cavities are shown in Fig 11. The cavities are quarter wave resonators (QWR) patterned after structures built for the low beta section of the INFN-Legnaro heavy ion linac. The cavities have a simple construction with

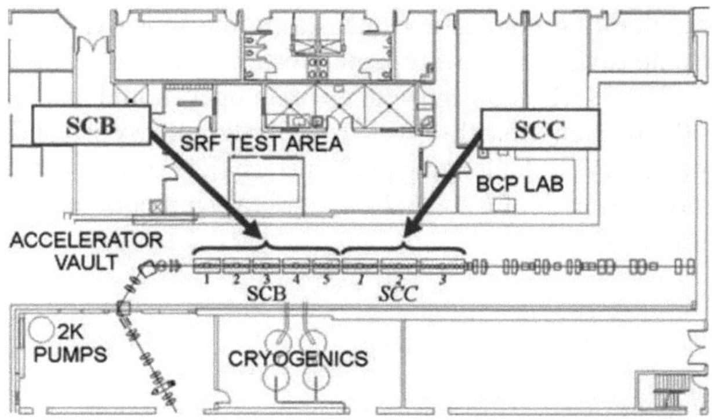

Fig. 10 Schematic representation of the installed SCB and SCC section of the ISAC-II superconducting linac

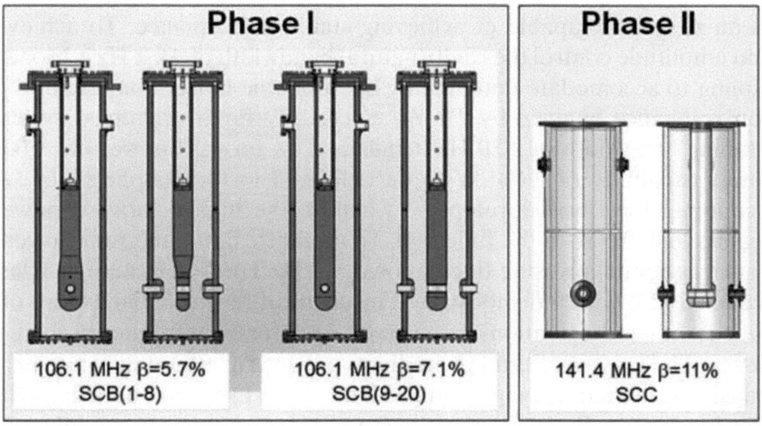

Fig. 11 Superconducting linac cavities of the SCB (*left*, Phase-I) and SCC (*right*, Phase-II)

a cylindrical shape, a rigid upper flange and an annular lower flange designed for mounting a removable tuning plate. The helium jacket is a cylinder of reactor grade niobium formed from two sheets and welded to the upper and lower flanges. A stainless steel flange is bolted to the upper niobium flange and sealed with an indium gasket to connect the cavity to the helium reservoir. A common outer conductor diameter of 180 mm is used for all cavities. The main difference between the Phase I and II cavities besides the frequency (and therefore the height) is that in Phase II the inner conductor beam port region is outfitted with a donut style drift tube. Each cavity is outfitted with a mechanical damper assembly supported from the stainless steel upper helium flange and limits microphonics to less than a few Hz rms.

Cavities are leak and pressure tested, degreased, chemically etched to remove ~100 μm from the surface then rinsed with high pressure 18 MΩ water in preparation for rf test. All cavities are characterized via cold test in a single cavity test cryostat prior to mounting in the cryomodule [23]. Characterizations involve measuring the operating frequency, measuring the cavity quality factor, Q, as a function of accelerating voltage and measuring the sensitivity to helium pressure fluctuations. Some cavities receive repeated tests depending on the initial performance.

The ISAC-II design goal is for an operational gradient of 6 MV/m across an 18 cm effective length with $P_{\mathrm{cav}} \leq 7$ W. The gradient corresponds to an acceleration voltage of 1.1 MV, a challenging peak surface field of $E_p = 30$ MV/m and $H_p = 60$ mT and a stored energy of $U_o = 3.2$ J and was a significant increase over other operating heavy ion facilities.

4.3 Towards higher gradient

Several design and hardware choices were made in an effort to reach the gradient goal. The high peak surface field demands clean rf surfaces. For simplicity of mechanical assembly a single vacuum space for cavity and thermal isolation is used but clean assembly methods and cavity rinsing are adopted. The large stored energy requires an rf system capable of achieving stable performance. To achieve stable phase and amplitude control the cavity natural bandwidth of ±0.1 Hz is broadened by overcoupling to accomodate detuning by microphonic noise. The required forward power on resonance is given by $P_f(W) \simeq \pi U_o \Delta f_{\frac{1}{2}}$ for overcoupled systems. The chosen tuning bandwidth of ±20 Hz demands a cw forward power of ~200 W and peak power capability of ~400 W to be delivered to the coupling loop. An LN_2 cooled coupling loop was developed to handle the higher forward power while releasing less than 0.5 W to the LHe [24]. To minimize detuning from slower perturbations such as helium pressure fluctuations (~ 2 Hz/Torr) a fast zero backlash tuner was developed [25] with a demonstrated mechanical response bandwidth of 30 Hz. Amplitude and phase regulation can be maintained for eigenfrequency changes of up to 60 Hz/sec. The large accelerating gradients produce a large rf defocussing. A linac lattice consisting of modules with a single high field (9 T) superconducting solenoid in the center is adopted. Beam diagnostics are positioned between modules at a waist in the beam envelope. Each diagnostic box contains a Faraday cup (FC) for beam intensity measurement and an 'x' and 'y' slit plate that gives profile information when scanned in front of the FC. The lattice is compatible with acceleration of multi-charge beams of $\Delta Q/Q \leq 7$ %. Also unique is the use of unshielded high field solenoids with added cancelling coils operating in close proximity to the cavities [26].

4.4 Cryomodule design and assembly

Each module has two main assemblies, the top assembly and the tank assembly [27]. The top assembly, shown in Fig. 12, includes the vacuum tank lid, the lid mu-metal and LN_2 shield, the cold mass and the cold mass support. All components are supported from the top plate; with the cold mass supported off a stainless steel strongback suspended from the lid through stainless steel struts.

The tank consists of the vacuum tank, the mu-metal liner and the LN_2 box insert. Both the top and bottom sub-components were assembled separately in a 'dirty assembly' area as a pre-assembly step. The sub-components were then disassembled, cleaned and delivered to the ISAC-II clean room for final assembly, alignment and testing. The tank is installed in the test pit that is serviced with an overhead crane.

Fig. 12 Cryomodule SCC1 assembly prior to the first cold test

After completion the top assembly is moved to the crane area and inserted in the tank assembly for cooldown and rf testing.

The cold mass elements are pre-aligned when warm to off-sets determined in the initial cooldowns. A Wire Position Monitor [28] system was used during initial studies to characterize the motion of the cold mass during cooldown. Optical targets inserted in the solenoid are used to align the cold mass with respect to the tank beam ports once thermalization is reached. After the final cold test the top assembly is opened to remove the targets then re-installed, pumped down and moved to the vault. Alignment in the vault is done with optical targets in the tank beam ports.

5 Cryogenics

The linac cryomodules are cooled by 4K LHe at 1.1 Bar [29, 30]. The measured static heat load for cryomodules SCB1–5 is 13 W each with 15W for SCC1–2 and 18W for SCC3. Each cryomodule consumes about 5 l/h of LN_2. ISAC-II is outfitted with two Linde TC-50 cold boxes each with a demonstrated refrigeration power of ~600 W. The demonstrated liquefaction rate for a single cold box based on a rising dewar level is 225 l/h. During normal operation the two cold boxes operate independent from each other. Each cold box produces LHe into one of two 1000 liter dewars. Cold helium distribution lines supplied by DEMACO deliver LHe from the dewars to a main 'trunk' manifold line separated by manual valves into Phase I and Phase II sections. The trunk line delivers LHe in parallel to each cryomodule from this helium supply 'trunk' line through variable supply valves and field joints. The cold return from the cryomodules comes back to the trunk cold return line through open/close valves and field joints. In particular the Phase I system delivers LHe to SCB1–3 and to the SRF development areas and the Phase II system delivers LHe to SCB4–5 and SCC1–3. In non-standard operation one system can cool the whole linac while the other one is serviced. During cooldown, when warmer than 20°K, the returning gas is sent back to the suction side of the compressor through 'warm return' piping and in-line vaporizers. Keep-cold sections with proportional valves join the trunk supply and the trunk cold return at each end. All supply and cold return piping is vacuum jacketed and except for the short feed line from supply valve to cryomodule is cooled with LN_2.

The estimated static loads on the distribution system are 140 W total including the dewar to trunk lines and trunk (70 W each for Phase I and Phase II), 12 W per stinger pair for each cryomodule giving a total of 145 W for Phase I and 206 W in Phase II. The active load for each cavity is 8 W.

5.1 Cooldown

The cavities are first baked at ~90° C for 48 h. LN_2 is then fed through the side-shields and the cold mass is cooled by radiation for 24 h to bring the average temperature to about 240 K before helium transfer.

Linac cooldown is done sequentially, one cryomodule at a time, to achieve a cavity cooling rate of ~100 K/h to mitigate the effects of Q-disease. This requires a LHe flow of ~100–150 l/h. It takes about 5 h to establish a 120 l inventory in the cryomodule and roughly 24 h to complete the bulk of the thermalization. A full cooldown takes a minimum of ten days with two days for the cold box, dewar and trunk line and one day each for the cryomodules. After each module is filled it remains at level even as the warm modules are cooled. A level probe in the cryomodule helium reservoir is used to regulate the variable supply valve during operation. An immersion heater in the main dewar is used to regulate the dewar level. With these two regulation systems the helium levels are maintained independent of the rf active load. Before linac tuning commences a refrigerator mode is selected to allow enough headroom in the dewar heater to accept the planned active load. As the cavities are turned on the dewar heater value decreases to maintain the dewar level.

5.2 Measurements

The rf cavities are initially conditioned with low and high power pulses to optimize performance where required. The cavities are set to the power limit of 7 W per cavity at critical coupling. The coupler is then moved to a position requiring a forward power of ~160 W for a coupling $\beta \sim 100$ that has been determined to provide sufficient rf bandwidth to maintain lock. The cavities are initially locked and left for 24 hours to test the operational stability and tuner performance.

Each cavity is turned on starting at the upstream end. The cavities are phased by measuring the beam energy for five different phases and fitting the data to a cosine profile to find 0°. The phasing detector consists of a gold foil to scatter ions and a silicon detector. All cavities are set to a sychronous phase of −25° for acceleration. The focusing solenoids and beamline optics are set to their theoretical settings as the acceleration progresses.

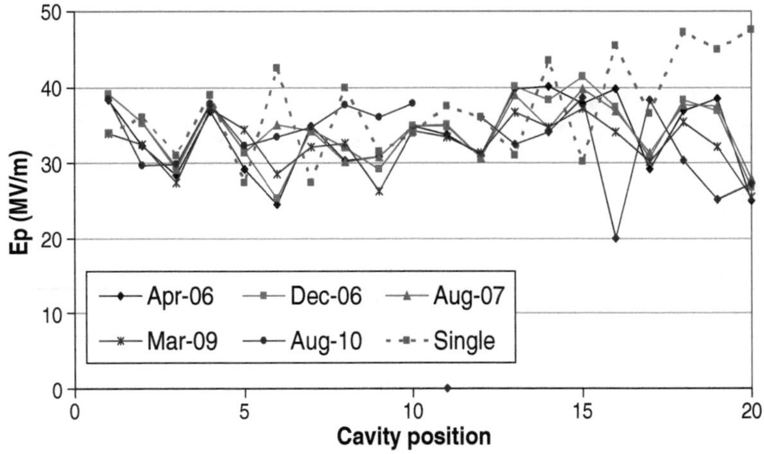

Fig. 13 SCB cavity performance over the first four years of operation as indicated by associated peak electric field, E_p at 7 W cavity power. *Dashed line* shows initial single cavity test results

5.3 Installed performance

During the initial Phase II commissioning a $^{16}O_{5+}$ beam (A/q = 3.2) from the ISAC off-line (stable) source was accelerated to 10.8 MeV/u. The final energy is equivalent to acceleration of a beam with A/q = 6 to 6.5MeV/u and as such demonstrates the ISAC-II goals on the first acceleration. This was achieved despite the fact that five cavities were not available for acceleration.

5.3.1 Phase I performance

The performance of the Phase I SCB cavities is monitored periodically typically during start-up following a shutdown. The linac is warmed up once per year for three months as part of the site maintenance shutdown. In addition several of the cryomodules have been vented (pump replacement) and some have been taken off line for disassembly to repair internal faults (rf cable, coupling loop faults). For characterization, cavity voltages are measured at critical coupling and with the design cavity power of 7 W. Figure 13 summarizes the performance over the first four years of operation for all twenty cavities. There is some fluctuation in individual cavity performance over time but the average values are consistent: the average gradients correspond to peak surface fields of 33.6, 34.2, 34.4, 32.5, and 33.2 MV/m chronologically for the five dates plotted in the figure spanning from April 2006 to Aug. 2010. For comparison the initial single cavity test results are plotted for the same conditions with an average performance of E_p = 37.1 MV/m at 7 W.

5.3.2 Phase II performance

The cavity performance on-line is summarized in Fig. 14 [31]. Shown are single cavity test results at 7 W (blue) and cavity performance on-line (red). Data taken from beam energy gain measurements during the first acceleration are also plotted (green). The average performance corresponds to peak surface fields of 32 MV/m at

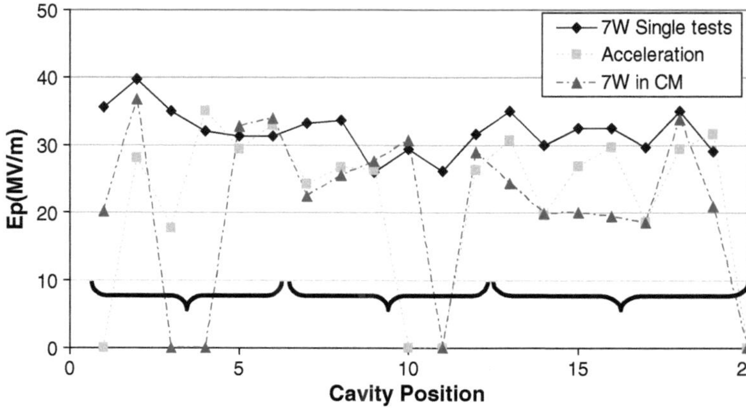

Fig. 14 Phase II (SCC) cavity performance. Shown are single cavity test results at 7 W (*blue*), performance during first acceleration at arbitrary power and cavity performance in situ (*red*). Brackets indicate SCC1, 2 and 3 range

7 W in single cavity tests, 26 MV/m for online tests at 7 W and 27 MV/m during first acceleration with higher cryogenics load. Brackets indicate the cryomodule SCC1, 2 or 3 where the cavity is mounted. In each of the on-line data sets four cavities are off. In the initial acceleration series these were attributed to multipacting and controls issues. Two of these cavities recovered during commissioning and two were found to have shorted cables internal to the cryomodule.

5.3.3 Beam quality

The transmission is ~100 % and the tuning is straightforward. Beam emittance measurements are taken with a standard slit and harp device. A sample measured with the Phase I linac is shown in Fig. 15. Here the vertical and horizontal emittance of a $^4He_{1+}$ beam is displayed for energies of 2.7 MeV/u and 6.8 MeV/u. The energies correspond to acceleration with the first cryomodule and with five cryomodules respectively. The measured vertical emittances of 2.3 π μm and 1.6 π μm correspond to normalized emittances of 0.17 π μm and 0.19 π μm respectively while the measured horizontal emittances of 2.3 π μm and 1.0 π μm correspond to normalized emittances of 0.17 π μm and 0.12 π μm respectively. Measurements from the Phase II linac agree with these values with all normalized emittances $\leq$0.2 π μm.

The longitudinal emittance is estimated using three time of flight monitors located downstream of the linac. This technique measures the time spread of the drifting beam at the three different locations. Due to the marginal separation between the monitors the measurement method is not very precise. Nevertheless the measurments indicate a longitudinal emittance of the beam that is around 1 keV/u·ns.

There is presently no buncher in the high energy beamline. However we have successfully demonstrated that for experiments that do not require the full energy, and hence have only a limited number of cavities 'on' starting from the linac's upstream end, a downstream cavity, normally 'off', can be tuned to manipulate the longitudinal phase space to provide either time focussed or energy focussed beams at the experiment [32]. Figure 16 shows the case with cavities 1–8 'on' and cavity 19 used as a buncher to provide a time focus at the downstream silicon detector.

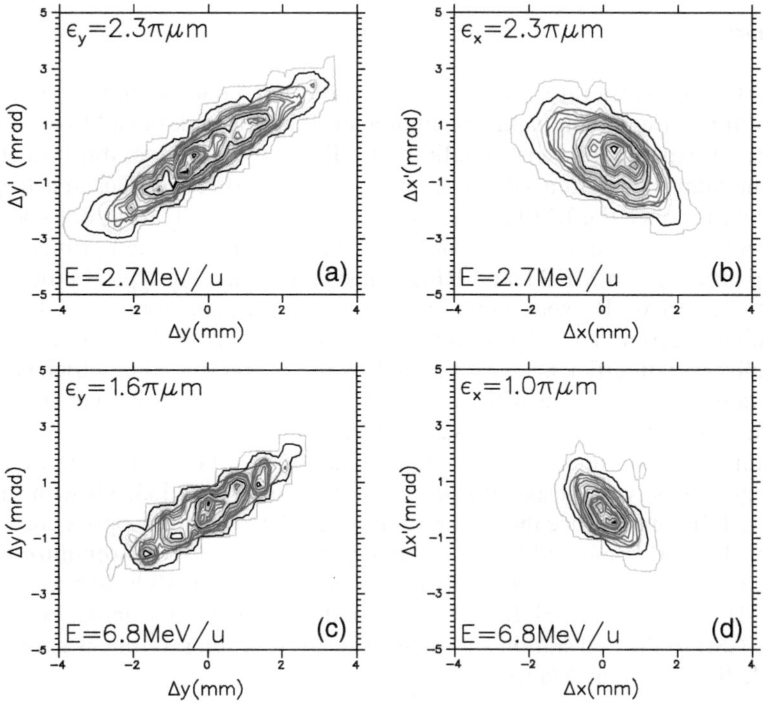

Fig. 15 Measured transverse emittance for a $^{4}He^{1+}$ beam at two different energies: after the first cryomodule at 2.7 MeV/u (*top*) and the fifth at 6.8 MeV/u (*bottom*)

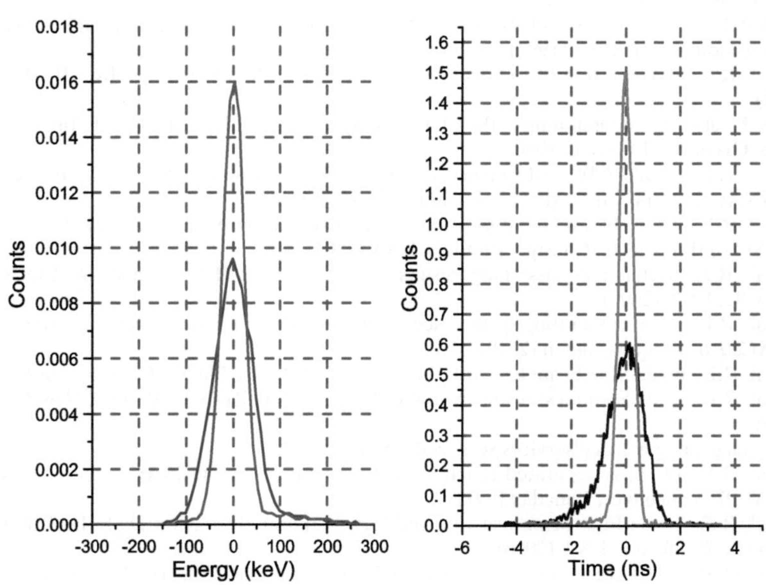

Fig. 16 Resulting longitudinal profiles after using cavity 19 to bunch in energy (*left*) and time (*right*)

6 Summary

The ISAC post-accelerator is a flexible heavy ion linac combining both room temperature and superconducting technologies. It has been designed to focus on two key areas of experimental investigation. The ISAC room temperature accelerators serve a medium energy area with full energy variability to key on nuclear astrophysics investigations. Since the RFQ can accelerate up to mass $A = 30$ astrophysics experiments with $A \leq 30$ can receive beam with high efficiency defined primarily by the stripping efficiency in MEBT and RFQ transmission yielding typical efficiencies of 20–40 %. The ISAC-II experimental program concentrates on nuclear physics near the Coulomb barrier. The 40 MV superconducting linac provides extremely flexible beam delivery with efficiencies dominated by the charge breeding efficiency in the charge state booster for ions with $A > 30$. A second stripping foil after the DTL can be added to achieve higher energies at slightly lower efficiency.

The tuning of the linac for low intensity radioactive beams is done first with a stable high intensity 'pilot' beam from an off-line ion source [33, 34] with the same A/q as the RIB beam. Once the tune is established the low energy optics are switched to accept the RIB beam and low intensity diagnostics are used to optimize the RIB yield at the end station. Recently high mass beams from the CSB have been delivered to ISAC-II experiments [35]. In this case a 'buddy beam' with a similar A/q is tuned from the CSB to the end station and then an accurate scaling routine is used to move the whole linac chain in A/q to the RIB setting.

References

1. Laxdal, R.E., et al.: RNB post-accelerator for ISAC at TRIUMF—present and future. Nucl. Phys. A **701**, 647c–650c (2002)
2. Schmor, P.W., et al.: Status of the TRIUMF ISAC-facility for accelerating radioactive beams. In: Proceedings of PAC97 (1997)
3. Laxdal, R.E.: Acceleration of radioactive ions. Nucl. Inst. Methods Phys. Res. B **204**, 400–409 (2003)
4. Ames, F., et al.: Commissioning of the ECRIS charge state breeder at TRIUMF. In: ECRIS 2010, p. 178. Grenoble, France (2010)
5. Poirier, R.L., et al.: CW performance of the TRIUMF 8 meter long RFQ for exotic ions. In: LINAC 2000, pp. 1023. Monterey, California, USA (2000)
6. Laxdal, R.E., et al.: Beam commissioning and first operation of the ISAC DTL at TRIUMF. In: PAC 2001, pp. 3942. Chicago, Illinoois, USA (2001)
7. Laxdal, R.E.: Initial commissioning results from the ISAC-II SC Linac. In: LINAC 2006. Knoxville, USA (2006)
8. Laxdal, R.E., et al. Operating experience of the 20 MV upgrade Linac. In: Proceedings of LINAC2010. Tsukuba, Japan (2010)
9. Laxdal, R.E., et al.: First beam test with the ISAC RFQ. In: LINAC98. Chicago, USA (1998)
10. Laxdal, R.E., et al.: Beam test results with the ISAC 35 MHz RFQ. In: PAC99. New York, USA (1999)
11. http://laacg.lanl.gov/laacg/services/serv_codes.phtml
12. Laxdal, R.E., et al.: A separated function drift-tube Linac for the ISAC project at TRIUMF. In: PAC97. Vancouver, Canada (1997)
13. Laxdal, R.E., et al.: First beam test with the ISAC separated function DTL. In: LINAC 2000. Monterey, California, USA (2000)
14. Wie, J., et al.: FRIB accelerator status and challenges. In: Linac 2012. Tel Aviv, Israel (2012)
15. Bernaudin, P.E.: Status of the SPIRAL-II superconducting LINAC. In: IPAC 2010. Kyoto, Japan (2010)

16. Berkovits, D.: Operational experience and future goals of the SARAF Proton/Deuteron Linac. In: LINAC 2012. Tel Aviv, Israel (2012)
17. Mosnier, A.: The IFMIF 5 MW Linacs. In: LINAC 2008. Victoria, Canada (2008)
18. Kadi, Y.: Status and future perspectives of the HIE-ISOLDE project at CERN. In: IPAC 2012. New Orleans, USA (2012)
19. Bollinger, L.M.: The Argonne Tandem-Linac accelerator system. IEEE Trans. Nucl. Sci. **NS-30**, 2065 (1983)
20. Fortuna, G., et al.: Status of ALPI and related developments of superconducting structures. In: LINAC96. Geneva, Switzerland (1996)
21. Takeuchi, S., Matsuda, M.: Status of the superconducting heavy-ion tandem-booster Linac at Jaeri. In: LINAC98. Chicago, USA (1998)
22. Laxdal, R.E.: Commissioning and early experiments with ISACII. In: PAC2007, pp. 2593. Albuquerque, New Mexico, USA (2007)
23. Laxdal, R.E., et al.: ISAC-II QWR cavity characterizations and investigations. Physica C **441**, 193–196 (2006)
24. Poirier, R., et al.: RF coupler design for the TRIUMF ISAC-II superconducting quarter wave resonators. In: LINAC 2004. Lubeck, Germany (2004)
25. Ries, T., et al.: A mechanical tuner for the ISAC II quarter wave superconducting cavities. IOn: PAC 2003. Portland, USA (2003)
26. Laxdal, R.E., et al.: Magnetic field studies in the ISAC-II cryomodule. Physica C **441**, 225–228 (2006)
27. Stanford, G., et al.: Engineering and cryogenic testing of the ISAC-II medium beta cryomodule. In: LINAC 2004. Lubeck, Germany (2004)
28. Rawnsley, W., et al.: A wire position monitor system for the ISAC-II cryomodule components alignment. In: LINAC 2004. Lubeck, Germany (2004)
29. Laxdal, R.E., et al.: Performance and early operating experience with the ISAC-II cryogenic system. In: LINAC 2006. Knoxville, USA (2006)
30. Sekachev, I., et al.: Recent operating experience for the ISAC-II SC-Linac cryogenic system at TRIUMF. In: Advances in Cryogenic Engineering: Transactions of the Cryogenic Engineering Conference—CEC, vol. 52. AIP Conference Proceedings, vol. 985, pp. 1580–1585 (2008)
31. Longuevergne, D., et al.: RF cavity performance in the ISAC-II superconducting heavy ion Linac. Tsukuba, Japan (2010)
32. Marchetto, M., et al.: Beam dynamics study on the ISAC-II superconducting Linac. In: LINAC 2006. Knoxville, USA (2006)
33. Jayamanna, K., et al.: Off-line ion source terminal for ISAC at TRIUMF. Rev. Sci. Instrum. **79**, 02C711 (2008)
34. Jayamanna, K., et al.: A multicharge ion source (Supernanogan) for the OLIS facility at ISAC/TRIUMF. Rev. Sci. Instrum. **81**(2), 02A331 (2010)
35. Marchetto, M., et al.: In flight ion separation using a Linac chain. In: LINAC 2012. Tel-Aviv, ISRAEL (2012)

DOI 10.1007/978-94-007-7963-1_9
Reprinted from *Hyperfine Interactions* Journal, DOI 10.1007/s10751-013-0885-7

High energy beam lines

M. Marchetto · R. E. Laxdal

Published online: 10 October 2013

Abstract The ISAC post accelerator comprises an RFQ, DTL and SC-linac. The high energy beam lines connect the linear accelerators as well as deliver the accelerated beams to two different experimental areas. The medium energy beam transport (MEBT) line connects the RFQ to the DTL. The high energy beam transport (HEBT) line connects the DTL to the ISAC-I experimental stations (DRAGON, TUDA-I, GPS). The DTL to superconducting beam (DSB) transport line connects the ISAC-I and ISAC-II linacs. The superconducting energy beam transport (SEBT) line connects the SC linac to the ISAC-II experimental station (TUDA-II, HERACLES, TIGRESS, EMMA and GPS). All these lines have the function of transporting and matching the beams to the downstream sections by manipulating the transverse and longitudinal phase space. They also contain diagnostic devices to measure the beam properties.

Keywords ISAC · ISAC-I · ISAC-II · MEBT · HEBT · DSB · SEBT · Pre-buncher · Chopper · Rotator · Stripping foil · Rebuncher

1 Introduction

The ISAC post-accelerator comprises a 35.4 MHz RF quadrupole (RFQ) [1] to accelerate beams of $A/q \leq 30$ from the source energy of 2 keV/u to 150 keV/u and a post stripper, 106.1 MHz variable energy drift tube linac (DTL) [2] to accelerate ions of $A/q \leq 7$ to a final energy between 0.117 MeV/u to 1.8 MeV/u. A 40 MV superconducting linac (SC-Linac) [3] further accelerates beam from 1.5 MeV/u to energies above the Coulomb barrier. All linacs operate cw to preserve beam

ISAC and ARIEL: The TRIUMF Radioactive Beam Facilities and the Scientific Program.

M. Marchetto (✉) · R. E. Laxdal
TRIUMF, 4004 Wesbrook Mall, Vancouver, BC V6T 2A3, Canada
e-mail: marco@triumf.ca

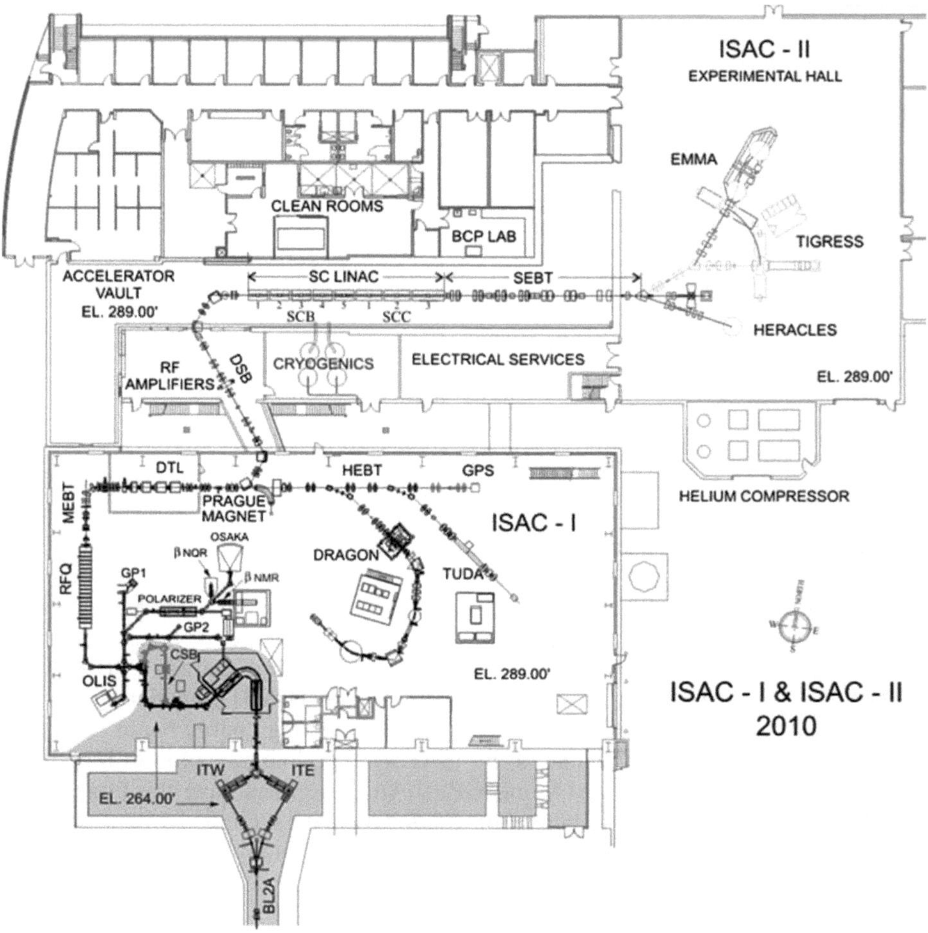

Fig. 1 ISAC facility overview

intensity. The continuous beam from the source(s) is prebunched before the RFQ at 11.8 MHz the third sub-harmonic of the RFQ.

The ISAC high energy beam lines connect the accelerators to each other plus deliver the beam to the experimental end stations (Fig. 1). The accelerators were installed over a period of time and the names of each section are historic. The RFQ and DTL are joined by the medium energy beam transport (MEBT) section. The RFQ and DTL are used in tandem to produce beams for the medium energy experimental area. The beam line between the exit of the DTL and the medium energy experimental stations (DRAGON and TUDA-I) is called the High Energy Beam Transport (HEBT).

The RFQ/DTL tandem is also used to produce beams at 1.5 MeV/u for injection into the SC-linac in ISAC-II. The beam is delivered to ISAC-II using an S-shaped transfer line called the DTL to superconducting beam transport line (DSB). The beam line between the SC-linac and the high energy experimental end stations

(TIGRESS, TUDA-II and EMMA) is called the superconducting energy beam transport (SEBT).

The ISAC high energy transport lines are composed of magnetic optical elements, specialized RF devices and diagnostic elements. Standard magnetic optical elements in the high energy beam lines are quadrupole lenses to focus the beam, magnetic dipoles to bend the beam and magnetic correctors (steerers) to adjust the beam angle in order to compensate for misalignment of the beam line elements. Diagnostic boxes are interspersed throughout the lines where required and are joined by evacuated beam tubes. Diagnostic elements include Faraday cups to measure beam intensities with sensitivities down to a few pico-Amperes, low intensity diagnostics (see below), profile monitors to monitor the shape and position of the beam and timing diagnostics to monitor the longitudinal characteristics of the beam bunches.

2 MEBT

The primary aim of the MEBT is to provide transport to a stripping foil to raise the charge state of the ion and then to prepare the beam for injection into the DTL. Increasing the charge state reduces the voltage required in the downstream accelerators at the expense of an emittance increase and a reduction in total beam intensity since multiple charge states are produced but only one is selected. The MEBT section is comprised of three sections; the matching section to the stripping foil, the charge selection section and the matching section to the DTL (see Fig. 2).

2.1 Matching to stripping foil

The ion charge is increased by stripping the beam through a 4 μg/cm^2 carbon foil. A carousel with ~30 foils is loaded periodically. Foil lifetime is mass dependent but typically is of the order of $200nA \cdot hours$. The beam is brought to both a spatial focus and a time focus at the stripping foil to reduce the emittance increase due to multiple scattering and energy straggling respectively. The spatial focus is produced using a set of five quadrupoles. The five quadrupoles (Q1–5) are skewed by 45 degrees with respect to the lab frame to accommodate the RFQ vane rotated orientation. A round beam image is created on the stripping foil to allow switching the downstream optics to the lab frame without emittance growth. A triple gap split ring buncher [4] (bunch rotator) positioned between Q2 and Q3 is used to produce a time focus (typically 1 ns) at the stripping foil. The 106 MHz device is the slightly modified prototype buncher for the DTL. The bunch rotator is tuned by using a coaxial 50Ω fast Faraday cup (FFC).

The multi-harmonic pre-buncher operates at the third sub-harmonic of the RFQ. This scheme generates a time spectrum at the exit of the RFQ with two main bunches 85 ns apart and two small satellites bunches in between the two main bunches each spaced by 28 ns.

Two choppers [5] are installed between Q3–Q4 to clean the time spectrum upon request. An 11.78 MHz chopper removes the satellite bunches to give 85 ns between bunches with a 5 % reduction in beam intensity. A 5.89 MHz chopper can eliminate every second main bunch to produce 170 ns between bunches with a 53 % reduction in beam intensity. The chopper consists of a series of two sets of plates located

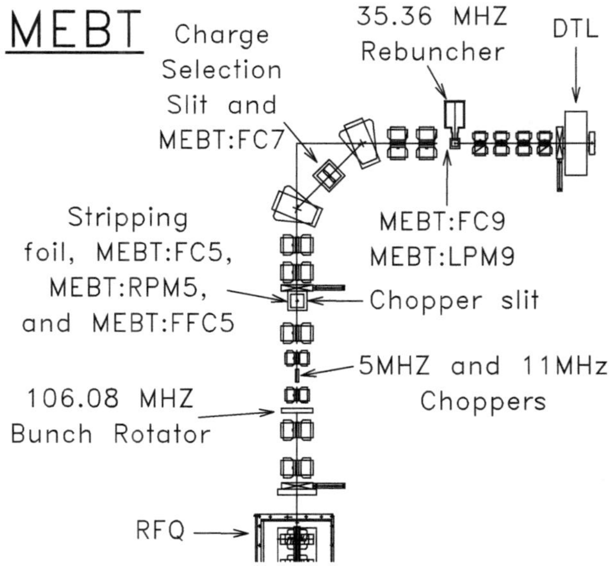

Fig. 2 Medium energy beam transport (MEBT) line

where horizontal divergence has been minimized followed by selecting slits near the stripping foil $\sim 90°$ phase advance downstream. Each plate pair has one plate driven at RF voltage (11.8 MHz and 5.9 MHz respectively) and the other compensating plate is DC biased to produce zero deflecting field at the base of the RF waveform to reduce transverse emittance growth. In the first mode the two 35.4 MHz side-bands in the pulse structure (Fig. 3d) are deflected at 11.8 MHz (Fig. 3a) yielding the time structure shown in Fig. 3e . In the second mode the side-bands plus every second main pulse are deflected by adding the $5.9MHz$ deflection (Fig. 3b) from the second set of plates to yield the combined deflecting field shown in Fig. 3c giving the time structure measured in Fig. 3f.

Two lumped circuits drive the RF voltage on the plates up to 7 kV each. A profile monitor at the chopper slit is used to record beam deflection and to optimize the chopper phase and then the dc bias is set to optimize transmission through the chopper slit. The beam intercepted by the chopper slit is measured during beam delivery and forms part of the on-line beam current monitoring system.

2.2 Charge selection section

The charge selection section is a mirror symmetric singly achromatic 90° bend section with QQDDQQ optics. A set of adjustable slits is positioned between dipoles at the dispersed focus to provide some energy selection. The stripping foil is used for ions with $A \leq 30$ so that charge states are separated by more than 15 %. The bending strength of the dipoles has an acceptance of $A/q \leq 7$. For masses with $A > 30$ the CSB must be used and since the stripping foil is inefficient at producing high charge states for heavier beams the CSB must produce beams with $A/q \leq 7$. This sets the mass range for efficient production to $\sim A < 150$.

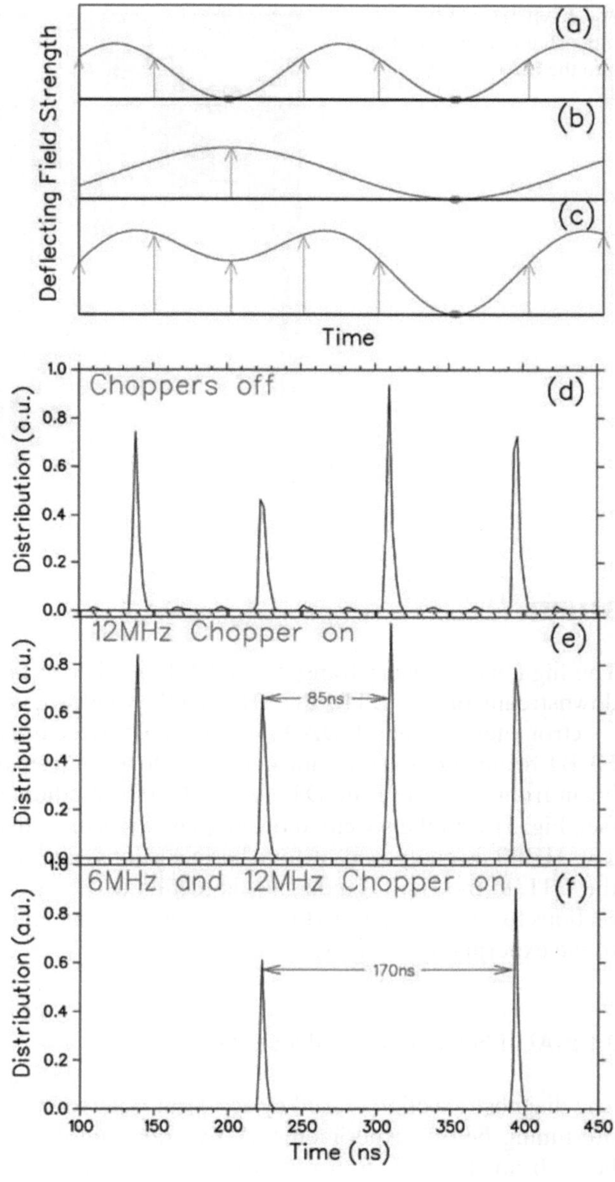

Fig. 3 **a** Output time structure from the RFQ is shown in **d**. The field produced by the 11.8 MHz chopper plate **a** produces a 85 ns time structure shown in **e**. The field produced by the 5.9 MHz plate in **b** is combined with the 11.8 MHz deflection to produce the combined deflection shown in **c** and generates the time structure given in **f**

2.3 DTL matching section

The DTL matching section matches the beam both longitudinally and trasversally for optimal acceptance by the DTL. Transverse matching is done by four quadrupoles. Longitudinal matching is done by a 35 MHz double gap spiral resonator called the MEBT rebuncher [6] (see Fig. 4). The goal is to produce a beam that is slightly converging in all planes at the entrance of DTL Tank1.

Fig. 4 MEBT 35 MHz rebuncher to match the beam into the ISAC-I DTL

3 HEBT

The high energy beam transport (HEBT) feeds two target stations 23 m and 30 m downstream of the DTL; the DRAGON windowless gas target and recoil mass spectrometer and the TUDA-I multi-purpose detector array. The primary aim of the HEBT beam line is to tune and diagnose the beam from the DTL and to transport the beam from the exit of the DTL to the ISAC-I medium energy experimental station (see Fig. 5) with the specified beam quality to satisfy the experimental requirement. The HEBT is composed of four basic sections; a section to match the beam from the DTL to the HEBT, a diagnostic section and bunching section, achromatic bend sections to deliver beams to the experiments and matching sections to focus the beam to the experimental targets.

3.1 ISAC-I beam tuning and analysis

The diagnostic section is used by accelerator personnel for beam commissioning and pre-tuning before experiments. A five quadrupole section matches the diverging beam from the DTL to a round beam focus before entering the beam diagnostic section. The beam energy and energy spread is analyzed in a 90° spectrometer with a radius of curvature of 1.5 m. A quadrupole triplet focuses beam at an object slit. A harp at the image slit measures the beam width with energy dispersion of 15 mm %. Energy is measured by correlating the field setting to optimize the beam through the image slit.

The beam composition can be analyzed at a phasing/purity monitor (HEBT:PSID5). The monitor is composed of a gold foil and an off-axis silicon detector (SiD). Ions experience Rutherford scattering in the gold foil and are energy analyzed in the SiD.

Once the energy, energy spread and composition of the beam are optimized the optics are changed and the beam is delivered to the experiment.

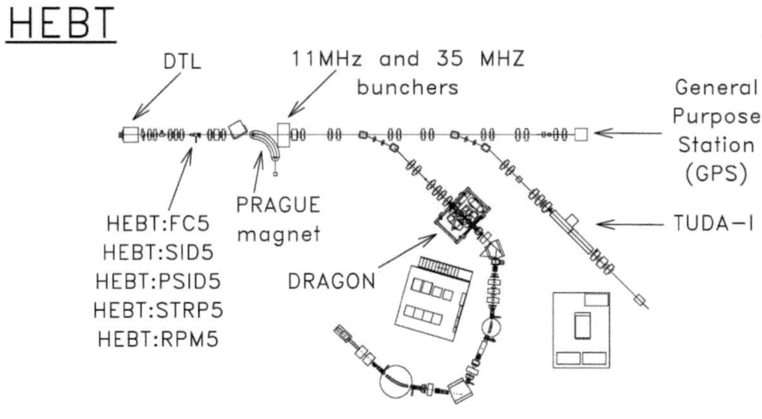

Fig. 5 The high energy beam transport (HEBT) line

Fig. 6 Beam time distributions measured on a fast Faraday cup near the DRAGON target for high-β buncher off and on

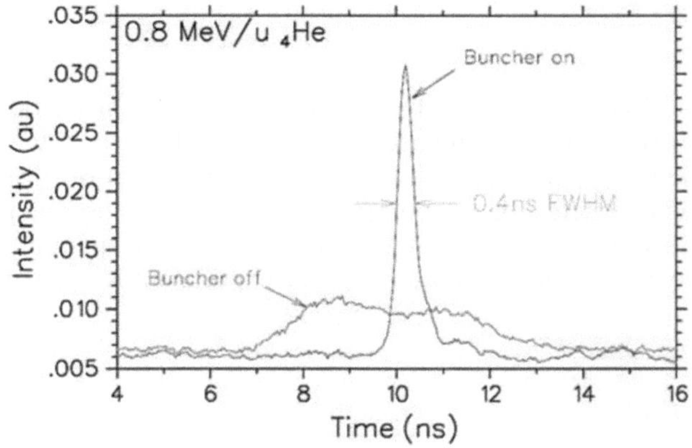

3.2 ISAC-I beam delivery mode

A low-β ($\beta_0 = 0.022$) 11.8 MHz rebuncher and a high-β ($\beta_0 = 0.032$) 35.4 MHz rebuncher [7] are positioned on either side of a double focus positioned 1 m downstream of the DTL. The 11.8 MHz rebuncher is a three gap structure driven by two lumped element circuits with up to 30 kV required on each drift tube. The 35.4 MHz buncher is a two-gap spiral device similar to the MEBT rebuncher with up to 170 kV required on the drift tube. For beams with $E \leq 0.4$ MeV/u the low beta buncher is used; for higher energies the high beta buncher is used. The longitudinal beam quality at the exit of the DTL is typically quantified at 0.4 % energy spread FWHM resulting in an unbunched time spread of several nanoseconds. The time profile is measured just upstream of each experimental station by mean of a FFC. The bunchers can be used to provide either a time focus or an energy focus at the experiment. Assuming the typical longitudinal π-emittance of $\sim 1\ \pi\mathrm{keV/u} \cdot \mathrm{ns}$ time spreads of 1 ns or less can be achieved while the energy spread can be reduced to less than 0.1 %. Figure 6 shows the beam time distribution close to the DRAGON target for the high-β buncher off and on.

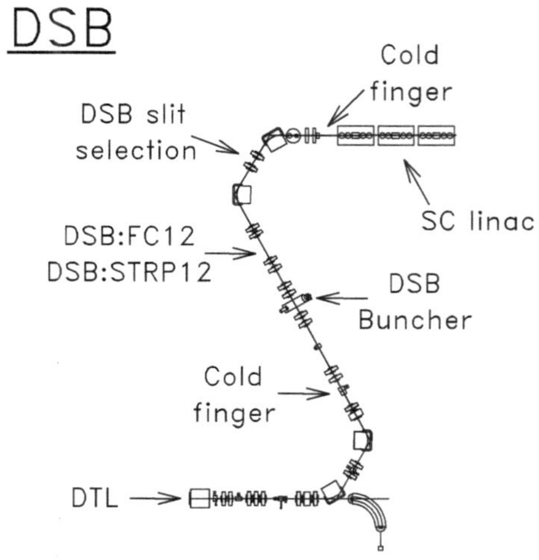

Fig. 7 The DTL to superconducting linac beam transport (DSB) line

Each experimental station has a doubly achromatic 45° bend section plus a four quadrupole matching section to adjust the size of the beam on target. Adjustable slits are positioned between each dipole pair at the dispersed focus to provide some energy selection. A four quadrupole section is used to match the beam to the target. The transverse π-emittance after the DTL is quantified at $\sim 0.2\ \pi\text{mm}\cdot\text{mrad}$. The typical beam spot required at the experiment is a round shape with 2 mm diameter.

4 DSB

The DSB line is an S-bend transport line with two 116° doubly achromatic bending sections that connect the DTL to the SC linac (see Fig. 7). The second achromatic bend contains a multi-slit collimator between dipoles at the dispersed focus to allow some energy and mass selection with a resolving power of $M/\Delta M \sim 500$. The two achromatic bend sections are joined by two four quadrupole sections; the first matches the beam into a buncher and the second matches the beam to the SC linac.

An optional stripping foil is positioned downstream of the second four quadrupole matching section to allow either an energy boost in the linac by increasing the charge or a charge dependent element selection to help purify the beam.

The beam is longitudinally matched into the SC linac using the 35 MHz DSB buncher. The buncher is a spiral double gap resonator similar to the MEBT rebuncher with $\beta_0 = 0.05$. The buncher is tuned to minimize the time spread at the entrance of the linac either by monitoring the beam on a FFC at the entrance to the SC-linac or tuning to a phase monitor downstream of the linac and scaling the buncher amplitude.

Two cold traps (cold finger) cooled at liquid nitrogen temperature are installed in the DSB. These are present in order to prevent trace contaminants (like oil) from migrating to the superconducting cavity from ISAC and help maintain the linac performance.

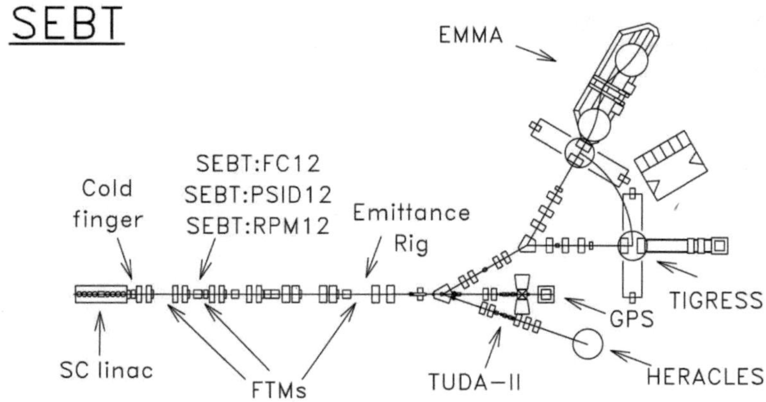

Fig. 8 The superconducting energy beam transport (SEBT) line

5 SEBT

The SEBT line connects the SC linac to the ISAC-II experimental stations (see Fig. 8). There are four experimental stations in ISAC-II; TUDA-II and HERACLES at −22°, TIGRESS after an achromatic dog-leg at 0° and EMMA after a doubly achromatic 45° bend.

The line is composed of a four quadrupole section to match the beam to the periodic 8 quadrupole transport section that delivers the beam to the bend sections at the end of the accelerator vault.

5.1 ISAC-II beam tuning and analysis

A silicon detector 4 m downstream from the linac monitors ions back scattered from a thin gold foil. The monitor is used for cavity phasing and energy measurement. A time of flight monitor also in the downstream beam line is used for more precise energy measurement. The monitor consists of three identical units spaced 2.8 m and 9.2 m apart respectively. Each unit consists of a biased wire inside a grounded can. A hole in the can allows the beam to pass. Electrons driven off the wire by the ions are accelerated through an aperture in the can to a micro-channel plate for timing information. The three timing signals are used to calculate the precise velocity of the beam to an accuracy of 0.1 %. The distance between monitors and the response and delay times of the monitors are calibrated using an on-line laser system. A slit and harp transverse emittance rig is used to measure the transverse emittance of the beam.

Presently no longitudinal beam tuning is performed in the SEBT beam line. A superconducting quarter wave buncher with $\beta_0 = 0.07$ is planned for installation at the end of the vault section to bunch the beam to the experimental station.

5.2 ISAC-II beam delivery mode

Since the SC linac is capable of accelerating ions above the Coulomb barrier there are limitations in the beam current that can be sent to the ISAC-II experimental

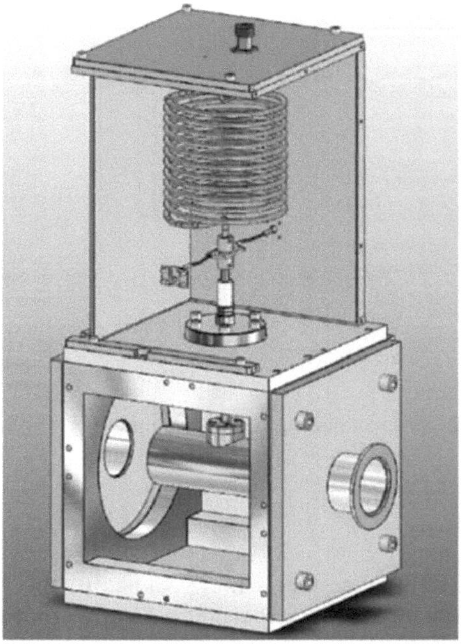

Fig. 9 A SolidWorks drawing of the monitor showing the pickup tube and helical resonator

stations in order to maintain the prompt radiation within regulatory limits. The deliverable current limits are energy and mass dependent and are defined in the operating license.

In order to deliver beam outside the accelerator vault the beam current has to be continuously monitored to be in compliance with the license. The current monitoring system [8] includes a partially intercepting monitor in MEBT and two non-intercepting monitors. The partially intercepting monitor is the ISAC-I chopper slit that is electrical isolated and intercepts the satellite bunches chopped by the MEBT chopper. The current from the satellite bunches is directly proportional to the total beam current. The two non-intercepting monitors are capacitive pickup tubes with lumped element RF resonating circuits (see Fig. 9). The pickup consists of a 12.7 cm long, 4.76 cm inner diameter copper plated aluminum tube through which the beam passes. The tube is housed in an 8 inches vacuum box and has a capacitance to ground of 5 pF. The signal is brought out through a robust vacuum feedthrough, a Ceramaseal 16941-01-CF, with a capacitance of 8.5 pF to a helical resonator.

The signal from each of the three monitors is compared in an EPICS interface with respect to a relative reference trip limit set according to the maximum allowed beam intensity. If any of the monitors reads back a current higher than 80 % of the relative trip limit for more than 5 seconds the current monitoring system removes the safety enable and the source Faraday cups and the SEBT beam blocker, that prevents the beam from exiting the accelerator vault, are driven in.

5.3 RIB delivery and low intensity diagnostics

Radioactive Ion Beams (RIB) tend to have low intensities making accelerator and beam line tuning impractical. The approach taken at TRIUMF is to use a 'pilot' stable beam from a stable off-line ion source [9, 10] (OLIS) of the same A/q as the desired radioactive beam. The stable beam is of high enough intensity, $5 - 100enA$, that tuning of the accelerator and beam lines is straightforward. The LEBT switchyard is then changed to accept beam from the on-line source and low intensity diagnostics at a few locations in the accelerator chain are used to optimize the RIB to the experiment. Diagnostics are either silicon detectors inserted directly into the beam $10 \leq counts \leq 1000$ or channeltrons $1000 \leq counts \leq 10^6$. The low intensity diagnostics are positioned just after the MEBT dipole to optimize the RFQ phase and the beam matching from the on-line source and in the SEBT vault section to optimize the RF phase between the RFQ and the DTL and the MEBT dipole section.

References

1. Poirier, R.L. et al.: CW performance of the TRIUMF 8 meter long RFQ for exotic ions. In: LINAC 2000, pp. 1023. Monterey, California, USA (2000)
2. Laxdal, R.E. et al.: Beam commissioning and first operation of the ISAC DTL at TRIUMF. In: PAC 2001, pp. 3942. Chicago, Illinoois, USA (2001)
3. Laxdal, R.E.: Commissioning and early experiments with ISACII. In: PAC 2007, pp. 2593. Albuquerque, New Mexico, USA (2007)
4. Bylinsky, Y.V. et al.: High power test of the ISAC triple gap buncher operating in CW mode. In: PAC 1999, pp. 893. New York, USA (1999)
5. Mitra, A.K. et al.: Design, test and commissioning of a dual frequency chopper for the TRIUMF ISAC facility. In: LINAC 2002, pp. 485. Gyeoungju, Korea (2002)
6. Mitra, A.K., Poirier, R.L.: A 35 MHz Spiral re-buncher cavity for the TRIUMF ISAC facility. In: PAC 1999, pp. 839. New York, USA (1999)
7. Mitra, A.K. et al.: RF test and commissioning of the low and high beta bunchers for the TRIUMF ISAC facility. In: EPAC 2002, pp. 2178. Paris, France (2002)
8. Marchetto, M. et al.: The ISAC-II current monitor system. In: LINAC 2010, pp. 623. Tsukuba, Japan (2010)
9. Jayamanna, K. et al.: Off-line ion source terminal for ISAC at TRIUMF. Rev. Sci. Instrum. **79**, 02C711 (2008)
10. Jayamanna, K. et al.: A multicharge ion source (Supernanogan) for the OLIS facility at ISAC/TRIUMF. Rev. Sci. Instrum. **81**(2), 02A331 (2010)

DOI 10.1007/978-94-007-7963-1_10
Reprinted from *Hyperfine Interactions* Journal, DOI 10.1007/s10751-013-0886-6

The experimental facilities at ISAC

J. Dilling · R. Krücken

Published online: 10 October 2013

Abstract The ISAC accelerator facility provides rare isotope beams in three different areas at low (20–60 keV), medium-high (up to 1.8 AMeV) or higher energies (up to 16 AMeV, depending on the mass-to-charge ratio of the isotopes). There is a corresponding suite of experimental installations on the floor, most of them permanently, which are uniquely matched to the requirements of the scientific goals as well as to the conditions as the arise at rare-beam facilities, like ISAC. An introduction to the three distinct experimental areas, given by the available energy of beams, is given.

Keywords ISAC radioactive beams · Post-acceleration · Instrumentation

1 Introduction

The rare-isotope beam program at TRIUMF started out in 1987 with an ISOL (Isotope Separator on-line) test facility, called TISOL (Triumf's Isotope Separator On-line), which paved the way to first experimental studies. Those studies included target and ion-source developments as well as experimental programs in nuclear astrophysics [1] and fundamental symmetries via beta-decay correlation studies [2, 3]. In the late 1990 the ISAC (Isotope Separator and Accelerator) facility was started and initially beams at low energy [4] and soon thereafter at medium-high energies [5] were being delivered. The post-accelerated beams were mostly used for nuclear astrophysics applications. In 2007 the ISAC II complex with further post-acceleration capacities was added [6], which provides now beam energies up to 16 AMeV using superconducting RF technology.

ISAC and ARIEL: The TRIUMF Radioactive Beam Facilities and the Scientific Program.

J. Dilling (✉) · R. Krücken
TRIUMF, 4004 Wesbrook Mall, Vancouver, BC V6T2A3, Canada
e-mail: JDilling@triumf.ca

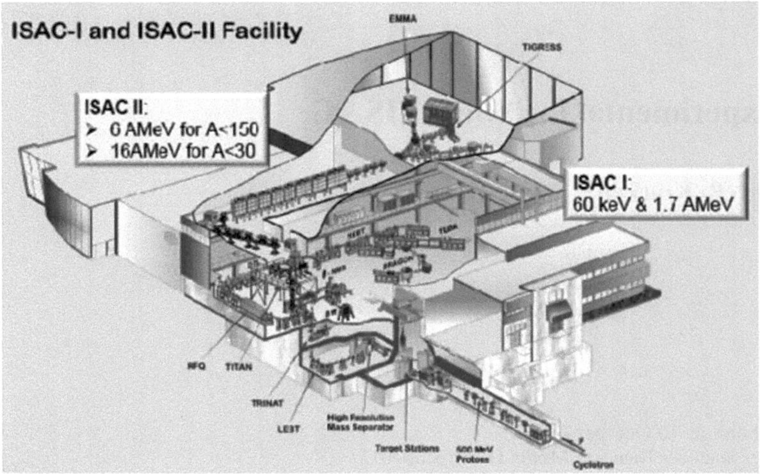

Fig. 1 The ISAC I and ISAC II experimental complex. ISAC I provides beam to experiments at low and medium-high energies. ISAC II delivers post-accelerated beam up to 6 AMeV or 16 AMeV, depending on the mass-to-charge ratio

The ISAC I and II experimental areas are shown in Fig. 1. The ISAC I area provides radioactive ion beams to two separate experimental areas, the low energy area with beams at 20–60 keV, and the post-accelerated area with beam of up to energies of 1.8 AMeV. The ISAC II facility can provide ion beams with further post-acceleration of energies up to either 6 AMeV (for isotopes with mass A up to 150 amu) or 16 AMeV (for lighter isotopes up to mass 30 amu). The experiemental program covered in both areas can be divided into the following categories:

- solid state interface studies and new material with βNMR and βNQR.
- nuclear astrophysics
- physics beyond the Standard Model and fundamental symmetries
- fundamental nuclear physics and structure studies

2 Experimental facilities in ISAC I

The experimental fcilities in ISAC I are organised into the low energy area, and the post-accelerated area. The low energy facilities (20–60 keV) follow these science goals and motivations: measurement of ground state properties of radioactive isotopes, scattering and reaction experiments for nuclear astrophysics, precision measurements for tests of fundamental symmetries and physics beyond the Standard Model. The ISAC I experimental facilities and their primary purpose are listed in the following:

Nuclear structure precision measurements:

- 8π gamma-ray spectrometer [4] (soon to be replaced by:)
- GRIFFIN [7] (Gamma-Ray Infrastructure For Fundamental Investigations of Nuclei) gamma-ray spectrometer for decay studies

- DESCANT [8] (Deuterated Scintillator Array for Neutron Tagging) neutron detector array to be coupled to GRIFFIN (and TIGRESS)
- TITAN [9, 10] (Triumf's Ion Trap for Atomic and Nuclear science) Penning trap mass spectrometer, in trap decay studies
- Decay tape station [11], precision half-live measurements
- OSAKA [12] beta decay spectroscopy of laser polarized nuclei

Measurement to test fundametal symmetries and physics beyond the Standard Model:

- TRINAT [13] (TRIumf's Neutral Atom Trap) laser trap for beta-decay correlation measurements
- TITAN-ec [14] EC decay branching ratios for $2\nu 2\beta$ intermitten isotopes
- RADON EDM [15], apparatus to search for atomic electric dipol moment in spin-polarized Rn isotopes
- MTV [16] (Mott polarimetry for T-Violation experiment) – Drift Chamber for precision measurement of Mott-scattering of electrons from polarized ^{8}Li

Material science studies and new materials from measurements of magnetic properties at surfaces and interfaces.

- βNMR facility [17], measurements of beta-decay assymetry of polarized ^{8}Li implanted in solid state material, and using high magnetic fields.
- βNQR facility [17], polarized ^{8}Li asymetry measurements at zero magnetic field

Nuclear astrophysics with medium-high energy radioactive (and stable) beams:

- DRAGON [18, 19] (Dectector of Recoils and Gamma Rays for Nuclear Astrophysics) a windowless gas target recoil separator for radiative capture reaction
- TUDA [20] (TRIUMF and UK Detector Array) for particle detection in nuclear reaction experiments for nuclear astrophysics
- TACTIC [20] (TRIUMF Annular Chamber for Tracking and Identification of Charged particles) active target systems for nuclear reaction experiments at very low rates
- DSL [21] (Doppler Shift Lifetime measurement apparatus) femto- to picosecond lifetime determination of excited nuclear states

All experimental facilities are specificly set up to match the demands of radioactive beam experiments with often very low intensities.

3 Experimental facilities in ISAC II

The ISAC II superconducting accelerator complex provides beams (depending on the achievable mass-to-charge ratio) of up to 16 AMeV. This allows reaction experiments at energies below and above the Coulomb barrier as well as Coulomb excitation measurements. Another area of interest is the Equation of State (EOS) for assymetric nuclei, such as very n-rich isotopes. The following experimental facilities are permanently installed in the ISAC II experimental hall.

- TIGRESS [22, 23] (**TRIUMF-ISAC Gamma-Ray Escape Suppressed Spectrometer**), high resolution segmented high purity Germanium gamma-ray detector

- EMMA [24, 25] (Electro-Magnetic Mass Analyser) recoil mass analyzer separator and detector
- HERACLES [26] (HEavy-ion Reaction Array for the Characterization of Light Excited Systems) multipurpose chamber for reaction studies like multifragmentation
- IRIS [27] (ISAC Charged Particles Spectroscopy Station) solid hydrogen target station with charged particle detector set-up for nuclear reaction studies at low intensities

In addition, temporaery installations are frequenty added at one of the general purpose stations at ISAC I or II. The experimental facilities are described in detail in the following contributions in this issue.

References

1. Buchmann, L., et al.: Phys. Rev. Lett. **70**, 726–729 (1993)
2. Behr, J.A., et al.: Phys. Rev. Lett. **79**, 375 (1997)
3. Trinzcek, M., et al.: Phys. Rev. Lett. **90**, 012501 (2001)
4. Ball, G.C., et al.: Phys. Rev. Lett. **86**, 1454 (2001)
5. Ruiz, C., et al.: Phys. Rev. C **65**, 042801 (2001)
6. Tanihata, I., et al.: Phys. Rev. Lett. **100**, 192502 (2008)
7. Svensson, C.E., Garnsworthy, A.B.: Springer Hyperfine Interaction (2013, this issue)
8. Garrett, G., et al.: Springer Hyperfine Interaction (2013, this issue)
9. Dilling, J., et al.: Int. J. Mass. Spec. **251**, 198 (2006)
10. Kwiatkowski, A.A., et al.: Springer Hyperfine Interaction (2013, this issue)
11. Ball, G.C., et al.: Springer Hyperfine Interaction (2013, this issue)
12. Shimoda, T., et al.: Springer Hyperfine Interaction (2013, this issue)
13. Gorelov, A., et al.: Phys. Rev. Lett. **94**, 142501 (2005)
14. Lennarz, A., et al.: Springer Hyperfine Interaction (2013, this issue)
15. Tardiff, E., et al.: Springer Hyperfine Interaction (2013, this issue)
16. Murata, J., et al.: Springer Hyperfine Interaction (2013, this issue)
17. Morris, G.D.: Springer Hyperfine Interaction (2013, this issue)
18. Hutcheon, D.A., et al.: NIM A **498**, 190–210 (2003)
19. Fallis, J.: Springer Hyperfine Interaction (2013, this issue)
20. Laird, A.M., et al.: Springer Hyperfine Interaction (2013, this issue)
21. Davids, B.: Springer Hyperfine Interaction (2013, this issue)
22. Svensson, C.E., et al.: J. Phys. G **31**, S1663 (2005)
23. Hackman, G., Svensson, C.E.: Springer Hyperfine Interaction (2013, this issue)
24. Davids, B., Davids, C.N.: NIM A **544**, 565 (2005)
25. Davids, D.: Springer Hyperfine Interaction (2013, this issue)
26. Patrick St-Onge, et al.: Springer Hyperfine Interaction (2013, this issue)
27. Kanungo, R.: Springer Hyperfine Interaction (2013, this issue)

DOI 10.1007/978-94-007-7963-1_11
Reprinted from *Hyperfine Interactions* Journal, DOI 10.1007/s10751-013-0887-5

TRINAT: measuring β-decay correlations with laser-trapped atoms

J. A. Behr · A. Gorelov · K. P. Jackson · M. R. Pearson · M. Anholm · T. Kong · R. S. Behling · B. Fenker · D. Melconian · D. Ashery · G. Gwinner

Published online: 10 October 2013

Abstract TRIUMF's neutral atom trap for β decay (TRINAT) has been used to measure the most accurate β-ν correlation. Plans include improving that measurement, and also carrying out a full program of spin-polarized β decay correlation measurements complementary to the decay of the neutron

Keywords Weak interactions · Laser cooling · Beta decay

ISAC and ARIEL: The TRIUMF Radioactive Beam Facilities and the Scientific Program.

J. A. Behr (✉) · A. Gorelov · K. P. Jackson · M. R. Pearson
TRIUMF, 4004 Wesbrook Mall, Vancouver, BC, Canada V6T 2A3
e-mail: behr@triumf.ca

M. Anholm · T. Kong
Department of Physics and Astronomy, University of British Columbia, Vancouver, BC, Canada V6T 1Z1

R. S. Behling · B. Fenker · D. Melconian
Cyclotron Institute, Texas A&M University, College Station, TX, 77843, USA

R. S. Behling
Department of Chemistry, Texas A&M University, College Station, TX, 77843, USA

B. Fenker · D. Melconian
Department of Physics, Texas A&M University, College Station, TX, 77843 USA

D. Ashery
School of Physics and Astronomy, Tel Aviv University, Tel Aviv, Israel

G. Gwinner
Department of Physics and Astronomy, University of Manitoba, Winnipeg, MB, Canada R3T 2N2

1 Motivation

Nuclear β decay correlation experiments helped establish the nature of the weak interaction, a theory with massive spin-1 bosons (partners of the photon) coupling only to left-handed leptons [1]. This standard model of the weak interaction makes definite predictions for the angular distribution of the two leptons produced with respect to each other, and with respect to the spin direction of the decaying nucleus.

Some of these distributions can be understood intuitively. For example, since the leptons and antileptons are emitted with opposite helicity in the standard model, in a nuclear spin^{parity} $I^\pi = 0^+ \rightarrow 0^+$ decay the β^+ and ν cannot be emitted back-to-back, because that would carry off one unit of spin. Writing the angular distribution $W[\theta] = 1 + a\,\frac{v}{c}\cos[\theta]$, that immediately implies $a = 1$ in the standard model. There are no corrections from isospin mixing, because this prediction only depends on the nuclear spin—there is a calculable correction of order parts per thousand from radiative emission of real bremsstrahlung-like photons [2]. So a measurement of a to high accuracy tests this standard model prediction, allowing sensitivity to interactions that produce leptons and antileptons with the same helicity, like the exchange of a spin-0 boson [2]. Similar predictions emerge for correlations with the nuclear spin, depending on the nature of the exchanged boson and the degree of handedness of the leptons produced [3, 4].

Such β-decay measurements can still assist particle physics by determining 1st-generation lepton-quark couplings to possible new exchange bosons, but must reach excellent accuracy ≤ 0.1 %. Using atom trap technology, the TRINAT collaboration is pursuing two experimental programs with the potential to reach this accuracy.

We are concentrating efforts on isobaric analog decays of ^{37}K and $^{38\text{m}}$K. In ^{37}K higher-order standard model effects are also small (e.g. $\leq$ 0.2 % for the β asymmetry [5]), and can be determined to good accuracy from the electromagnetic moments of the nuclei by the conserved vector current hypothesis, itself a standard model prediction [6]. In the I^π 3/2$^+$ $\rightarrow$ 3/2$^+$ decay of ^{37}K, the ratio of spin-flipping to spin preserving operators (Gamow-Teller/Fermi) is also needed for the standard model predictions, and can be determined from reliable measurements of the lifetime and branching ratio [7].

If successful at 0.001 accuracy, our correlation experiments would be complementary with constraints from radiative π decay [8] and indirect effective field theory constraints from $\pi \rightarrow e\ \nu$ [9]. They would begin to allow sensitivity to R-parity conserving supersymmetric left-right sfermion mixing, which can produce effects as large as 0.001 in nuclear β decay [10].

2 Techniques

A magneto-optical trap (MOT) can be thought of as a damped harmonic oscillator. The damping is produced by laser beams from all 6 directions tuned slightly lower in frequency than an atomic resonance, so that atoms moving in any direction absorb more photons from the Doppler-shifted opposing beam and slow down. The position-sensitive force is trickier, and its demonstration was a large part of the Nobel Prizes [11] awarded for laser cooling and trapping. The Zeeman shift from a 10 G/cm gradient magnetic field produced by anti-Helmholtz coils flips sign at the origin, changing the probability of the atoms to absorb different handedness of circularly polarized light from opposing beams. In TRINAT, the atoms produced by the ISAC/TRIUMF isotope separator are first collected with 0.001 efficiency in a MOT, and are then transferred with 75 % efficiency to a 2nd MOT to escape backgrounds from untrapped atoms [12]. Achieved trapping lifetimes are ≈15 s. For more details see [13] and references therein.

Confined in a 1 mm-sized cloud, the nuclei undergo beta decay, producing three products: a β^+, a ν, and a recoiling daughter nucleus. The daughter nucleus has less than 500 eV kinetic energy and would stop in a nm of material, but is unperturbed by the small B field and has transmuted to an element unperturbed by the resonant laser light. Trap temperatures of order milliKelvin produce negligible effects on the recoil velocities. In β^+ decay about 15 % of the recoils are charged, which after collection and acceleration by a uniform electric field of 1 kV/cm are efficiently detected by a micro-channel plate (MCP). The β^+ is detected in coincidence in δE-E telescopes using double-sided silicon strip detectors backed by plastic scintillator. The resulting time-of-flight (TOF) of the recoil with respect to the β^+, together with its position readout by delay-line anode, lets us reconstruct its momentum. Together with the β^+ momentum, this yields the momentum and angular distribution of the ν (see [13] for details).

The atom trap methods provide many unique tools to determine experimental systematics. Kinematic redundancy in most of the data set lets us make many consistency tests; for example, the construction of the β momentum from other observables tests our simulations of the detector response [13]. We measure the cloud dimensions and temperature by photoionizing a small fraction of the atoms and measuring their TOF and position on the same MCP that measures the nuclear recoils. Detection of shakeoff atomic electrons in coincidence provides a high-efficiency TOF trigger to determine the recoil moment and suppresses possible backgrounds from untrapped atoms for singles β^+ measurements [14].

We have learned to spin-polarize nuclei and measure the polarization in situ of the decaying species by atomic methods independent of the nuclear decays. This involves the following sequence with a 1 kHz repetition rate: a) turn off the trap and its perturbing magnetic and laser fields; b) optically pump the atoms to close to 100 % spin polarization with a single on-resonance circularly polarized laser beam [15] and count; c) re-trap the same atoms to minimize their expansion.

Recently we have harnessed an "AC MOT" [16] to achieve better and faster extinction of the trap B fields, which otherwise hurt the spin polarization. The MOT B field is sinusoidally varied, in phase with flipping the circular polarization of the MOT laser light. The result is a similar confining potential, but the trap can be switched off when the eddy currents from the stainless steel chamber are near zero. An offline version of this AC MOT has produced trap lifetimes similar to those of the DC MOT with adequate duty cycles. The MOT B field in the full apparatus (Fig. 1) has been reduced to 1 % in 100 μs, four times better than previously achieved, with further improvements in progress [17]. Together with use of 316LN and 316L stainless steel, careful welds, and a chamber with twice the diameter, it is hoped the polarization of ^{37}K achieved before (97 ± 2 % [15]) can be improved to the ≥99.6 % achieved in ^{41}K in a pyrex chamber. Optical pumping beams will also shine from both directions so they can be balanced better to minimize a false asymmetry from movement of the atoms during optical pumping.

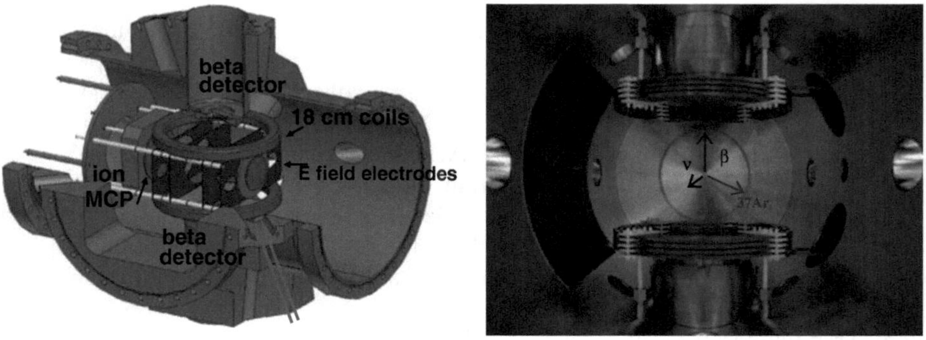

Fig. 1 (*Left*) Chamber design for spin-polarized ^{37}K experiments, showing AC MOT coils inside the vacuum, β detectors top and bottom, 80 mm ion MCP, and 40 mm atomic electron MCP facing it. The *red arrows* show the direction of an optical pumping beam as it travels to the mirror in front of the top β detector to be reflected directly downward—its twin opposing beam and the four horizontal MOT beams are not shown. The vertical 766.5 nm MOT beam has the same path as the 770 nm optical pumping beam, an unusual geometry leaving more space for efficient beta detection. The beams are combined together by an angle-tuned bandpass optical filter that passes one frequency while reflecting the other. The glassy carbon electrodes (low-Z to minimize β scattering, semiconductor to minimize eddy currents) provide an open enough geometry for all the beams—this geometry will change for a more uniform field for ^{38m}K. (*Right*) The assembled in-vacuum MOT B field coils with re-entrant flanges for β detectors, looking along the ion detection axis without the MCP assembly installed. The split copper shields minimize eddy currents while shielding detectors from inductive heating. The horizontal MOT vacuum ports are visible. Momentum vectors for the products and one of two possible spin polarization directions are shown

Detector improvements include delay-line anode MCP readout to improve the momentum resolution, developed for a direct massive particle search [18]. Potentials of 10 kV are needed, necessitating care with inductance and impedance matching of high-voltage feedthroughs to cleanly bring out MCP timing and delay-line anode signals while minimizing pulse distortion. Larger β detectors will improve solid angle. Data acquisition for the larger number of strip detector channels will be done by waveform digitizers, and this is also being explored for the delay-line anode signals.

3 Experiments

3.1 Spin-polarized experiments

We have published a measurement of the ν asymmetry B_ν at 3 % accuracy [15]. The goals of our spin correlation program (TRIUMF experiment S1188) include a simultaneous measurement of β singles asymmetry A_β and recoil singles asymmetry A_{recoil} at part per thousand sensitivity, and eventually B_ν at 0.3 %. E.g., from the detailed dependence of the recoil asymmetry A_{recoil} on recoil momentum, a statistical sensitivity to 4-fermi tensor interactions [8] $C_T + C_T'$ of 0.002 can be reached in one week of counting. Other physics, including right-handed currents in left-right models and 2nd-class currents [5], can be simultaneously constrained by the other observables. This program would provide similar but complementary sensitivity to the new physics probed by the community's measurement of neutron β-decay.

3.2 β-ν correlation in pure Fermi decay

We also would improve our best limits on general scalar interactions coupling to the first generation of particles by measuring the β-ν correlation in the pure Fermi decay of ^{38m}K [2]. The complete angular acceptance for recoils we have developed will minimize key systematic errors in our upgraded experiment, with a goal of reaching 0.1 % accuracy in a and possibly b_{Fierz} (the interference term that is constrained by the endpoint energy dependence of the $0^+ \to 0^+$ decay Q-values). We determine systematic errors for this experiment from statistics-limited kinematic observables that are independent of the angular correlation (see TRIUMF S1070 proposal).

3.3 Time-reversal

Given our ability to polarize nuclei and measure β and recoil momenta, we could design a trap to efficiently measure the time-reversal violating correlation [3, 4]

$$D\mathbf{I} \cdot \frac{\mathbf{p}_\beta}{E_\beta} \times \frac{\mathbf{p}_\nu}{E_\nu} \overset{t \to -t}{\rightarrow} -D\mathbf{I} \cdot \frac{\mathbf{p}_\beta}{E_\beta} \times \frac{\mathbf{p}_\nu}{E_\nu} = D\mathbf{I} \cdot \frac{\mathbf{p}_\beta}{E_\beta} \times \frac{\mathbf{p}_{\mathbf{recoil}}}{Q - E_\beta} \tag{1}$$

Since this correlation flips sign with time, it is zero if time-reversal is a good symmetry (with known correction from 'final state' effects of 0.0002 p_{recoil}/Q [19]). We have a conceptual design for a dedicated geometry with 20 % β efficiency (that the polarized light comes at 90° to the β detectors helps) so we could achieve 2×10^{-4} statistical error/week. The best measurement in the decay of the neutron was recently published by emiT, achieving $2 \mathrm{x} 10^{-4}$ sensitivity [20]. However, a recent general effective field theory calculation has constrained D to be an order of magnitude smaller from other experiments, including the electric dipole moment of the neutron and atomic parity violation [21]. Since possible contributions to time-reversal violation from various new physics sources could accidentally cancel, measurement of D could still be complementary. We plan a feasibility study of time-reversal systematics in our present ^{37}K geometry.

References

1. Commins, E.D., Bucksbaum, P.H.: Weak Interactions of Leptons and Quarks. Cambridge University Press (1983)
2. Gorelov, A., Melconian, D., Alford, W.P., Ashery, D., Ball, G., Behr, J.A., Bricault, P.G., D'Auria, J.M., Deutsch, J., Dilling, J., Dombsky, M., Dubé, P., Fingler, J., Giesen, U., Glück, F., Gu, S., Häusser, O., Jackson, K.P., Jennings, B., Pearson, M., Stocki, T.J., Swanson, T.B. Trinczek, M.: Scalar interaction limits from the β-ν correlation of trapped radioactive atoms. Phys. Rev. Lett. **94**, 142501 (2005)
3. Jackson, J.D., Treiman, S.B., Wyld, H.W.: Possible tests of time reversal invariance in beta decay. Phys. Rev. **106**, 517 (1957)
4. Jackson, J.D., Treiman, S.B., Wyld, H.W.: Coulomb corrections in allowed beta transitions. Nucl. Phys. **4**, 206 (1957)
5. Melconian, D., Behling, R.S., Behr, J.A., Ashery, D., Gorelov, A., Jackson, K.P., Kong, T., Mehlman, M., Pearson, M.R., Shidling, P.: The β^+ decay of ^{37}K as a multifaceted probe of fundamental physics. Second international Ulaanbaatar conference on nuclear physics and applications. AIP Conf. Proc. **1342**, 53 (2011). doi:10.1063/1.3583167
6. Holstein, B.R.: Recoil effects in allowed beta decay: the elementary particle approach. Rev. Mod. Phys. **46**, 789 (1974)

7. Shidling, P.D., Behling, R.S., Hardy, J.C., Iacob, V.E., Mehlmann, M., Melconian, D., Mroeder, B.T., Stephens, H.: Precise lifetime measurement of ^{37}K. Bull. Am. Phys. Soc. B10.004 (2011)
8. Bhattacharya, T., Cirigliano, V., Cohen, S.D., Filipuzzi, A., Gonzalez-Alonso, M., Graesser, M.L., Gupta, R., Lin, H.-W.: Probing novel scalar and tensor interactions from (ultra)cold neutrons to the LHC. Phys. Rev. D **85**, 054512 (2012)
9. Campbell, B.A., Maybury, D.W.: Constraints on scalar couplings from $\pi^{\pm} \to l^{\pm}\nu_l$. Nucl. Phys. B **709**, 419 (2005)
10. Profumo, S., Ramsey-Musolf, M.J., Tulin, S.: Supersymmetric contributions to weak decay coefficients. Phys. Rev. D **75**, 075017 (2007)
11. Chu, S., Cohen-Tannoudji, C., Phillips, W.: Nobel prize lectures. Rev. Mod. Phys. **70**, 685, 707, 721 (1998)
12. Swanson, T.B., Asgeirsson, D., Behr, J.A., Gorelov, A., Melconian, D.: Efficient transfer in a double magneto-optical trap system. J. Opt. Soc. Am. B **15**, 2641 (1998)
13. Behr, J.A., Gwinner, G.: Standard model tests with trapped radioactive atoms. J. Phys. G, Nucl. Part. Phys. **36**, 033101 (2009)
14. Pitcairn, J.R.A., Roberge, D., Gorelov, A., Ashery, D., Aviv, O., Behr, J.A., Bricault, P.G., Dombsky, M., Holt, J.D., Jackson, K.P., Lee, B., Pearson, M.R., Gaudin, A., Dej, B., Höhr, C., Gwinner, G., Melconian, D.: Tensor interaction constraints from beta-decay recoil spin asymmetry of trapped atoms. Phys. Rev. C **79**, 015501 (2009)
15. Melconian, D., Behr, J.A., Ashery, D., Aviv, O., Bricault, P.G., Dombsky, M., S. Fostner, A. Gorelov, S. Gu, V. Hanemaayer, Jackson, K.P., Pearson, M.R., Vollrath, I.: Measurement of the neutrino asymmetry in the decay of laser-cooled, polarized ^{37}K. Phys. Lett. B **649**, 370 (2007)
16. Harvey, M., Murray, A.J.: Cold atom trap with zero residual magnetic field: the ac magneto-optical trap. Phys. Rev. Lett. **101**, 173201 (2008)
17. Anholm, M., Norton, H., Anderson, R.M.A., Theriault, O., Donohue, J., Behr, J.A.: The construction and properties of an AC-MOT. Bull. Am. Phys. Soc. DAMOP C2.00008 (2012)
18. Kong, T., Gorelov, A., Höhr, C., Wiebe, T., Chatwin-Davies, A., Berman, A., Behling, S., Ashery, D., Gwinner, G., Pearson, M.R., Behr, J.A.: Specific mass shift measurements in rubidium by Doppler-free two-photon transitions. J. Phys. B Atomic Mol. Opt. Phys. **44**, 215004 (2011)
19. Holstein, B.R.: Tests for T invariance in allowed nuclear β decay. Phys. Rev. C **5**, 1529 (1972)
20. Mumm, H.P., Chupp, T.E., Cooper, R.L., Coulter, K.P., Freedman, S.J., Fujikawa, B.K., Garcia, A., Jones, G.L., Nico, J.S., Thompson, A.K., Trull, C.A., Wilkerson, J.F., Wietfeldt, F.E.: New limit on time-reversal violation in beta decay. Phys. Rev. Lett. **107**, 102301 (2011)
21. Ng, J., Tulin, S.: CP violation in beta decay and electric dipole moments. Phys. Rev. D **85**, 033001 (2012)

DOI 10.1007/978-94-007-7963-1_12
Reprinted from *Hyperfine Interactions* Journal, DOI 10.1007/s10751-013-0888-4

The 8π spectrometer

A. B. Garnsworthy · P. E. Garrett

Published online: 13 November 2013

Abstract The 8π spectrometer installed at the TRIUMF-ISAC radioactive beam facility provides a powerful and versatile detection system for radioactive decay, allowing measurements of γ rays and conversion electrons with high resolution, tagging on β particles, and fast-timing measurements. The facility supports a wide program of research in the fields of nuclear structure, nuclear astrophysics and fundamental symmetries with low-energy radioactive beams.

Keywords Gamma-ray spectroscopy · Beta decay

1 Introduction

Radioactive decay measurements provide access to excited states in nuclei in an environment which does not suffer from many common effects associated with in-beam reaction studies such as Doppler broadening of lineshapes, Coulomb-excitation of target chamber materials or target-backings, or X-ray and delta electron production which act to degrade spectral quality. Determination of parent half-lives, daughter decay schemes and level properties, including spin assignments, transition multipolarities, transition rates, and log ft values, and even β-decay Q-values are all possible. With the event-by-event identification of exotic nuclei produced in relativistic fragmentation reactions, and the intense yields of radioactive isotopes

ISAC and ARIEL: The TRIUMF Radioactive Beam Facilities and the Scientific Program.

A.B. Garnsworthy (✉)
TRIUMF, Vancouver, BC VT6 2A3, Canada
e-mail: garns@triumf.ca

P. E. Garrett
Department of Physics, University of Guelph, Guelph, ON N1G 2W1, Canada
e-mail: pgarrett@physics.uoguelph.ca

available at ISOL facilities such as TRIUMF-ISAC, β-decay studies have re-emerged as a powerful probe of nuclear structure phenomena across the nuclear chart.

The 8π γ-ray spectrometer began operation at the Chalk River Laboratory, Canada, in 1985 focused on in-beam reaction experiments investigating high-spin nuclear structure. In 1997, the spectrometer was moved to the Lawrence-Berkeley Laboratory, also for in-beam studies, until its move to TRIUMF in 2000. At ISAC, it has been reconfigured for use with stopped radioactive beams and, with the addition of several ancillary detector sub-systems, is now a very sensitive and powerful instrument for radioactive decay studies [1–4].

2 Facility details

The radioactive beam from ISAC is implanted into a tape at the focus of the 8π detector array. Experiments are performed in a cycling mode where each cycle consists of typically; a tape move, background measurement, beam implantation, and a decay period. The timing and control of this cycle is achieved by a CAMAC-based timing and sequence Jorway module. This allows for long-lived daughter nuclei and isobars to be removed from sight of the detectors, behind a Pb wall, into the tape box.

The 8π spectrometer incorporates 20 Compton-suppressed, hyper-pure germanium (HPGe) detectors. The array photo-peak efficiency for single events is ~1 % at 1.3 MeV. The SCintillating Electron-Positron Tagging ARray (SCEPTAR) consists of 20 plastic scintillators to detect coincident β particles emitted in the decay of the parent nucleus. When used in coincidence with the HPGe detectors, it dramatically improves the signal-to-background ratio of the resultant spectrum. In this way, the radioactive decay of beams delivered at a rate of just 2–3 ions/s has been achieved [5]. The Di-pentagonal Array for Nuclear Timing Experiments (DANTE) [6] has 10 fast scintillators of either BaF_2 or $LaBr_3(Ce)$. These detectors have excellent timing response to γ rays and can therefore be used to measure the lifetimes of excited states in the picosecond to nanosecond range. The downstream half of SCEPTAR can be replaced by a single fast-plastic scintillator for use in β-γ fast-timing measurements. PACES (Pentagonal Array for Conversion Electron Spectroscopy) consists of 5 lithium-drifted silicon detectors located inside the upstream hemisphere of the vacuum chamber and is cryogenically cooled. They are used to detect internal conversion electrons (e^-) as well as α particles emitted from the ground states of the actinide beams now available at ISAC.

The data acquisition system of the 8π spectrometer is separated into 4 data streams representing the detector sub-systems outlined above. These 4 streams each use analogue electronics to process energy and timing signals. Trigger-processing logic can process nearly any desired combination of coincidence and singles signals, and a master trigger can be issued separately to each data stream to initiate readout. Signals are digitized in CAMAC modules with control via the LeCroy FERA standard, and the data are passed to VME memory units for ultimate writing to disk. This architecture provides an extremely versatile platform for triggering and readout, as well as allowing for high data throughput (up to 12 Mb/s). The DAQ also enables precision measurements to be performed, including precise tracking of

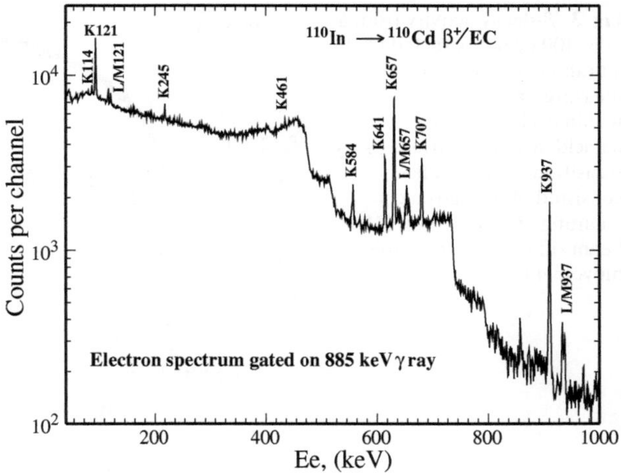

Fig. 1 Portion of the electron spectrum obtained in coincidence with the 885-keV $4^+ \rightarrow 2^+$ γ-ray transition in the decay of ^{110}In→^{110}Cd

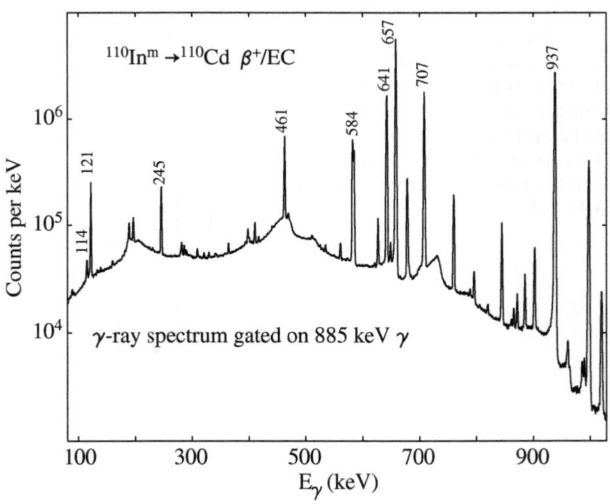

Fig. 2 Portion of the γ-ray spectrum obtained in coincidence with the 885-keV $4^+ \rightarrow 2^+$ γ-ray transition in the decay of ^{110}In→^{110}Cd

pile-up factors, dead-time determination on an event-by-event basis, and singles *vs.* coincidence event flagging.

3 Experimental performance

The physics program utilizing the 8π spectrometer is focused on two main areas; precision measurements of super-allowed β decay to provide fundamental tests of the description of electroweak interactions in the Standard Model [3, 7–13], and investigations of a wide variety of nuclear structure phenomena such as the validity of vibrational models to describe the Cd isotopes or the decay properties and level structures of neutron-rich nuclei [1, 2, 5, 14–18]. The sensitivity of the spectrometer can be demonstrated with some examples of recent experiments.

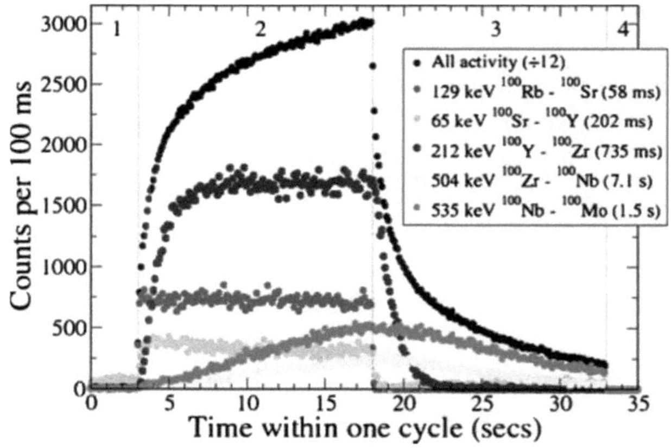

Fig. 3 β-decay activity from a $A = 100$ beam extracted from a uranium carbide target measured as a function of time within each cycle, in coincidence with various γ-ray transitions. Each cycle consisted of background counting (*3 s*), beam-on (*15 s*), beam-off (*15 s*), and a tape movement

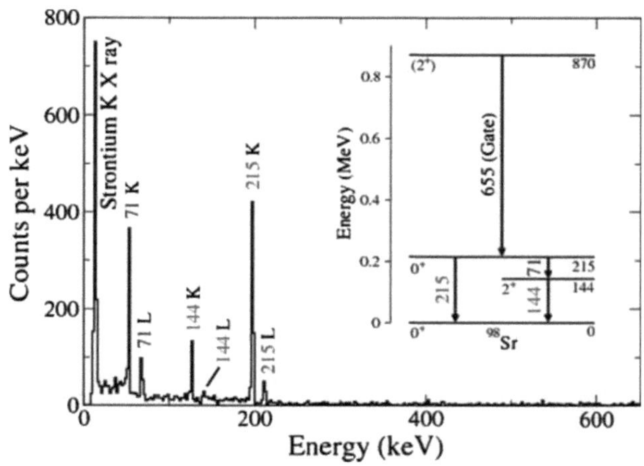

Fig. 4 PACES spectrum in coincidence with a 655 keV γ ray detected in the 8π HPGe. The conversion electrons from the lowest 3 transitions in ^{98}Sr (As shown in the partial level scheme [19]) are clearly visible along with the K-shell X ray from Sr at 14 keV

Shown in Figs. 1 and 2 are portions of coincident e^- and γ-ray spectra obtained in a recent experiment investigating the decay of (predominately) ^{110}In$\rightarrow$^{110}Cd. The spectra have a condition of prompt coincidence with the 885-keV $4_1^+ \rightarrow 2_1^+$ transition. The mass 110 beam was dominated by the ^{110}In(gs) at a rate of 10^7 ion/s, and the 3.5 d experiment resulted in background-subtracted γ-γ matrix containing 850×10^6 events. With both the high-statistics γ-γ and e^- data, a detailed decay scheme is being developed (Diaz Varela et al., unpublished; Jigmeddorj et al., unpublished).

Figures 3 and 4 display data obtained with beams from a recent uranium carbide target, with mass selection of $A = 100$ and $A = 98$, respectively. In Fig. 3, the β activity in coincidence with various γ-ray transitions as a function of time is shown, demonstrating the distinction that can be made between γ-rays emitted in the decay of species with different half-lives. Figure 4 displays a portion of the e^- spectrum in coincidence with the $(2^+) \rightarrow 0_2^+$ transition in ^{98}Sr. The relative intensities of the 215-keV and 71-keV transitions allow the determination of the $E0$ branching ratio – vital information for understanding the nature of the low-lying 0^+ excitations in the neutron-rich Sr isotopes.

The 8π will continue operation until the end of 2013 when the facility will be significantly upgraded to the Gamma-Ray Infrastructure For Fundamental Investigations of Nuclei (GRIFFIN) spectrometer [20]. The factor of $\simeq 300$ improvement in γ-γ coincidence efficiency, coupled to a very-high-throughput data acquisition system, will enable β decay experiments of unparalleled sensitivity to both weak decay branches, and at the extreme of low beam intensities.

Acknowledgements The authors would like to thank A. Close, A. Diaz Varela, and B. Jigmeddorj for assistance in the preparation of this article. Operational funding for the 8π spectrometer and its ancillary detector systems is provided by the Natural Sciences and Engineering Research Council of Canada (NSERC) and TRIUMF.

References

1. Garrett, P.E. et al.: Nucl. Instrum. Meth. **B261**, 1084 (2007)
2. Garrett, P.E. et al.: Acta Phys. Pol. **B38**, 1169 (2007)
3. Zganjar, E.F. et al.: Acta Phys. Pol. **B38**, 1179 (2007)
4. Garnsworthy, A.B.: In: Garrett, P.E. (ed.) Proc. 14'th Int. Sym. Capt. Gamma-Ray Spect. Rel. Topics. World Scientific, Singapore (2012)
5. Mattoon, C.M. et al.: Phys. Rev. **C75**, 017302 (2007)
6. Cross, D. et al.: JINST **6**, 08008 (2011)
7. Hyland, B. et al.: Phys. Rev. Lett. **97**, 102501 (2006)
8. Grinyer, G.F. et al.: Nucl. Instrum. Meth. **A579**, 1005 (2007)
9. Grinyer, G.F. et al.: Phys. Rev. **C76**, 025503 (2007)
10. Leach, K.G. et al.: Phys. Rev. Lett. **100**, 192504 (2008)
11. Finlay, P. et al.: Phys. Rev. **C78**, 025502 (2008)
12. Finlay, P. et al.: Phys. Rev. Lett. **106**, 192504 (2011)
13. Dunlop, R. et al.: In: Garrett, P.E. (ed.) Proc. 14'th Int. Sym. Capt. Gamma-Ray Spect. Rel. Topics. World Scientific, Singapore (2012)
14. Smith, M.B. et al.: Phys. Rev. **C68**, 031302(R) (2003)
15. Sarazin, F. et al.: Phys. Rev. **C70**, 031302(R) (2004)
16. Chakrawarthy, R.S. et al.: Phys. Rev. C **73**, 024306 (2006)
17. Green, K.L. et al.: Phys. Rev. **C80**, 032502(R) (2009)
18. Mattoon, C.M. et al.: Phys. Rev. **C80**, 034318 (2009)
19. Singh, B., Hu, Z.: Nucl. Data Sheets **98**, 355 (2003)
20. Svensson, C.E., Garnsworthy, A.B.: The GRIFFIN Spectrometer, This Edition

DOI 10.1007/978-94-007-7963-1_13
Reprinted from *Hyperfine Interactions* Journal, DOI 10.1007/s10751-013-0889-3

The GRIFFIN spectrometer

C. E. Svensson · A. B. Garnsworthy

Published online: 10 October 2013

Abstract Gamma-Ray Infrastructure For Fundamental Investigations of Nuclei (GRIFFIN) is an advanced new high-efficiency γ-ray spectrometer being developed for use in decay spectroscopy experiments with low-energy radioactive ion beams provided by TRIUMF's Isotope Separator and Accelerator (ISAC-I) radioactive ion beam facility. GRIFFIN will be comprised of sixteen large-volume clover-type high-purity germanium (HPGe) γ-ray detectors coupled to custom digital signal processing electronics and used in conjunction with a suite of auxiliary detection systems. This article provides an overview of the GRIFFIN spectrometer and its expected performance characteristics.

Keywords Gamma-Ray spectrometer · HPGe clover detectors · Radioactive ion beams · Digital electronics

1 Introduction

The Isotope Separator and Accelerator (ISAC) facility [1] at TRIUMF in Vancouver, Canada produces high-quality beams of radioactive ions via the isotope separation on-line (ISOL) technique with primary production targets driven by a 500 MeV proton beam from TRIUMF's main cyclotron with a current up to 100 μA. In addition to ISAC's post-acceleration capabilities [2], low-energy (~30 keV) beams of radioactive isotopes with masses up to $A = 238$ can be transported directly to a variety of experimental facilities, including the 8π γ-ray spectrometer [3, 4] and its

ISAC and ARIEL: The TRIUMF Radioactive Beam Facilities and the Scientific Program.

C. E. Svensson (✉)
Department of Physics, University of Guelph, Guelph, ON, Canada N1G 2W1
e-mail: sven@uoguelph.ca

A. B. Garnsworthy
Science Division, TRIUMF, 4004 Wesbrook Mall, Vancouver, B.C., Canada V6T 2A3

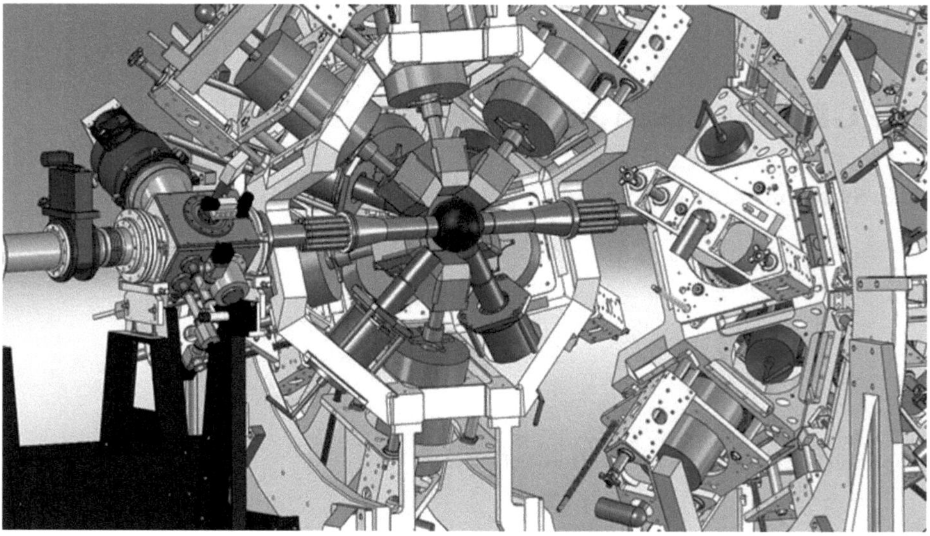

Fig. 1 SolidWorks design of the GRIFFIN spectrometer on its low-energy beamline at ISAC-I. One hemisphere of GRIFFIN is withdrawn to reveal the inner vacuum chamber, which will accommodate a number of GRIFFIN auxiliary detection systems

auxiliary detection systems, which have served as the primary decay spectroscopy facility at ISAC-I for the past decade. The Advanced Rare IsotopE Laboratory (ARIEL) [5] project is currently under construction at TRIUMF and will provide both a simultaneous multi-user radioactive beam capability at ISAC and access to unprecedented intensities of neutron-rich radioactive beams through actinide production targets driven by both a new superconducting electron linear accelerator and a second high-intensity proton beam. In order to fully exploit the new opportunities for nuclear structure, nuclear astrophysics, and fundamental symmetries research that will be provided by ARIEL, the current 8π facility for decay spectroscopy research at ISAC-I will be replaced by the advanced new high-efficiency GRIFFIN spectrometer described in this article.

2 GRIFFIN detector and spectrometer design

GRIFFIN will consist of sixteen clover-type detectors, each comprised of four 90 mm long, ~40 % relative efficiency, n-type high-purity germanium (HPGe) crystals packed in a four-leaf clover geometry. The mechanical dimensions of the GRIFFIN HPGe crystals and clover detectors have been designed to be identical to those used in the TIGRESS [6, 7] array at ISAC-II and, as in TIGRESS, the outer edges of the HPGe crystals will be tapered at 22.5 degrees over the first 30 mm of their length to enable close-packing of neighbouring clover detectors in the GRIFFIN array. The segmentation of the outer electrical contacts of the TIGRESS detectors [8, 9] used for γ-ray Doppler shift corrections in experiments with accelerated radioactive beams, however, will not be present in the GRIFFIN detectors, which will be used in decay spectroscopy experiments with stopped low-energy beams.

As shown in Fig. 1, the GRIFFIN HPGe detectors will be close-packed to fill 16 of the 18 square faces of a rhombicuboctahedral geometry. One of the remaining square faces will be used for the entry of the low-energy radioactive ion beam from ISAC-I, which will be implanted at the center of the array in the same in-vacuum moving tape collector system currently used with the 8π Spectrometer [4]. The other square face will be used for the exit of the in-vacuum tape, allowing long-lived daughter and contaminant activities to be moved into a collection box shielded from the HPGe detectors by a lead wall. As shown in Fig. 1, the 8 triangular faces of the rhombicuboctahedron will also be available for mounting GRIFFIN auxiliary detection systems, including the BaF_2 and $LaBr_3$ fast-timing γ-ray detectors currently used with the 8π Spectrometer [4], or 8π HPGe detectors themselves, which have 0.5 mm Be entrance windows and can efficiently trigger on photons below 10 keV in those experiments for which low-energy detection efficiency is crucial.

In the close-packed geometry, the front faces of the GRIFFIN HPGe detectors will be located 110 mm from the center of the array, comfortably accommodating the existing central vacuum chamber of the 8π Spectrometer, which has an outer radius of 89.4 mm. The existing in-vacuum auxiliary detection systems of the 8π [4], including the 20 plastic scintillator β detectors of the 4π Scintillating Electron Positron Tagging Array (SCEPTAR), the 5 liquid nitrogen cooled Si(Li) conversion electron detectors of the Pentagonal Array of Conversion Electron Spectrometers (PACES), which also serve as α particle and proton detectors, and the 1π zero degree scintillator (ZDS) β detector for fast timing experiments, will all be immediately available for use in GRIFFIN experiments. The gap between the inner vacuum chamber and the GRIFFIN HPGe detectors can optionally be filled by spherical shells of Delrin with thicknesses up to 20 mm in order to stop energetic β particle from directly hitting the HPGe detectors, while minimizing bremsstrahlung production through the use of a low-Z stopping material. By removing the 4 downstream HPGe clover detectors, it will also be possible to couple the new 70-element Deuterated Scintillator Array for Neutron Tagging (DESCANT) [10] neutron detector array to GRIFFIN, providing a powerful β-delayed n-γ coincidence detection capability. Finally, the mechanical support structure for GRIFFIN shown in Fig. 1 has been designed for full compatibility with the future addition of modular bismuth germanate (BGO) Compton suppression shields of a similar design to those used in the TIGRESS array [11].

3 GRIFFIN data acquisition system

The data acquisition (DAQ) system for GRIFFIN will take advantage of state-of-the-art digital electronics based on the TIGRESS DAQ System [12]. The main design aims for the system are to facilitate the requirements of the two themes of the physics program for nuclear structure and precision fundamental symmetries measurements; to operate at high data throughput with each HPGe crystal operating at rates up to 50 kHz, and to achieve a level of accountability and deadtime/event traceability consistent with half-life and branching ratio measurements with precisions better than 0.05 %. Signal processing will be performed with digitizers designed and built at the Université de Montréal and TRIUMF. The front-end digitizers will use 14-bit precision operating at 100 MHz for HPGe, Si(Li), BaF_2 and $LaBr_3$ energy signals and at 1 GHz for SCEPTAR/ZDS and DESCANT. The fast coincident timing for the BaF_2 and $LaBr_3$ scintillators will be performed in conventional analogue

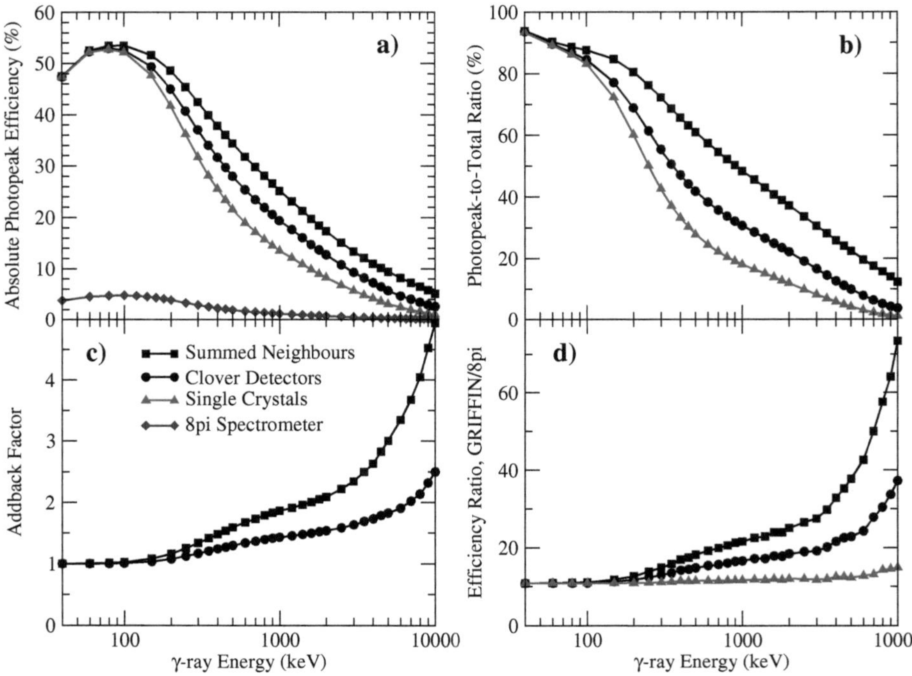

Fig. 2 GEANT4 simulations of GRIFFIN performance metrics: **a** absolute photopeak efficiency, **b** photopeak-to-total ratio, **c** addback factor, and **d** ratio of photopeak efficiency to that of the current 8π Spectrometer, as a function of γ-ray energy. GRIFFIN results are shown treating individual HPGe crystals as separate detectors (*triangles*), summing the energies deposited in the four crystals within a clover detector (*circles*), and summing energies deposited in neighbouring clover detectors (*squares*), while the photopeak efficiencies of the 8π Spectrometer are indicated by diamonds in panel **a**

electronics and the output signal of the TAC modules digitized to be considered in filtering decisions. All events will be pushed to filtering logic where user-selectable coincidence conditions must be satisfied before the data is committed to storage.

4 GRIFFIN performance simulations

Detailed Monte Carlo studies of GRIFFIN performance have been carried out with the GEANT4 simulation package. Figure 2 shows results as a function of γ-ray energy treating the individual HPGe crystals as separate detectors (triangles), summing the energies deposited in the crystals of each clover detector (circles), and summing energies deposited in neighbouring clover detectors in the close-packed array (squares). As shown in Fig. 2a, the absolute photopeak efficiency of GRIFFIN exceeds 50 % in the 50–150 keV energy range, in the "clover" mode is 19 % at 1 MeV and 2.6 % at 10 MeV, while in the "summed neighbour" mode is 25 % at 1 MeV and remains 5.1 % at 10 MeV. The large gains in photopeak efficiency achieved by adding the γ-ray energies deposited in neighbouring crystals, either within the same clover detector or in neighbouring detectors, are displayed in Fig. 2c as the "addback" factors, which are the clover and summed neighbour photopeak efficiencies divided by those of the single crystal mode. For the clover detectors, the addback factor is 1.43 at 1 MeV and 2.5 at 10 MeV, while for the summed neighbour mode it is 1.86 at 1 MeV and 4.9 at 10 MeV, indicating that the close packing of the GRIFFIN detectors, with their flush tapered outer edges over the first 30 mm of HPGe crystal depth, effectively provides an additional addback factor in low γ-ray multiplicity β decay experiments that equals the usual clover addback factor for 1 MeV γ rays and significantly exceeds it for γ rays above 5 MeV.

The photopeak-to-total ratios for GRIFFIN are shown in Fig. 2b and for 1 MeV γ rays increase from 18 % for single crystals, to 31 % for the clover detectors, and 48 % for the summed neighbour mode. Finally, Fig. 2d displays the ratios of the GRIFFIN photopeak efficiencies to those of the current 8π Spectrometer at ISAC-I (shown as diamonds in Fig. 2a for reference). For low-energy γ rays, GRIFFIN is a factor of 11 more efficient than the 8π Spectrometer due to the increased solid angle coverage by HPGe, and this factor is relatively constant in the single crystal mode, increasing to a factor of 15 for 10 MeV γ rays. The addback factors of the clover and summed neighbour modes, however, provide large gains with increasing γ-ray energy. The GRIFFIN clover mode is predicted to be 17 times more efficient than the 8π Spectrometer at 1 MeV and 37 times more efficient at 10 MeV, while the summed neighbour mode is 22 times more efficient at 1 MeV and more than 70 times more efficient than the 8π Spectrometer for 10 MeV γ rays. In addition to these very large efficiency gains for high-energy γ rays, it is also important to note that many GRIFFIN experiments will be carried out in γ-γ coincidence mode, for which the relevant figure of merit is the square of the photopeak efficiency. For typical γ-ray energies in the 1–2 MeV range, the clover mode of GRIFFIN will thus provide approximately 300 times the γ-γ coincidence efficiency of the current 8π Spectrometer, while the summed neighbour mode of GRIFFIN will have almost 500 times the γ-γ efficiency of the 8π. These enormous gains in high-energy and γ-γ coincidence efficiency provided by GRIFFIN will enable both high-precision branching ratio measurements for complex high Q-value β decays and detailed spectroscopic studies of the most exotic beams produced by the new ARIEL facility with intensities below 0.01 ions/s.

5 Summary and outlook

GRIFFIN has been designed to provide the excellent γ-ray energy resolution characteristic of HPGe detectors, the very high photopeak detection efficiency required for experiments with the most exotic radioactive beams, high rate capabilities for experiments with intense radioactive beams closer to stability, and the rigorous control of electronic deadtimes required for high-precision fundamental symmetries measurements, such as the RnEDM experiment [13]. First experiments with $\sim$10 GRIFFIN HPGe clover detectors at ISAC-I are planned for 2014, while the full 16-detector GRIFFIN spectrometer will be operational in 2015. As is the case with the current 8π Spectrometer, GRIFFIN will operate in conjunction with a suite of auxiliary detection systems, including β, conversion electron, neutron, low-energy photon, and fast-timing γ-ray detectors. The combination of the state-of-the-art GRIFFIN spectrometer, this powerful suite of auxiliary detectors, and the intense beams of radioactive ions from ISAC, and its major upgrade to ARIEL, will provide a wide range of new opportunities for nuclear structure, nuclear astrophysics, and fundamental symmetries research.

Acknowledgements The authors would like to thank the entire GRIFFIN collaboration for their ongoing contributions to the GRIFFIN program, and D. Brennan and E. Rand in particular for assistance in the preparation of this article. GRIFFIN is funded jointly by the Canada Foundation for Innovation (CFI), TRIUMF and the University of Guelph. This work has been partially supported by the Natural Sciences and Engineering Research Council of Canada (NSERC). TRIUMF receives funding via a contribution agreement through the National Research Council (NRC) of Canada.

References

1. Dilling, J., Krücken, R., Ball, G.C.: The experimental facilities at ISAC. Hyperfine Interact. (2013). doi:10.1007/s10751-013-0886-6
2. Laxdal, R., Marchetto, M.: The ISAC post-accelerator. Hyperfine Interact. (2013). doi:10.1007/s10751-013-0884-8
3. Svensson, C.E., et al.: Radioactive beam experiments with large Gamma-Ray detector arrays. Nucl. Instrum. Meth. Phys. Res. B **204**, 660–665 (2003)
4. Garnsworthy, A.B., Garrett, P.E.: The 8π Spectrometer. Hyperfine Interact. (2013). doi:10.1007/s10751-013-0888-4
5. Dilling, J., Krücken, R., Merminga, L.: ARIEL overview. Hyperfine Interact. (2013). doi:10.1007/s10751-013-0906-6
6. Svensson, C.E., et al.: TIGRESS: TRIUMF-ISAC Gamma-Ray escape-suppressed spectrometer. J. Phys. G **31**, S1663–1668 (2005)
7. Hackman, G., Svensson, C.E.: The TRIUMF-ISAC Gamma-Ray escape suppressed spectrometer, TIGRESS. Hyperfine Interact. (2013). doi:10.1007/s10751-013-0905-7
8. Scraggs, H.C., et al.: TIGRESS highly segmented high-purity Germanium clover detector. Nucl. Instrum. Meth. Phys. Res. A **543**, 431–440 (2005)
9. Svensson, C.E., et al.: Position sensitivity of the TIGRESS 32-Fold segmented HPGe detector. Nucl. Instrum. Meth. Phys. Res. A **540**, 348–360 (2005)
10. Garrett, P.E.: DESCANT—the deuterated scintillator array for neutron tagging. Hyperfine Interact. (2013). doi:10.1007/s10751-013-0891-9
11. Schumaker, M.A., et al.: Measured and simulated performance of compton-suppressed TIGRESS HPGe clover detectors. Nucl. Instrum. Meth. Phys. Res. A **570**, 437–445 (2007)
12. Martin, J.-P., et al.: The TIGRESS DAQ/Trigger system. IEEE Trans. Nucl. Sci. **55**, 84–90 (2008)
13. Tardiff, E.R., et al.: The Radon EDM apparatus. Hyperfine Interact. (2013). doi:10.1007/s10751-013-0898-2

DOI 10.1007/978-94-007-7963-1_14
Reprinted from *Hyperfine Interactions* Journal, DOI 10.1007/s10751-013-0890-x

Precision nuclear β-decay half-life measurements

Gordon C. Ball

Published online: 11 October 2013

Abstract Precise measurements of the intensities of superallowed Fermi β decays between 0^+ analog states provide a demanding test of the electroweak standard model. At ISAC, nuclear β-decay lifetimes are measured with high-precision (<0.05 %) by direct β counting using a 4π continuous flow gas proportional counter coupling to a fast tape transport system. This facility has provided the most precise half-life measurements for several superallowed β emitters including: $^{26}Al^{m}$, $^{38}K^{m}$, ^{62}Ga and ^{74}Rb.

Keywords Superallowed fermi β decay · Half-life · Direct β counting

1 Introduction

Precise measurements of superallowed $0^+ \rightarrow 0^+$ β^+ transitions have led to increasingly precise tests of several key ingredients in the electroweak standard model, specifically: the conserved vector current (CVC) hypothesis, the absence of scalar currents, and the unitarity of the Cabibbo-Kobayashi-Maskawa (CKM) matrix. At the present time, from the world data on superallowed decays [1] it has been determined that the vector coupling constant has the same value for all 13 precisely measured transitions to within ± 0.013 %. These data can also be used to set an upper limit on a possible scalar current at 0.2 % of the vector current, and to confirm CKM unitarity to ± 0.06 %.

The *ft* value that characterizes any β transition depends on three measured quantities: the total transition energy, the half-life of the parent state; and the branching ratio for the particular transition of interest. The transition energy is

ISAC and ARIEL: The TRIUMF Radioactive Beam Facilities and the Scientific Program.

G. C. Ball (✉)
TRIUMF, Vancouver, BC VT6 2A3, Canada
e-mail: ball@triumf.ca

required to determine the statistical rate function, f, while the half-life and branching ratio combine to yield the partial half-life, t. The ISAC facility is ideally suited to precision nuclear β-decay studies and we now have experimental facilities capable of measuring all three quantities, namely high precision half-lives using the β-counting nuclear lifetime facility, high precision branching ratio and lifetime measurements using the 8π gamma-ray spectrometer and high precision mass measurements using the TITAN penning trap mass spectrometer. The β-counting facility is described in this paper. The other two facilities are described elsewhere in the ISAC Laboratory Portrait.

2 Facility details

The high-precision β-decay half-life measurements are carried out by direct β-counting using a technique established previously in experiments elsewhere (see Ref. [2] for more details). Although the measurements are simple in principle, great care must be taken to achieve the required precision (<0.05 %). A schematic diagram of the apparatus is shown in Fig. 1. The low-energy, ~30 keV radioactive ion beams from ISAC are implanted into the 25-mm-wide aluminized mylar tape of a fast tape-transport system [3] modified to operate at 4 m/s. After a collection period of up to 4 half-lives, the ISAC beam is interrupted and the samples are moved out of the vacuum chamber through two stages of differential pumping, finally being positioned in the center of a 4π continuous gas-flow proportional β counter [2, 4]. At the highest tape transport speed, sample counting can begin ~100 ms after the collection period. The signals from the 4π counter are then multiscaled for about 25 half-lives and the data stored. This collect-move-count cycle is repeated continuously until sufficient statistics are accumulated.

A 1 MHz ± 2 Hz laboratory clock is used to provide a time standard for the experiment which is controlled by a Jorway 221, twelve channel timing and sequence module. The signals from the 4π detector are sent to an Ortec 572 timing filter amplifier and then passed to a discriminator, the output from which is split and sent to two fixed-width non-extendable Lecroy 222 gate generators. The durations of the gates are set to different times, typically 3 and 4 μs, both being significantly longer than any series dead time in the circuit preceding it and each thus establishing a single dominant dead time in its branch. The gate signals from each branch are then routed simultaneously to two different and completely independent CAMAC multichannel scaler (MCS) modules. The dead times in both branches of the electronics are measured with an uncertainty of ±0.01 μs using the source-plus-pulser technique [6]. By comparing the dead-time corrected results from both branches, we are able to check for any systematic dead-time effects.

To test for possible systematic errors associated with the other key equipment parameters the cycles are grouped into separate runs, each with different settings for these parameters. The detector bias voltage is altered within the plateau region (typically 2600–2900 V); the lower-level-discriminator threshold is varied by a factor of ~3; the fixed 3 and 4 μs dead times are interchanged and the dwell times of the MCS modules are also varied by ~10–20 %. The run-to-run consistency in the measured half-life values then provides a sensitive test for any unwanted dependence on these parameters.

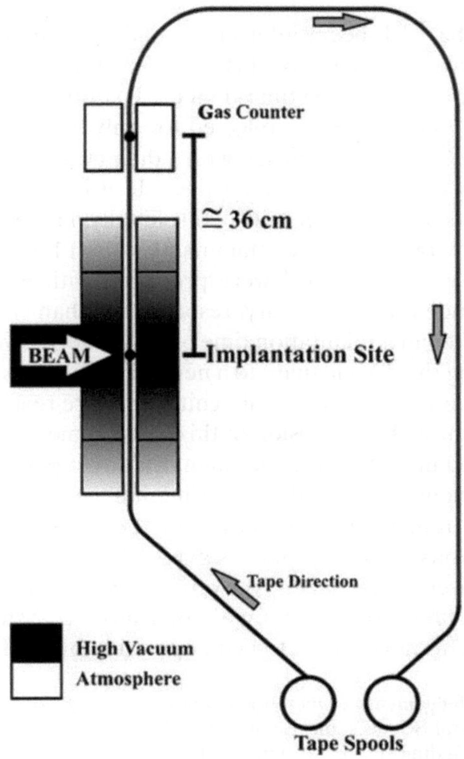

Fig. 1 Schematic of the 4π continuous gas flow proportional counter and fast tape transport system used in the β-counting facility [5]

At ISAC the radioactive ion beams from the source are mass analyzed by a low-resolution preseparator followed by a high acceptance mass analyzer with a resolving power of $M/\Delta M = 1000$. This is more than sufficient to eliminate any nuclides with a different mass number from the one of interest, but not neighbouring isobars. Often, the intensity of these isobaric contaminants are too small to be determined by fitting the measured decay curve but can still result in an unknown systematic error in these high-precision measurements. As a result, the upper limit on the possible contribution from such isobars is monitored throughout the experiment with a large volume HPGe-detector located just outside of the $4\pi\beta$ counter.

3 Experimental program

The β-counting nuclear lifetime facility has been used to determine the half-lives of several short-lived superallowed β emitters including $^{26}Al^m$, t 1/2 = 6346.54 ± 0.76 ms [7], $^{38}K^m$, t 1/2 = 924.46 ± 0.14 ms [8], ^{62}Ga, t 1/2 = 116.100 ± 0.025 ms [9] and the first high precision measurement for ^{74}Rb, t 1/2 = 64.761 ± 0.031 ms [10].

Each measurement presented its own unique challenges. One of the first experiments carried out at ISAC was the half-life measurement for the supperallowed β-emitter ^{74}Rb. The precision of this measurement (0.05 %) was limited by the ^{74}Rb beam intensity available during the initial operation of ISAC. The statistical uncertainty in this measurement could be reduced by more than a factor of two given the ^{74}Rb beam intensity available today. For the short-lived β-emitter ^{38}K^{m}, the yield of the long-lived (t 1/2 = 7.6 min) ground state of the same nucleus, ^{38}K, was approximately 50 times higher. To optimize the ratio of initial activities in favor of ^{38}K^{m}, samples were collected for only 0.3 s and, in addition, the proton beam on the ISAC target was pulsed with a duty cycle of 6–11 %; that is, the beam was on for only 4 s of every 30–60 s. As a result, it was possible to increase the ratio of initial activities for ^{38}K^{m}/^{38}K to 60–80:1. Finally, in our recent measurement of the half-life of ^{26}Al^{m}, two isobaric contaminants ^{26}Na (t 1/2 = 1.07128(25) s [5] and ^{26}Al^{g} (t 1/2 = 7.17(24) $\times$ 10^{5} yrs [11] were produced with intensities of one and two orders of magnitude higher intensity, respectively, than ^{26}Al^{m} which had a yield 10^{6}/s. In this case, after an implantation time of 6–14 s we allowed the samples to "cool" for 26–34 s reducing the ^{26}Na activity to a negligible level. As a result, we were able to follow the decay for over 4 orders of magnitude before reaching the background limit of the 4π gas counter. The precision of this half-life measurement, 0.012 %, is the highest ever achieved in a single measurement for any superallowed β-emitter and represents the experimental limit of this experimental technique.

One fundamental limitation of the present β-counting facility is the diffusion of gaseous radioisotopes such as ^{18}Ne and molecular ^{12}C^{14}O implanted in the aluminized mylar tape. To overcome this limitation (Zganjar, private communication) is presently designing a tape transport system that will use thick Al tape. We expect this new system to be in operation by the fall of 2012.

Acknowledgements Operational funding for the β-counting nuclear lifetime facility is provided by the Natural Sciences and Engineering Research Council of Canada (NSERC). TRIUMF receives federal funding via a contribution agreement through the National Research Council of Canada.

References

1. Hardy, J.C., Towner, I.S.: Phys. Rev. **C79**, 055502 (2009)
2. Koslowsky, V.T., et al.: Nucl. Inst. Methods Phys. Res. **A401**, 289 (1997)
3. Hagberg, E., et al.: Nucl. Phys. **A571**, 555 (1994)
4. Koslowsky, V.T., et al.: Nucl. Phys. **A405**, 29 (1983)
5. Grinyer, G.F., et al.: Phys. Rev. **C71**, 044309 (2005)
6. Baerg, A.P.: Metrologia **1**(3), **131** (1965)
7. Finlay, P., et al.: Phys. Rev. Lett. **106**, 032501 (2011)
8. Ball, G.C., et al.: Phys. Rev. **C82**, 045501 (2010)
9. Grinyer, G.F., et al.: Phys. Rev. **C77**, 015501 (2008)
10. Ball, G.C., et al.: Phys. Rev. Lett. **86**, 1454 (2001)
11. Geppart, C., et al.: Nucl. Phys. **A746**, 631c (2004)

DOI 10.1007/978-94-007-7963-1_15
Reprinted from *Hyperfine Interactions* Journal, DOI 10.1007/s10751-013-0891-9

DESCANT – the deuterated scintillator array for neutron tagging

P. E. Garrett

Published online: 15 October 2013

Abstract The DESCANT array is a new device for neutron detection based on deuterated liquid scintillator. It has been designed to be coupled with the TIGRESS and GRIFFIN γ-ray spectrometers to enable neutron tagging in fusion-evaporation reactions, and β-delayed neutron studies.

Keywords Gamma-ray spectroscopy · Beta decay

1 Introduction

The study of neutron-rich nuclei is at the forefront of nuclear structure research, but is made difficult by the limitation of methods by which they can be produced and observed. Studies of neutron-rich nuclei using reactions, such as fusion-evaporation reactions that have formed the backbone of techniques to study nuclear structure on the proton-rich side of the valley of stability, will have the tendency of the compound system to emit multiple neutrons during the cooling phase, driving it back towards stability. Reactions that produce the compound system in a rather cool state will likely be widely used to study the most neutron-rich species. Charged-particle detection, recoil detection, and neutron detection from compound systems will be used, with some detection methods offering clear advantages over others in any particular reaction. For some reactions, for example, the recoil will not be sufficiently energetic or constrained within the acceptance cone of a recoil device like the EMMA spectrometer [1]. Furthermore, there will be a definite need to characterize the emitted particles, including neutrons, and their multiplicity in the reaction. Therefore, there is a definite need for sensitive neutron detector arrays that can be coupled with powerful γ-ray spectrometers.

ISAC and ARIEL: The TRIUMF Radioactive Beam Facilities and the Scientific Program.

P. E. Garrett (✉)
Department of Physics, University of Guelph, Guelph, ON N1G 2W1, Canada
e-mail: pgarrett@physics.uoguelph.ca

Fig. 1 The DESCANT neutron-detector array coupled to the TIGRESS γ-ray spectrometer. The downstream lampshade of TIGRESS is removed and replaced by DESCANT. The DESCANT solid angle is 1.08π sr covered by 70 individual detector units, with a target-to-detector distance of 50 cm

In addition to the need of neutron detection in fusion-evaporation studies, a neutron detector array will also be used in reaction studies, such as those involving halo nuclei like ^{11}Li. Small-angle correlations between neutrons emitted in reactions of ^{11}Li with various targets contain structural information about the halo neutrons. However, in typical neutron detector arrays these small opening-angle events must be rejected due to the large amount of detector-to-detector scatterings that are always present. A method that could successfully distinguish between multiple scattering events and true high-multiplicity events would offer a tremendous advantage.

A sensitive neutron detector array can also prove invaluable in β-decay studies of neutron-rich systems. In neutron-rich nuclei, the β-decay Q-values become very large and it is possible to populate final states in the daughter nucleus above the neutron separation energy. The neutron-emission probability, P_n, is vital information not only for nuclear structure studies, but also for detailed understanding of nuclear astrophysical scenarios, as well as applications in nuclear engineering and advanced fuel-cycle reactor design.

In order to meet these needs, the Deuterated Scintillator Array for Neutron Tagging, DESCANT, is begin constructed. Coupled to a high-efficiency γ-ray spectrometer, like TIGRESS [2] or GRIFFIN [3], DESCANT will enable spectroscopy with neutron-rich beams, accelerated to several tens of keV for implantation for β-decay studies, or above the Coulomb barrier for reaction studies.

2 DESCANT design

The DESCANT spectrometer is designed to be coupled with both the TIGRESS γ-ray spectrometer [2], and the future GRIFFIN spectrometer [3]. It will replace the forward "lampshade" of 4 clover-type HPGe detectors (with their BGO suppression shields), and will occupy a solid angle of 1.08π sr with the maximum angle of 65.5°

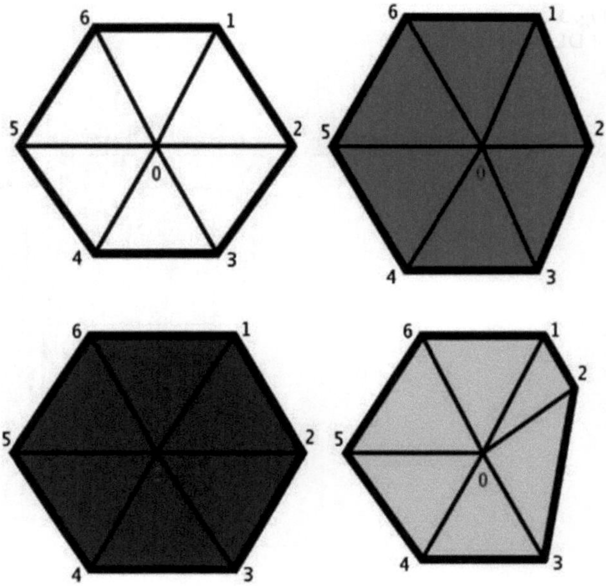

Fig. 2 The four basic shapes of the DESCANT detectors. For scale, the length of the bisector from point 2 to 5 on the White detector is 14.3 cm

degrees with respect to the initial beam direction. The target-to-detector distance is 50 cm, and the individual detector cans are 15 cm thick. When fully loaded, DESCANT contains 70 individual neutron detectors, as pictured in Fig. 1 where it is coupled to the TIGRESS array. DESCANT is designed so that the inner ring, and its adjacent ring, of detectors surrounding the beam line can be removed to facilitate larger auxiliary detector systems that may be placed downstream of the target. The 70 detector units contain liquid deuterated scintillator, BC537, and were fabricated by the Bicron division of St. Gobain.

The geometry of DESCANT is based on 3 basic shapes of detectors; Red, White, and Blue. However, in order to maximize the solid angle coverage, the outer ring of detectors, which normally would be based on the White shape, were modified to have a truncated shape, resulting in the Green and Yellow detectors that are mirror images of each other. The shapes of the front faces of the detectors are shown in Fig. 2, and the actual detectors are shown in Fig. 3.

Signals from the DESCANT detectors will be digitized by custom-built 1 GHz waveform digitizers manufactured by Instrumentation Services of the University of Montréal. On-board digital signal processing will integrate the pulse to yield the total charge, determine the event time via a constant-fraction algorithm, and perform neutron-γ discrimination. The new digitizers are designed to be fully integrable into the TIGRESS [4] and GRIFFIN data acquisition systems.

3 Neutron detector performance

DESCANT uses a deuterated-benzene liquid scintillator, BC-537, from St. Gobain. Due to the asymmetric nature of *n*-*d* scattering in the center of mass, unlike *n*-*p* scattering that is isotropic, the pulse-height information from the deuterated scin-

Fig. 3 The 5 detectors of DESCANT

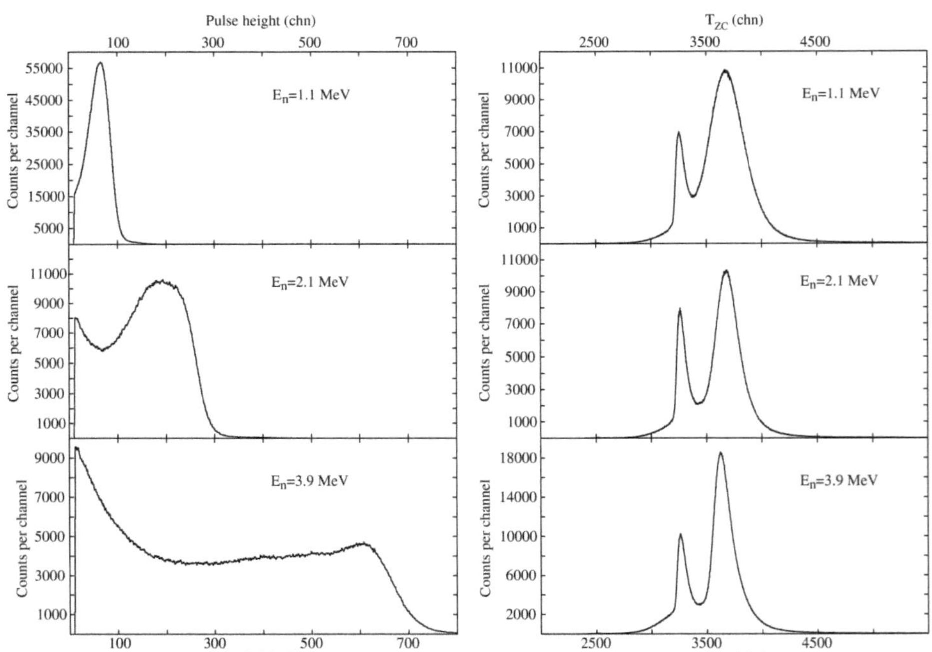

Fig. 4 Pulse height (*left*) and $n\gamma$ discrimination (*right*) data from a DESCANT detector obtained with mono-energetic neutrons of 1.1 MeV (*top*), 2.1 MeV (*middle*), and 3.9 MeV (*bottom*)

tillator contains useful information on the initial neutron energy. While deuterated detectors have been used in active-target experiments, and as γ-ray detectors in the presence of large thermal neutron fluxes, they had not been used in a large-scale array of detectors for fast neutrons.

The DESCANT detectors have a time resolution, measured with a ^{60}Co source, of slightly better than 1 ns FWHM, and an energy resolution of $\simeq$ 25 % measured by the width of the Compton edge from sources of ^{22}Na, ^{137}Cs, and ^{60}Co. Changes in pulse height position, energy, and time resolution as a function of count rate and front-face position show less than a 2 % variation.

Shown on the left of Fig. 4 are examples of pulse-height spectra obtained for monoenergetic neutron energies of 1.1 MeV, 2.1 MeV, and 3.9 MeV, incident on a DESCANT detector. The neutrons were obtained via the $^{3}H(p, n)^{3}He$ reaction at the University of Kentucky accelerator facility. A condition on the $n\gamma$ discrimination shown on the right in Fig. 4 to select the neutrons (the right-most peak) based on their zero-cross-over time was made. Note that the pulse-height spectrum has a prominent peak-like structure, especially for low neutron energies. By combining both the pulse height information and the TOF, DESCANT should be able to differentiate multiple-scattered from true multiple-neutron events in the array better than detectors using hydrogen-based scintillators.

The DESCANT array is currently in the assembly stage with an operational status projected in the spring 2014. The first commissioning experiment will utilize stable beam in order to explore the capacity of the device to handle widely varying count rates, and different neutron-multiplicity events.

Acknowledgements The author would like to thank L. Bianco, V. Bildstein, B. Hadinia, and J. Wong for assistance in the preparation of this article, S.W. Yates and M. McEllistrem for assistance in making measurements at the University of Kentucky, and TRIUMF for the design and fabrication of the DESCANT support structure. Construction funds for DESCANT were provided by the Canadian Foundation for Innovation, the Ontario Ministry for Research and Innovation, and TRIUMF.

References

1. Davids, B., Davids, C.N.: Nucl. Instrum. Meth. **A544**, 565 (2005)
2. Svensson, C.E., Amaudruz, P., Andreoiu, C., Andreyev, A., Austin, R.A.E., Ball, G.C., Bandyopadhyay, D., Boston, A.J., Chakrawarthy, R.S., Chen, A.A., Churchman, R., Drake, T.E., Finlay, P., Garrett, P.E., Grinyer, G.F., Hackman, G., Hyland, B., Jones, B., Kanungo, R., Maharaj, R., Martin, J.P., Morris, D., Morton, A.C., Pearson, C.J., Phillips, A.A., Ressler, J.J., Roy, R., Sarazin, F., Schumaker, M.A., Scraggs, H.C., Smith, M.B., Starinsky, N., Valiente-Dobón, J.J., Waddington, J.C., Watters, L.M.: J. Phys. **G31**, S1663 (2005)
3. Svensson, C.E., Garnsworthy, A.B.: Hyper. Int., This edition
4. Martin, J.P., Mercier, C., Starinski, N., Pearson, C.J., Amaudruz, P.A.: IEEE Trans. Nucl. Sci. **55**, 88 (2008)

DOI 10.1007/978-94-007-7963-1_16
Reprinted from *Hyperfine Interactions* Journal, DOI 10.1007/s10751-013-0892-8

TITAN: An ion trap facility for on-line mass measurement experiments

A. A. Kwiatkowski · C. Andreoiu · J. C. Bale · T. Brunner · A. Chaudhuri · U. Chowdhury · P. Delheij · S. Ettenauer · D. Frekers · A. T. Gallant · A. Grossheim · G. Gwinner · F. Jang · A. Lennarz · T. Ma · E. Mané · M. R. Pearson · B. E. Schultz · M. C. Simon · V. V. Simon · J. Dilling

Published online: 10 October 2013

Abstract Precision determinations of ground state or even isomeric state masses reveal fingerprints of nuclear structure. In particular, at the limits of existence for very neutron-rich or -deficient isotopes, one can extract detailed information about

ISAC and ARIEL: The TRIUMF Radioactive Beam Facilities and the Scientific Program.

A. A. Kwiatkowski (✉) · T. Brunner · A. Chaudhuri · U. Chowdhury · P. Delheij · S. Ettenauer · A. T. Gallant · A. Grossheim · A. Lennarz · E. Mané · M. R. Pearson · B. E. Schultz · M. C. Simon · V. V. Simon · J. Dilling
TRIUMF, 4004 Wesbrook Mall, Vancouver, BC V6T 2A3, Canada
e-mail: aniak@triumf.ca

C. Andreoiu · J. C. Bale · T. Ma
Department of Chemistry, Simon Fraser University, Burnaby, BC, Canada

T. Brunner
Physik Department E12, Technische Universität München, München, Germany

U. Chowdhury · G. Gwinner
Department of Physics and Astronomy, University of Manitoba, Winnipeg, MB, Canada

S. Ettenauer · A. T. Gallant · F. Jang · J. Dilling
Department of Physics and Astronomy, University of British Columbia, Vancouver, BC, Canada

D. Frekers · A. Lennarz
Institut für Kernphysik, Westfälische Wilhelms-Universität, Münster, Germany

V. V. Simon
Max-Planck-Institut für Kernphysik, Heidelberg, Germany

V. V. Simon
Ruprecht-Karls-Universität, Heidelberg, Germany

T. Brunner
Present Address:
Department of Physics, Stanford University, Stanford, CA, USA

nuclear structure from separation energies or binding energies. Mass measurements are important to uncover new phenomena, to test new theoretical predictions, or to refine model approaches. For example, the $N = 28$ shell has proven more stable than previously expected; however, the predicted new "magic" number at $N = 34$ in the K and Ca isotopes has yet to be confirmed experimentally. For these neutron-rich nuclei, the inclusion of three-body forces leads to significantly better predictions of the ground-state mass. Similarly, halo nuclei present an excellent application for ab-initio theory, where ground state properties, like masses and radii, test our understanding of nuclear structure. Precision mass determinations at TRIUMF are carried out with the TITAN (TRIUMF's Ion Traps for Atomic and Nuclear science) facility. It is an ion-trap setup coupled to the on-line facility ISAC. TITAN has measured masses of isotopes as short-lived as 9 ms (almost an order of magnitude shorter-lived than any other Penning trap system), and it is the only one with charge breeding capabilities, which allow us to boost the precision by almost 2 orders of magnitude. We recently made use of this feature by measuring short-lived, proton-rich Rb-isotopes, up to ^{74}Rb while reaching the 12+ charge state, which together with other improvements led to an increase in precision by a factor 36.

Keywords Ion traps • Mass spectrometry and Mass measurements • Nuclear structure • Nuclear shell model • Nuclear astrophysics and fundamental interactions • Highly charged ions • Charge breeding • Buffer gas cooling

1 Introduction

Ion traps offer a unique environment in which to study a single ion or an ensemble: 3D confinement in a small volume, well-defined electromagnetic fields, and long observations times. Ion traps are versatile tools used in many scientific disciplines spanning forensics to metrology [1]. At TRIUMF, several ion traps have been implemented for TRIUMF's Ion Traps for Atomic and Nuclear science (TITAN) facility [2, 3]. Each trap serves a distinct purpose. At the heart of the TITAN research program is Penning trap mass spectrometry, currently the most precise method of mass measurements. Precision can be boosted by a significant factor by charge breeding the singly charged ions delivered from ISAC (Isotope Separator and Accelerator) [4] in the electron beam and ion trap (EBIT) to a charge state Q. As the EBIT was constructed with optical access, decay spectroscopy can be performed on trapped ions as well. To reduce the beam emittance from the EBIT, a Cooler Penning Trap (CPET) is under construction. The radio-frequency quadrupole (RFQ) trap accumulates, cools, and bunches the beam in general preparation for the subsequent TITAN ion traps or for the collinear laser spectroscopy experiment [5]. This article will focus on the TITAN mass measurement program and the ion traps used therein.

The atomic mass represents the sum of neutrons, protons, and electrons and encompasses all interactions among them. It is unique to each nuclide and provides insight into nuclear structure [6]. A cornerstone of the TITAN mass measurement program has been the measurement of light, neutron-rich isotopes, including several halo nuclei [7–13]. In these very loosely bound systems, the wavefunction of the valence nucleons extends far beyond classical expectations. Close to the limits of nuclear existence and with relatively few nucleons, halo nuclei act as an excellent test of nuclear models, particularly the *ab initio* models [14]. Surveys of the mass landscape indicate changes in nuclear structure. Far from the so-called valley of stability, the nuclear shell structure often deviates from expectation. Shells may be quenched, and new, more stable shell gaps appear, leading to new "magic numbers" [15–17], for example, in oxygen isotopes. Recently, TITAN confirmed the $N = 28$ shell closure in neutron-rich potassium and calcium isotopes [18]. Accurate knowledge of shell behavior could be essential to calculate elemental abundances; mass indicates the energetically-allowed pathway for nucleosynthesis in e.g. supernovae or x-ray bursts [19]. Measurements of neutron-rich Rb and Sr isotopes in the $A \approx 100$ region yield important clues to the r-process [20]. Accurate and precise masses (or Q-values) are also key to precision tests of the unitarity of the Cabibbo-Kobayashi-Maskawa (CKM) matrix [21]. For this purpose, TITAN measured the mass of ^{74}Rb [22], a superallowed β-emitter. From its $\mathcal{F}t$-value together with the twelve other well-known $\mathcal{F}t$-values, one can extract V_{ud} of the CKM matrix. With the highest Z of all superallowed β-emitters [21], ^{74}Rb represents a favorable case to distinguish between conflicting nuclear model predictions of the isospin-breaking correction δ_C, which scales as Z^2. The measurement also merits attention as the first Penning trap mass measurement of a charge-bred, short-lived ($T_{1/2} < 100$ ms) nuclide.

2 The TITAN facility

Penning trap mass spectrometry requires low energy, cooled, and pure ion beams. The ISAC facility [4] delivers a continuous beam of radioactive ions to TITAN with a typical transport energy of ≈ 20–40 keV. A schematic of TITAN is shown in Fig. 1. The efficiency of Penning trap mass spectrometry can be improved by operating with bunched beams. To this end, the ISAC beam first enters the linear gas-filled, radio-frequency quadrupole (RFQ) trap [23, 24], where it is accumulated, cooled, and bunched. Upon ejection from the RFQ, the ion bunches can be sent to either the EBIT for charge breeding before the mass measurement or directly to the measurement Penning trap (MPET). In the near future, the CPET will be installed just upstream of the MPET (green arrow in Fig. 1) to provide additional cooling.

2.1 The RFQ beam cooler and buncher trap

The RFQ [23, 24] is a linear Paul trap filled with helium gas held at a pressure of $\approx 2\times10^{-2}$ mbar. The singly charged ions (SCI) collide with the inert buffer gas and thereby lose kinetic energy by transfer to the room-temperature gas, which is constantly exchanged. A photograph of the RFQ is shown in Fig. 2. The transverse confinement is supplied by an RF potential on four half-circular rods. The RF potential is driven with a square wave in order to provide deeper transverse confinement [23] than with the typical sine wave [25]. The rods are segmented to allow one to apply a long drag field and the electrostatic axial trap for beam accumulation. The RFQ serves both TITAN and the laser spectroscopy facilities; the latter is located further along the ISAC beamline (see Fig. 1). Therefore, the ions can be extracted from the RFQ either in the forward (TITAN) direction or in the reverse (laser spectroscopy or injection) direction. This capability is unique to the TITAN RFQ system. After forward extraction from the RFQ, the kinetic transport energy is adjusted to about 2 keV, and the ion bunches are sent to the switchyard where they can be directed towards the MPET for immediate mass measurement or to the EBIT for charge breeding.

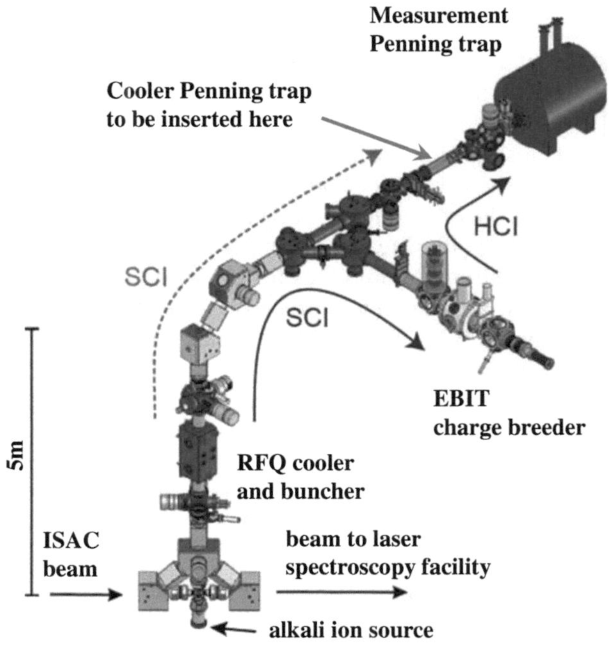

Fig. 1 Schematic of the TITAN facility. Beam is delivered from ISAC from the lower left and transported to the RFQ. The beam can be extracted in the forward (TITAN) direction or in the reverse (laser spectroscopy) direction. Downstream of the RFQ is the beam switchyard to the EBIT or the MPET. The EBIT is used for either charge breeding or electron capture branching ratio measurements. Singly charged ions (SCI) from the RFQ or highly charged ions (HCI) from the EBIT are delivered to the MPET, where the high precision mass measurement is performed. Between the switchyard and the MPET, at the *green arrow*, CPET will be installed to provide additional cooling for the highly charged ions

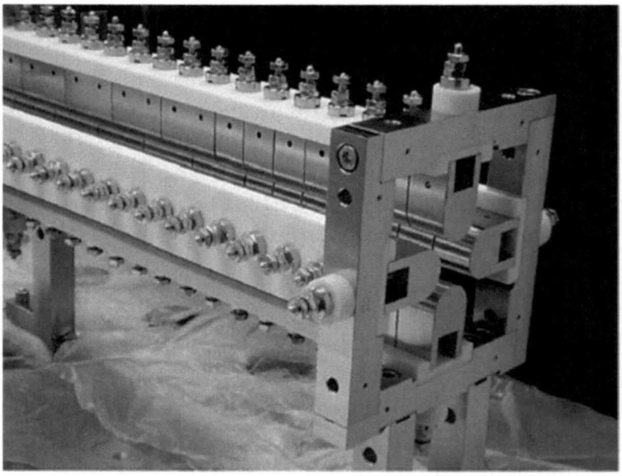

Fig. 2 Photograph of the RFQ cooler and buncher trap. The trap is composed of four rods, each segmented into 24 pieces

2.2 Charge breeding with EBIT

Penning trap mass spectrometry relies on the measurement of the cyclotron frequency, $2\pi\nu_c = q/m \cdot B$, where q is the charge, m is the mass, and B is the magnetic field strength. The precision follows

$$\frac{\delta m}{m} \propto \frac{m}{qBT_{\text{meas}}\sqrt{N}} \tag{1}$$

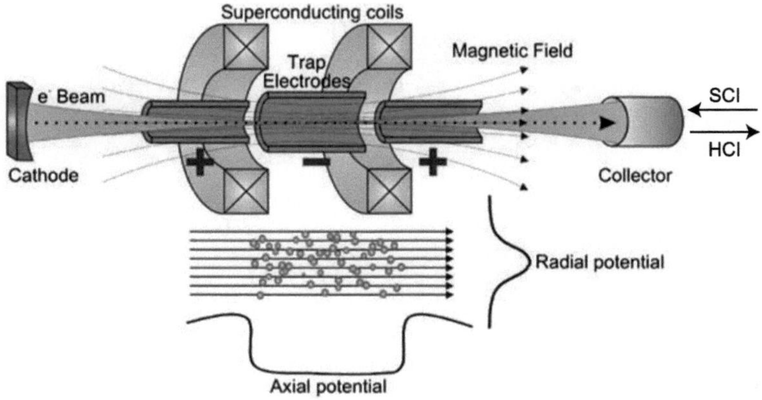

Fig. 3 Schematic of the EBIT. An electrostatic potential confines the ions axially and the space charge of the electron beam radially. Ions are bred to higher charge states by impact with the electron beam. Hitherto the highest charge states achieved are neon-like $^{84}Kr^{26+}$ and $^{71}Ge^{22+}$ [29]. Figure courtesy of Hjalmar Bruhns, Max Planck Institute for Nuclear Physics (Bruhns, private communication)

where N is the number of ions and T_{meas} the measurement time, typically limited by the half-life $T_{1/2}$ [26]. It can be seen that the precision increases with the charge q of the ion (assuming no significant losses due to the charge breeding). The EBIT [27, 28] initially traps singly charged ions and subsequently generates highly charged ions (HCI) by impact with an electron beam in a 6 T Helmholtz-style magnetic field. The strong magnetic field leads to a strong compression of the electron beam, leading to a high electron current density (ECD). The space charge of the electron beam provides radial confinement, and the high ECD leads to rapid charge breeding. An electrostatic potential axially confines the ions. Figure 3 shows a schematic of the EBIT, which was designed and assembled at the Max-Planck-Institute for Nuclear Physics (MPI-K) in Heidelberg. The final charge state distribution depends on the electron beam current, the energy, and the breeding time. Electron beams up to 500 mA at 70 keV may be used, and the design values are for currents as large as 5 nA. The highest charge state achieved and measured in the Penning trap thus far is $Q = 26+$ in stable nuclides and $Q = 22+$ in unstable species, creating neon-like $^{84}Kr^{26+}$ and $^{71}Ge^{22+}$ [29]. Unfortunately, the boost in charge state carries a price: ionization of background gas, additional contamination, increased beam emittance, lower transmission, and losses due to radioactive decay when preparation and breeding times exceed the species' half-lives. To remove the contaminant ions and charge states, a Bradbury-Nielsen gate [30] has been installed in the transfer beamline and allows one to select ions by their time of flight (TOF), which is proportional to their mass-to-charge ratio. The need for additional cooling will be addressed by the installation of CPET.

2.3 CPET for cooling of HCI

To match the emittance of the EBIT to the acceptance of the MPET, a cooler Penning trap [31–33] is under construction. Two cooling schemes are under consideration, electron and proton cooling. The former has been successfully implemented

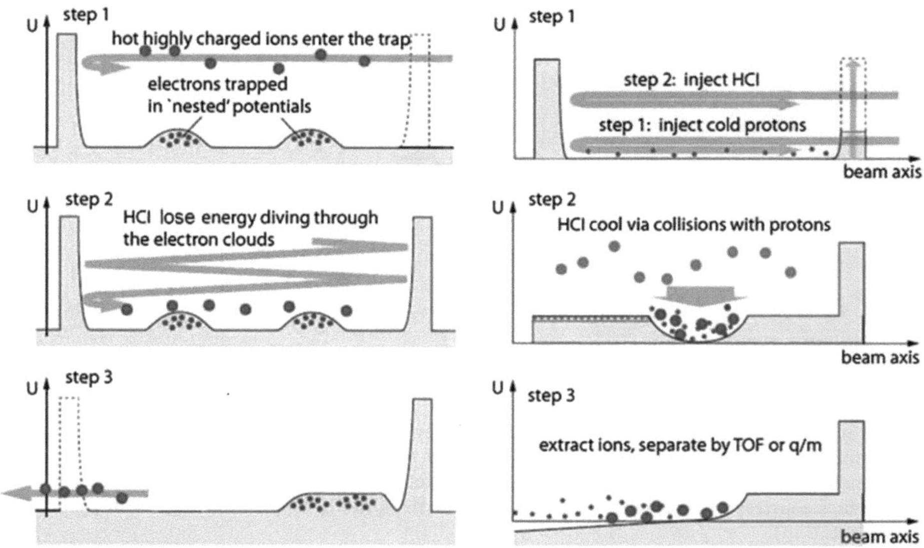

Fig. 4 Cartoons of two cooling schemes proposed for CPET. Electron cooling (*left*) relies on electrons self-cooling by synchrotron radiation but risks electron-ion recombination. Proton cooling (*right*) could lead to faster cooling although it requires a large quantity of initially cool protons

Fig. 5 Photograph of the CPET trapping electrodes mounted in its aluminum holder

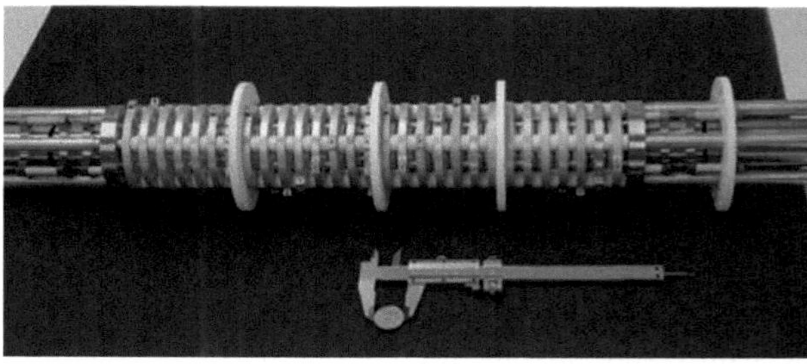

for anti-proton trapping [34] and anti-hydrogen production [35]. Electrons are injected into the Penning trap and undergo self-cooling by synchrotron radiation in the strong magnetic field ($B = 7$ T). A window of opportunity exists to trap, to cool, and to extract the HCI before electron-ion recombination becomes dominant. To trap simultaneously the electrons and the HCI, a nested Penning trap is required as illustrated in the left column of Fig. 4. The right column of Fig. 4 depicts the proton cooling scheme. It circumvents possible recombination and facilitates the trapping scheme. It requires a relatively large quantity of initially cool protons, which will be supplied from a plasma source. As the protons are ejected from CPET with the HCI, they must be selectively removed from the beam before it reaches the MPET. Additionally, the HCI can be evaporatively cooled with the protons. The key advantage of proton cooling is the significant phase-space overlap of protons ($A/Q = 1$) and HCI ($A/Q \approx 2$), which potentially leads to faster cooling.

Both cooling schemes will be investigated at TITAN. For sub-second synchrotron cooling of the electrons a 7 T superconducting, solenoidal magnet will be used.

Fig. 6 Photograph of the precision measurement Penning trap (MPET). The hyperbolic trap was manufactured from oxygen-free copper and gold-plated. A CAD $1 coin is shown for scale

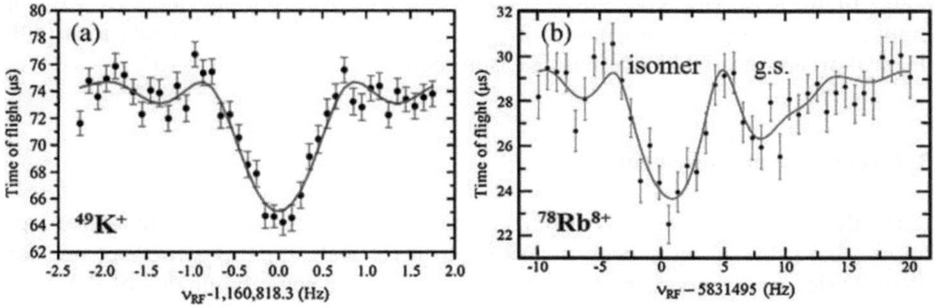

Fig. 7 A TOF-ICR resonance of **a** $^{49}K^+$ and **b** $^{78}Rb^{8+}$. The center of the dip occurs at the cyclotron frequency $2\pi \nu_c = q/m \cdot B$. $^{78}Rb^{8+}$ was trapped in both the isomer and ground states, resulting in two dips in the *resonance curve*. The *solid curve* is a fit of the theoretical line shape [39] to the data

The elongated HCI cloud and nested trapping potential necessitate an extended trapping volume. The homogeneity of the magnetic is constant within $\delta B/B = 10^{-5}$ over a region 10 cm long and within $\delta B/B = 10^{-3}$ over a region 40 cm long. Most of the cylindrical Penning trap electrodes have been manufactured and assembled. CPET will be tested offline, along with its optics and diagnostics, and installed in the beamline after successful tests. A photograph of the trapping electrodes is shown in Fig. 5.

2.4 High precision mass spectrometry in the MPET

For the mass measurement, the SCI or HCI bunches are trapped in the MPET [11, 36]. The hyperbolic Penning trap, pictured in Fig. 6, is housed in a 3.7 T superconducting, solenoidal magnet. The magnetic field provides radial confinement and the electrostatic quadrupole axial confinement. To determine the mass, its cyclotron frequency $2\pi \nu_c = q/m \cdot B$ is measured using the time-of-flight ion cyclotron resonance (TOF-ICR) method [37, 38]. Figure 7 depicts a typical resonance scan, where the RF-excitation frequency is varied and the TOF is recorded. The center of the resonance scan coincides with the cyclotron frequency. The charge state Q can be determined by the ion's TOF from the RFQ or EBIT to MPET. To calibrate the magnetic field B, the cyclotron frequency of a well known stable nuclide is measured before and after the radioactive ion's, and it is linearly interpolated to the measurement time of the radioactive ion. Therefore, the primary experimental result is the ratio of cyclotron frequencies

$$R = \frac{\nu_{\mathrm{ref}}}{\nu_c} = \frac{m}{m_{\mathrm{ref}}} \cdot \frac{q_{\mathrm{ref}}}{q} \tag{2}$$

where $\nu_{c(\mathrm{ref})}$, $m_{(\mathrm{ref})}$, and $q_{(\mathrm{ref})}$ refer to the cyclotron frequency, mass, and charge of the radioactive (reference) ion. To extract the atomic mass from the ratio, the electron binding energy must be included. For SCI, it is typically negligible compared to the statistical uncertainty; however for HCI, the total binding energy can be comparable or larger than the statistical uncertainty as the binding energy increases with charge state. In the case of $^{71}\mathrm{Ge}^{22+}$ and $^{84}\mathrm{Kr}^{26+}$, the electron binding energies are not empirically known and are based on precise atomic theoretical calculations [40]. Typical precisions of TITAN mass measurements are $\delta m/m \approx 10^{-9}$-$10^{-7}$.

3 Recent mass measurements

Since the commissioning of TITAN in 2007, 12 elements and 44 radioactive isotopes (see Table 1) have been measured to investigate nuclear structure, nuclear astrophysics, and fundamental interactions. Below are some highlights of the mass measurement research program.

3.1 Measurements of halo nuclei

In very loosely bound nuclei, the wavefunctions of the valence nucleons may extend far beyond what is expected classically [44]; this is interpreted as the valence nucleons orbiting a tightly bound core, giving rise to the name halo nuclei. As known neutron halo nuclei are relatively simple systems (the heaviest known is $^{19}\mathrm{C}$) at the limits of nuclear existence, their unusual nuclear properties act as stringent tests for nuclear models. TITAN has taken advantage of the excellent production yield of these nuclides at ISAC-TRIUMF and has measured the one-neutron halo $^{11}\mathrm{Be}$ [9], the two-neutron halos $^{6}\mathrm{He}$ [13] and $^{11}\mathrm{Li}$ [8], and the four-neutron halo $^{8}\mathrm{He}$ [7]. With the exception of $^{11}\mathrm{Be}$, the TITAN mass values deviated 2σ or more from the accepted literature value in Atomic Mass Evaluation 2003 [45]. A summary is shown in Fig. 8 and given in [12]. With a half life $T_{1/2} = 8.8$ ms, $^{11}\mathrm{Li}$ is the shortest-lived nuclide measured with Penning trap mass spectrometry.

3.2 Resurgence of the $N = 28$ shell closure

TITAN undertook a mass measurement campaign in the $N \approx 28$ region to ascertain the persistence of the nuclear shell closure in neutron-rich K and Ca [18]. While the TITAN-measured mass values of $^{44,47,50}\mathrm{K}$ agreed with the AME'03 [45] values, 7σ and 10σ deviations were found for $^{48,49}\mathrm{K}$. Our findings strengthen the $N = 28$ shell closure in K by 1 MeV and align the nuclear structure trend in K with that observed in Ca, Sc, and Ti. A comparison of empirical findings to several theoretical calculations indicates that the nuclear structure in the region is not fully understood. The prediction of new magic numbers at $N = 32$, 34 has led us to measure $^{51}\mathrm{K}$ and $^{50-52}\mathrm{Ca}$ [43] and to plan measurements beyond $N = 34$. Thus far, theoretical

Table 1 A table of radioactive nuclides measured at the TITAN facility listing the element, isotope, charge state, and references

Ele.	Isotope (A)	Q	Ref.
He	6	1+	[13]
	8	1+	[7, 13, 41]
Li	8, 9, 11	1+	[8]
Be	9–11	1+	[9]
	12	1+	[10]
Na	29–31	1+	[42]
Mg	30–34	1+	[42]
Al	29, 30, 32	1+	[42]
K	44	4+	[18]
	47–50	1+	[18]
	51, 52	1+	[43]
Ca	49, 50	1+	[18]
	51, 52	1+	[43]
Ga	74	8+	[29]
Ge	71	22+	[29]
Rb	74, 75	8+	[22]
	76	12+	[22]
	78	8+	[22]
	94, 97, 98	15+	[20]
Sr	94, 97–99	15+	[20]

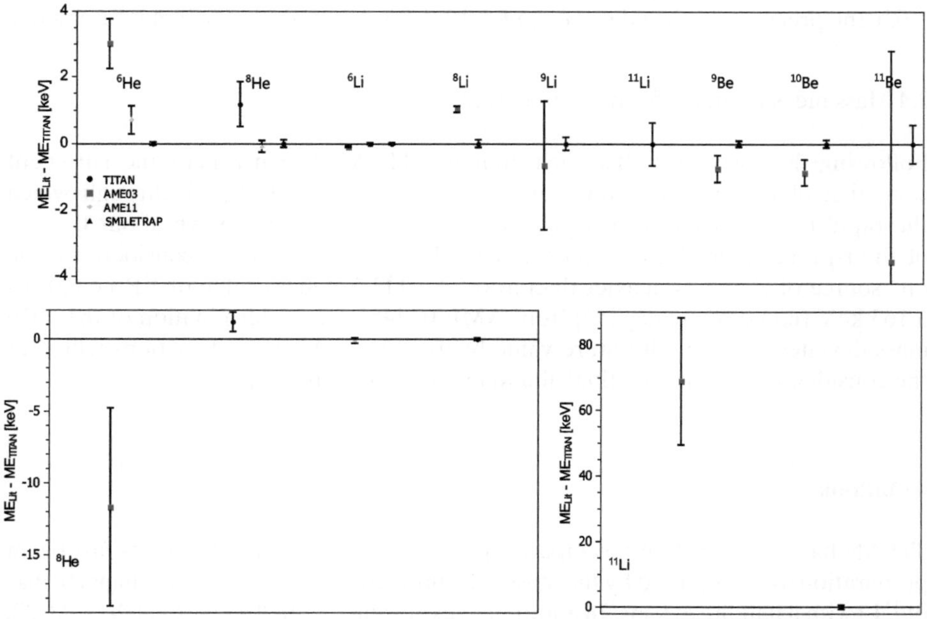

Fig. 8 Comparison of TITAN-measured mass excesses to values in literature for light, neutron-rich nuclides. Values for the Atomic Mass Evaluation [45, Audi and Meng, private communication] are taken from the version prior to the TITAN measurement

predictions have achieved better agreement with experiment when three-body forces are included [46].

3.3 Probing the CVC hypothesis

The Conserved Vector Current (CVC) hypothesis posits that the $\mathcal{F}t$ value of superallowed β-emitters, that is the corrected decay strength, is constant. From this $\mathcal{F}t$ value, the V_{ud} matrix element of the Cabibbo-Kobayashi-Maskawa (CKM) matrix can be calculated [21]. In the Standard Model, the CKM matrix must be unitary, and of the matrix elements V_{ud} contributes with the largest fraction to this test. In order to calculate the individual $\mathcal{F}t$, various corrections must be incorporated, including an isospin-symmetry breaking correction δ_C, which scales as Z^2. The superallowed β-emitter with the largest Z is ^{74}Rb, which has been the subject of a measurement campaign at TRIUMF: the mass at TITAN [22], the charge radius at the laser spectroscopy facility [5], and the branching ratio at the 8π spectrometer [47]. The charge radius measurement helped to refine calculations of δ_C. The TITAN mass measurement reduced the uncertainty in the Q_{EC}-value to 4 keV. Thus, the uncertainty in δ_C and Q_{EC} contribute equally to that of $\mathcal{F}t$.

The TITAN ^{74}Rb experiment used the EBIT to boost the ^{74}Rb charge state as well as of other proton-rich Rb isotopes to $Q = 8+$ to $12+$ and, hence, to reduce the uncertainty in the mass determination. This marks the first charge breeding of short-lived nuclides for Penning trap mass spectrometry. Also noteworthy is the half life of ^{74}Rb, $T_{1/2} = 65$ ms, which is considered short lived for Penning trap mass spectrometry. Typically, shorter half lives compromise the precision achieved; however, charge breeding ^{74}Rb compensates for the shorter half life. In the case of ^{74}Rb, the precision achieved was $\delta m/m \approx 10^{-7}$, a value typical of SCI measurements.

3.4 Mass measurements for nuclear astrophysics

Following the successful ^{74}Rb measurement, TITAN began a mass measurement campaign of Rb and Sr isotopes in the vicinity of A $\approx$ 100 [20], through which the rapid neutron capture or r-process passes [48]. Little is known about the site of the r-process, which may occur in Type II supernovae, and is considered to be one source of elements heavier than iron. The TITAN mass value of ^{97}Rb confirms a 163 keV (6σ) discrepancy [49] from AME'03 [45]. Our determination of the ^{98}Rb mass deviates from the literature value by 97 keV (2σ). These deviations will alter the considered path and the final abundances of the r-process.

4 Outlook

TITAN has capitalized on the superior production yields at ISAC and its fast beam preparation as exemplified by the mass measurements of the halo nuclei, in particular of ^{11}Li with a half life of 9 ms. In addition to exploring the shell closures at $N = 28, 32, 34$ in K and Ca isotopes and other aspects of nuclear structure, the TITAN research program also probes physics beyond the Standard Model and nuclear astrophysics. Key to many experiments is TITAN's charge breeding capabilities.

TITAN is the only Penning trap mass spectrometer at a rare isotope facility to include charge breeding capabilities. Measurement of HCI enhances precision relative to SCI, and improvements are underway to expand the charge breeding capabilities of the EBIT. Additionally, better isomeric resolution via charge breeding [50] is under investigation. The imminent installation of CPET will further improve mass measurements of HCI by reducing the beam emittance from the EBIT. Another upgrade to the TITAN facility will be a multi-reflection time-of-flight mass spectrometer [51] to improve beam purity in the Penning trap. The instrument is under development at the Justus Liebig University Giessen.

Acknowledgements We would like to thank the TRIUMF staff, in particular M. Good, the TRILIS group, and beam development group. This work has been supported by the Natural Sciences and Engineering Research Council (NSERC) of Canada and the National Research Council (NRC) of Canada. S.E. acknowledges support from the Vanier CGS, T.B. from the Evangelisches Studienwerk e.V. Villigst, and V.V.S. from the Deutsche Studienstiftung.

References

1. Blaum, K.: Phys. Rep. **425**(1), 1 (2006). doi:10.1016/j.physrep.2005.10.011
2. Dilling, J., Bricault, P., Smith, M., Kluge, H.J.: Nucl. Instr. Meth. **B204**, 492 (2003). doi:10.1016/S0168-583X(02)02118-3
3. Dilling, J., Baartman, R., Bricault, P., Brodeur, M., Blomeley, L., Buchinger, F., Crawford, J., Crespo López-Urrutia, J., Delheij, P., Froese, M., Gwinner, G., Ke, Z., Lee, J., Moore, R., Ryjkov, V., Sikler, G., Smith, M., Ullrich, J., Vaz, J.: Int. J. Mass Spec. **251**(2–3), 198 (2006). doi:10.1016/j.ijms.2006.01.044
4. Dombsky, M., Bishop, D., Bricault, P., Dale, D., Hurst, A., Jayamanna, K., Keitel, R., Olivo, M., Schmor, P., Stanford, G.: Rev. Sci. Instr. **71**(2), 978 (2000). doi:10.1063/1.1150364
5. Mané, E., Voss, A., Behr, J.A., Billowes, J., Brunner, T., Buchinger, F., Crawford, J., Dilling, J., Ettenauer, S., Levy, C., Shelbaya, O., Pearson, M.: Phys. Rev. Lett. **107**(21), 212502 (2011). doi:10.1103/PhysRevLett.107.212502
6. Lunney, D., Pearson, J.M., Thibault, C.: Rev. Mod. Phys. **75**, 1021 (2003)
7. Ryjkov, V., Brodeur, M., Brunner, T., Smith, M., Ringle, R., Lapierre, A., Ames, F., Bricault, P., Dombsky, M., Delheij, P., Lunney, D., Pearson, M., Dilling: J. Phys. Rev. Lett. **101**(1), 012501 (2008). doi:10.1103/PhysRevLett.101.012501
8. Smith, M., Brodeur, M., Brunner, T., Ettenauer, S., Lapierre, A., Ringle, R., Ryjkov, V., Ames, F., Bricault, P., Drake, G., Delheij, P., Lunney, D., Sarazin, F., Dilling, J.: Phys. Rev. Lett. **101**(20), 202501 (2008). doi:10.1103/PhysRevLett.101.202501
9. Ringle, R., Brodeur, M., Brunner, T., Ettenauer, S., Smith, M., Lapierre, A., Ryjkov, V., Delheij, P., Drake, G., Lassen, J.: Phys. Lett. **B675**(2), 170 (2009). doi:10.1016/j.physletb.2009.04.014
10. Ettenauer, S., Brodeur, M., Brunner, T., Gallant, A.T., Lapierre, A., Ringle, R., Pearson, M.R., Delheij, P., Lassen, J., Lunney, D., Dilling, J.: Phys. Rev. **C81**(2), 024314 (2010). doi:10.1103/PhysRevC.81.024314
11. Brodeur, M., Brunner, T., Champagne, C., Ettenauer, S., Smith, M., Lapierre, A., Ringle, R., Ryjkov, V.L., Audi, G., Delheij, P., Lunney, D., Dilling, J.: Phys. Rev. **C80**(4), 044318 (2009). doi:10.1103/PhysRevC.80.044318
12. Brodeur, M., Brunner, T., Ettenauer, S., Gallant, A.T., Simon, V.V., Smith, M., Lapierre, A., Mané, E., Ringle, R., Ryjkov, V.L., Bacca, S., Delheij, P., Lunney, D., Pearson, M., Dilling, J.: Hyperfine Interact. **199**(1–3), 167 (2011). doi:10.1007/s10751-011-0311-y
13. Brodeur, M., Brunner, T., Champagne, C., Ettenauer, S., Smith, M., Lapierre, A., Ringle, R., Ryjkov, V., Bacca, S., Delheij, P., Drake, G., Lunney, D., Schwenk, A., Dilling, J.: Phys. Rev. Lett. **108**(5), 052504 (2012). doi:10.1103/PhysRevLett.108.052504
14. Bacca, S., Schwenk, A., Hagen, G., Papenbrock, T.: Eur. Phys. J. **A42**(3), 553 (2009). doi:10.1140/epja/i2009-10815-5
15. Otsuka, T., Fujimoto, R., Utsuno, Y., Brown, B.A., Honma, M., Mizusaki, T.: Phys. Rev. Lett. **87**(8), 082502 (2001). doi:10.1103/PhysRevLett.87.082502

16. Kanungo, R., Nociforo, C., Prochazka, A., Aumann, T., Boutin, D., Cortina-Gil, D., Davids, B., Diakaki, M., Farinon, F., Geissel, H., Gernhäuser, R., Gerl, J., Janik, R., Jonson, B., Kindler, B., Knöbel, R., Krücken, R., Lantz, M., Lenske, H., Litvinov, Y., Lommel, B., Mahata, K., Maierbeck, P., Musumarra, A., Nilsson, T., Otsuka, T., Perro, C., Scheidenberger, C., Sitar, B., Strmen, P., Sun, B., Szarka, I., Tanihata, I., Utsuno, Y., Weick, H., Winkler, M.: Phys. Rev. Lett. **102**(15), 152501 (2009). doi:10.1103/PhysRevLett.102.152501
17. Janssens, R.V.F.: Nature **459**, 1069 (2009)
18. Lapierre, A., Brodeur, M., Brunner, T., Ettenauer, S., Finlay, P., Gallant, A., Simon, V., Delheij, P., Lunney, D., Ringle, R., Savajols, H., Dilling, J.: Phys. Rev. **C85**(2), 024317 (2012). doi:10.1103/PhysRevC.85.024317
19. Wiescher, M., Schatz, H., Champagne, A.: Phil. Trans. R. Soc. **A356**, 2105 (1998). doi:10.1098/rsta.1998.0265
20. Simon, V., Brunner, T., Chowdhury, U., Eberhardt, B., Ettenauer, S., Gallant, A., Mané, E., Simon, M., Delheij, P., Pearson, M., Audi, G., Gwinner, G., Lunney, D., Schatz, H., Dilling, J.: Phys. Rev. **C85**(6), 064308 (2012). doi:10.1103/PhysRevC.85.064308
21. Hardy, J.C., Towner, I.: Phys. Rev. **C79**(5), 055502 (2009). doi:10.1103/PhysRevC.79.055502
22. Ettenauer, S., Simon, M.C., Gallant, A.T., Brunner, T., Chowdhury, U., Simon, V.V., Brodeur, M., Chaudhuri, A., Mané, E., Andreoiu, C., Audi, G., Crespo López-Urrutia, J.R., Delheij, P., Gwinner, G., Lapierre, A., Lunney, D., Pearson, M.R., Ringle, R., Ullrich, J., Dilling, J.: Phys. Rev. Lett. **107**(27), 272501 (2011). doi:10.1103/PhysRevLett.107.272501
23. Smith, M., Blomeley, L., Delheij, P., Dilling, J.: Hyperfine Interact. **173**(1-3), 171 (2006). doi:10.1007/s10751-007-9554-z
24. Brunner, T., Smith, M., Brodeur, M., Ettenauer, S., Gallant, A., Simon, V., Chaudhuri, A., Lapierre, A., Mané, E., Ringle, R., Simon, M., Vaz, J., Delheij, P., Good, M., Pearson, M., Dilling, J.: Nucl. Instr. Meth. **676**, 32 (2012). doi:10.1016/j.nima.2012.02.004
25. Dawson, P.H. (ed.): Quadrupole Mass Spectrometry and its Applications, 1st edn. Elsevier, Amsterdam (1976)
26. Bollen, G.: Nucl. Phys. A **693**(1–2), 3 (2001). doi:10.1016/S0375-9474(01)00353-0
27. Sikler, G., Crespo López-Urrutia, J.R., Dilling, J., Epp, S., Osborne, C.J., Ullrich, J.: Eur. Phys. J. **A25**(S1), 63 (2005). doi:10.1140/epjad/i2005-06-072-6
28. Lapierre, A., Brodeur, M., Brunner, T., Ettenauer, S., Gallant, A., Simon, V., Good, M., Froese, M., Crespo López-Urrutia, J., Delheij, P., Epp, S., Ringle, R., Schwarz, S., Ullrich, J., Dilling, J.: Nucl. Instr. Meth. **A624**(1), 54 (2010). doi:10.1016/j.nima.2010.09.030
29. Frekers, D., Simon, M.C., Andreoiu, C., Bale, J.C., Brodeur, M., Brunner, T., Chaudhuri, A., Chowdhury, U., Crespo López-Urrutia, J.R., Delheij, P., Ejiri, H., Ettenauer, S., Gallant, A.T., Gavrin, V., Grossheim, A., Harakeh, M.N., Jang, F., Kwiatkowski, A.A., Lassen, J., Lennarz, A., Luichtl, M., Ma, T., Macdonald, T.D., Mané, E., Robertson, D., Schultz, B.E., Simon, V.V., Teigelhöfer, A., Dilling, J.: Penning-trap Q-value determination of the ^{71}Ga(nu,e^{-})^{71}Ge reaction using threshold charge breeding of on-line produced isotopes. Phys. Lett. B **722**(4–5), 233–237 (2013). doi:10.1016/j.physletb.2013.04.019
30. Brunner, T., Mueller, A., O'Sullivan, K., Simon, M., Kossick, M., Ettenauer, S., Gallant, A., Mané, E., Bishop, D., Good, M., Gratta, G., Dilling, J.: Int. J. Mass Spec. p. S1387380611003770 (2011). doi:10.1016/j.ijms.2011.09.004
31. Ryjkov, V.L., Blomeley, L., Brodeur, M., Grothkopp, P., Smith, M., Bricault, P., Buchinger, F., Crawford, J., Gwinner, G., Lee, J., Vaz, J., Werth, G., Dilling, J.: Eur. Phys. J. A **25**(S1), 53 (2005). doi:10.1140/epjad/i2005-06-122-1
32. Ke, Z., Shi, W., Gwinner, G., Sharma, K., Toews, S., Dilling, J., Ryjkov, V.L., the TITAN Collaboration: Hyperfine Interact. **173**(1–3), 103 (2006). doi:10.1007/s10751-007-9548-x
33. Simon, V.V., Chowdhury, U., Delheij, P., Dilling, J., Eberhardt, B., Gwinner, G.: J. Phys.: Conference Series **312**(5), 052024 (2011). doi:10.1088/1742-6596/312/5/052024
34. Rolston, S.L., Gabrielse, G.: Hyperfine Interact. **44**(1–4), 233 (1989). doi:10.1007/BF02398673
35. Andresen, G. and the ALPHA Collaboration: Phys. Rev. Lett. **100**(20), 203401 (2008). doi:10.1103/PhysRevLett.100.203401
36. Brodeur, M., Ryjkov, V., Brunner, T., Ettenauer, S., Gallant, A., Simon, V., Smith, M., Lapierre, A., Ringle, R., Delheij, P., Good, M., Lunney, D., Dilling, J.: Int. J. Mass Spec. **310**, 20 (2012). doi:10.1016/j.ijms.2011.11.002
37. Gräff, G., Kalinowsky, H., Traut, J.: Zeitschrift für Physik A: Atoms and Nuclei **297**(1), 35 (1980). doi:10.1007/BF01414243
38. Bollen, G., Moore, R.B., Savard, G., Stolzenberg, H.: J. Appl. Phys, **68**(9), 4355 (1990). doi:10.1063/1.346185

39. König, M., Bollen, G., Kluge, H.J., Otto, T., Szerypo, J.: Int. J. Mass Spec. Ion Process. **142**(1–2), 95 (1995). doi:10.1016/0168-1176(95)04146-C
40. Rodrigues, G., Indelicato, P., Santos, J., Patté, P., Parente, F.: At. Data Nucl. Data Tables **86**(2), 117 (2004). doi:10.1016/j.adt.2003.11.005
41. Brodeur, M., Brunner, T., Ettenauer, S., Lapierre, A., Ringle, R., Brown, B., Lunney, D., Dilling, J.: Phys. Rev. Lett. **108**(21), 212501 (2012). doi:10.1103/PhysRevLett.108.212501
42. Chaudhuri, A., Andreoiu, C., Brunner, T., Chowdhury, U., Ettenauer, S., Gallant, A.T., Gwinner, G., Kwiatkowski, A.A., Lennarz, A., Lunney, D., Macdonald, T.D., Schultz, B.E., Simon, M.C., Simon, V.V., Dilling, J.: Extinction of the $N = 20$ neutron-shell closure for ^{32}Mg examined by direct mass measurements. Phys. Rev. C (submitted to)
43. Gallant, A., Bale, J., Brunner, T., Chowdhury, U., Ettenauer, S., Lennarz, A., Robertson, D., Simon, V., Chaudhuri, A., Holt, J., Kwiatkowski, A., Mané, E., Menéndez, J., Schultz, B., Simon, M., Andreoiu, C., Delheij, P., Pearson, M., Savajols, H., Schwenk, A., Dilling, J.: Phys. Rev. Lett. **109**(3), 032506 (2012). doi:10.1103/PhysRevLett.109.032506
44. Tanihata, I., Hamagaki, H., Hashimoto, O., Shida, Y., Yoshikawa, N., Sugimoto, K., Yamakawa, O., Kobayashi, T., Takahashi, N.: Phys. Rev. Lett. **55**(24), 2676 (1985). doi:10.1103/PhysRevLett.55.2676
45. Audi, G., Wapstra, A., Thibault, C.: Nucl. Phys. A **729**(1), 337 (2003). doi:10.1016/j.nuclphysa.2003.11.003
46. Holt, J., Otsuka, T., Schwenk, A., Suzuki, T.: J. Phys. G. **39**(8), 085111 (2012). doi:10.1088/0954-3899/39/8/085111
47. Piechaczek, A., Zganjar, E.F., Ball, G.C., et al.: Phys. Rev. C **67**, 051305 (2003)
48. Seeger, P.A., Fowler, W.A., Clayton, D.D.: Astrophys. J. Suppl. **11**, 121 (1965)
49. Rahaman, S., Hager, U., Elomaa, V.V., et al.: Eur. Phys. J. A **32**, 87 (2007)
50. Gallant, A.T., Brodeur, M., Brunner, T., Chowdhury, U., Ettenauer, S., Simon, V.V., Mané, E., Simon, M., Andreoiu, C., Delheij, P., Gwinner, G., Pearson, M., Ringle, R., Dilling, J.: Phys. Rev. C **85**(4), 044311 (2012). doi:10.1103/PhysRevC.85.044311
51. Plass, W.R., Dickel, T., Petrick, M., et al.: Eur. Phys. J. S. T. **150**, 367 (2007)

DOI 10.1007/978-94-007-7963-1_17
Reprinted from *Hyperfine Interactions* Journal, DOI 10.1007/s10751-013-0893-7

Electron-capture branching ratio measurements of odd-odd intermediate nuclei in double-beta decay at the TITAN facility

A. Lennarz · T. Brunner · C. Andreoiu · A. Chaudhuri · U. Chowdhury · P. Delheij · J. Dilling · S. Ettenauer · D. Frekers · A. T. Gallant · A. Grossheim · F. Jang · A. A. Kwiatkowski · T. Ma · E. Mané · M. R. Pearson · B. E. Schultz · M. C. Simon · V. V. Simon

Published online: 11 October 2013

Abstract A novel technique to measure electron-capture branching ratios is being introduced, where the TITAN ion traps and the ISAC radioactive beam facility at TRIUMF are the central components. The technique will be applied to the intermediate odd-odd nuclei in double-beta decay. The decay properties of these nuclei will constrain theoretical models dealing with the evaluation of the nuclear matrix elements for both the 2ν and the 0ν mode. The present setup and its potential for measuring extremely low branching ratios at low instrumental backgrounds is described.

ISAC and ARIEL: The TRIUMF Radioactive Beam Facilities and the Scientific Program.

A. Lennarz (✉) · T. Brunner · A. Chaudhuri · U. Chowdhury · P. Delheij · J. Dilling · S. Ettenauer · D. Frekers · A. T. Gallant · A. Grossheim · F. Jang · A. A. Kwiatkowski · E. Mané · M. R. Pearson · B. E. Schultz · M. C. Simon · V. V. Simon
TRIUMF, 4004 Wesbrook Mall, Vancouver, BC V6T 2A3, Canada
e-mail: lennarz@triumf.ca

T. Brunner
Stanford University, Stanford, CA 94305, USA

C. Andreoiu · F. Jang · T. Ma
Department of Chemistry, Simon Fraser University, Burnaby, BC V5A 1S6, Canada

U. Chowdhury
University of Manitoba, Winnipeg, MB R3T 2N2, Canada

J. Dilling · S. Ettenauer · A. T. Gallant
University of British Columbia, Vancouver, BC V6T 1Z1, Canada

D. Frekers · A. Lennarz
Westfälische Wilhelms-Universtität, 48149 Münster, Germany

V. V. Simon
Max-Planck Institut für Kernphysik, 69117 Heidelberg, Germany

Keywords In-trap decay spectroscopy · Double-beta decay

1 Introduction

The TRIUMF TITAN ion-trap facility [1, 2] consists of a unique combination of ion traps dedicated to precision mass measurements. Additionally, it offers the opportunity to perform in-trap X-ray and γ-ray spectroscopy on radioactive isotopes stored in the center of the spectroscopy ion trap (also used as the Electron Beam Ion Trap (EBIT)).

One particular project deals with the measurement of the electron-capture branching ratios (ECBRs) of short-lived ($T_{1/2} < 1$ h) intermediate nuclei in double-beta ($\beta\beta$) decays [3]. The ECBRs are important for evaluating the nuclear matrix elements involved in the $\beta\beta$ decay for both decay processes, the two-neutrino ($2\nu\beta\beta$) and, to a lesser extent, the neutrinoless ($0\nu\beta\beta$) decay. The latter is of special interest, as its detection immediately signals the Majorana character of neutrinos [4].

The $2\nu\beta\beta$ decay is allowed in the Standard Model, where in the case of the $\beta^-\beta^-$-mode two electrons and two electron-antineutrinos are emitted [5]:

$$(Z, A) \overset{\beta^-\beta^-}{\rightarrow} (Z+2, A) + 2e^- + 2\bar{\nu}_e. \tag{1}$$

Out of the 35 isotopes which can undergo $\beta^-\beta^-$ decay, the half-lives have been determined for ^{48}Ca, ^{76}Ge, ^{82}Se, ^{96}Zr, ^{100}Mo, ^{116}Cd, ^{128}Te, ^{130}Te, ^{130}Ba, ^{150}Nd, ^{238}U [5], and more recently for ^{136}Xe [6]. The decay conserves the lepton number and is therefore consistent with the Standard Model (SM).

The connection between decay rate $\Gamma^{2\nu}_{(\beta^-\beta^-)} = T^{-1}_{1/2}$ and the nuclear matrix element is given by the most widely used Primakoff-Rosen approximation [7]:

$$\Gamma^{2\nu}_{(\beta^-\beta^-)} = G^{2\nu}(Q, Z)\left|M^{(2\nu)}_{\mathrm{DGT}}\right|^2, \tag{2}$$

where $G^{(2\nu)}(Q, Z)$ is the phase-space factor depending on the decay Q value and the atomic charge Z of the nucleus [8]. The $2\nu\beta\beta$ decay nuclear matrix element (NME)

$$M^{(2\nu)}_{\mathrm{DGT}} = \sum_m \frac{M^{\mathrm{GT}^+}_m \cdot M^{\mathrm{GT}^-}_m}{\frac{1}{2}Q + E(1^+_m) - E_0}, \tag{3}$$

which includes all the nuclear physics involved in the decay, is the sum of two consecutive single β decay matrix elements, each of Gamow-Teller (GT) type, between the initial and an intermediate state and between that intermediate state and the final state. Since the GT operator has no spatial dependence, the matrix element describes a $0^+ \rightarrow 1^+ \rightarrow 0^+$ decay path for a $0^+ \rightarrow 0^+$ initial to final ground-state to ground-state transition. The energy denominator in (3) contains the Q value of the decay and the energy difference $(E_x(1^+_m) - E_0)$ between the mth intermediate 1^+ state and the initial ground state. The fourth-power dependence on the weak axial-vector coupling constant g_A is included in $G^{(2\nu)}(Q, Z)$.

The situation is rather different for the $0\nu\beta\beta$ decay. Since in this case the neutrino enters as a virtual particle, the momentum constraints pertaining to the $2\nu\beta\beta$ case are lifted and almost all intermediate states may contribute to the nuclear matrix element. The understanding of the nuclear physics involved in the $0\nu\beta\beta$ case requires

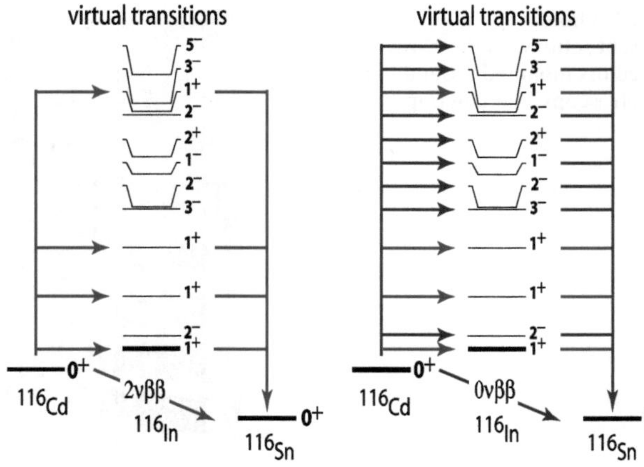

Fig. 1 Schematic drawing of the two $\beta\beta$ decay modes in ^{116}Cd showing the decay paths through excited states of the intermediate nucleus ^{116}In. The states are only indicative. Transitions with $J^{\pi}=1^{+}$ are marked with *red arrows*

a strong concerted effort between theory and experiment. Figure 1 sketches these two cases for the ^{116}Cd as an example.

The experimental search for the $0\nu\beta\beta$ decay process is presently the only viable method which can reveal the Majorana nature of neutrinos. Moreover, if the nuclear matrix elements were known to a reasonably high degree of confidence, the decay would also allow one to place a value on the neutrino mass. A recent paper by Escuderos et al. [9] shows the present situation of the NMEs, where they have grouped the calculated quantities for a large number of isotopes as they were determined from various model calculations: the Quasi-Particle Random Phase Approximation (QRPA) [10–14], the Shell Model (SM) [15–19], the angular momentum Projected-Hartree-Fock-Bogoliubov method (PHFB) [20–22] and the Interacting Boson Model (IBM-2) [23]. Despite the fact that significant progress has been made in theoretical calculations of the NME, the results still display large differences between different models. Of course the $0\nu\beta\beta$ NME are significantly more complex, since many more transitions besides the GT transitions have to be taken into account. Still, the GT driven matrix elements play an important role. In particular, the strength of the lowest lying GT transitions provide an important input to almost all theoretical calculations cited above.

The GT type matrix elements can be readily determined experimentally, either through charge-exchange reactions [24–32] or for ground-state transitions through EC or single β decay of the intermediate odd-odd nuclei [33]. Further, the EC process and the β^{-} decay provide information about the ground-state properties of the nuclear wave function, which is an important ingredient and a benchmark for theoretical calculations dealing with the neutrinoless decay.

2 The TITAN ECBR measurement setup

In most of the $\beta\beta$ decay cases EC branches are poorly known or not known at all, because the EC process is usually suppressed by several orders of magnitude compared to the β^{-} decay counterparts owing to its much lower decay Q-value. Traditionally the EC/β branching ratios have been obtained by producing radioactive isotopes through irradiation of a suitable target or, in the case of EC, by deposition

Fig. 2 Photograph of the current setup with the X-ray detectors mounted around the spectroscopy Penning trap

of a radioactive beam onto an appropriate and movable tape, followed by the measurement of the K-shell X-rays. The major concern and limiting factors in these latter experiments are beam/target contaminants, X-ray absorption in the carrier material and high backgrounds induced by the many orders of magnitude more intense β decays. To overcome these limitations, a unique method of measuring the ECBRs of the intermediate odd-odd nuclei in $\beta\beta$ decay has recently been developed at TITAN. The novel feature of the setup is the use of a Penning trap [34], where the ions are stored backing-free in a strong magnetic field. Therefore there is no isobaric contamination in the investigated sample as it is the case for the tape station technique. Two superconducting coils in a Helmholtz configuration create the magnetic field of up to 6 T, which confines the ions in the radial plane, and additional electric fields trap them in the axial plane. The presence of the magnetic field offers the additional advantage that electrons from the β^- decay are directed on axis out of the trap and away from the X-ray detectors. Detectors with separate dedicated functions at different locations can therefore be installed for either X-ray or β^- detection. For the latter a silicon surface barrier detector (SSB) [35–37] was installed at the exit of the trap to monitor the β decay electrons for normalization purposes.

One may note that the present ECBR project is yet another important research application, which the TITAN Penning trap setup offers on top of its use as an EBIT in the well established TITAN mass measurement program [38] with singly charged ions like ^{8}He or ^{10,11}Be [39, 40] or highly charged ions [41] created through charge-breeding.

For the K-shell X-ray detection following the EC process seven specially designed X-ray detectors (3 lithium drifted silicon (Si(Li)) detectors and 4 high-purity germanium (HPGe) detectors) have been mounted perpendicular to the beam axis around

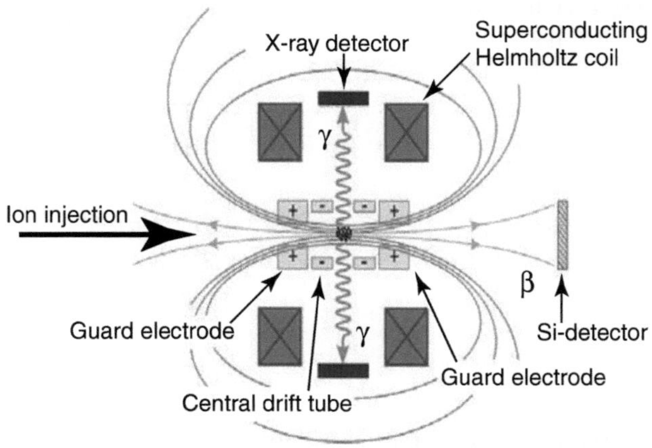

Fig. 3 Schematic drawing of the EC branching ratio measurement setup [37]. The X-rays from decaying ions inside the trap are detected in the Si(Li) and HPGe detectors, while the electrons are guided out of the trap by the magnetic field to an SSB detector

Table 1 Shown are the half-lives and $K_{\alpha_{1,2}}$ X-ray energies for the electron-capture decays of the intermediate odd-odd nuclei in $\beta\beta$ decay

Mother nucleus	Transition	Daughter nucleus	$K_{\alpha_{1,2}}$ [keV]	$T_{1/2}$
^{100}Tc	$1^+ \rightarrow 1^+$	^{100}Mo	17.5	15.8 s
^{110}Ag	$1^+ \rightarrow 1^+$	^{110}Pd	21.2	24.6 s
^{114}In	$1^+ \rightarrow 1^+$	^{114}Cd	25.3	71.9 s
^{116}In	$1^+ \rightarrow 1^+$	^{116}Cd	21.7	14.1 s
^{82m}Br	$2^- \rightarrow 1^+$	^{82}Se	11.2	6.1 min
^{128}I	$1^+ \rightarrow 1^+$	^{128}Te	27.5	25.0 min
^{76}As	$2^- \rightarrow 1^+$	^{76}Ge	9.9	26.2 h

the trap center in close geometry (see Fig. 2). The exact detector positions are at 0° (Si(Li)) at the bottom of the trap, at ± 45° (HPGe), ± 90° (Si(Li)) and at ± 135° (HPGe). In order to minimize absorption of the X-rays on their way towards the detectors, ultra-thin (0.25 mm thickness, 6" and 8" diameter), high-purity Be windows have been installed as vacuum barriers and heat shields. A schematic view of the spectroscopy Penning trap setup is shown in Fig. 3.

The advantage of the TITAN-EC program is given by the combination of pure and intense ISAC beams, no ion interaction with any surrounding materials, a spatial separation of X-rays and β^- particles and the high efficiency of the X-ray detectors at close distances (225 mm) from the center of the trap. All of these are important for a low background measurement, which is essential for the successful determination of the ECBRs with an uncertainty of approximately 10 %. Table 1 lists the seven odd-odd intermediate nuclei that will be investigated by the TITAN-EC experiment.

An example for one of the nuclei is presented in Fig. 4, which displays the ^{116}In decay scheme showing both, the EC and the β^- decay into the ^{116}Cd and ^{116}Sn daughter nuclei. The present data situation on ^{116}In is particularly unsatisfactory, since the EC branching ratio of ^{116}In into ^{116}Cd measured by Bhattacharya et al. [33] is in conflict by a factor of 2–3 with the result obtained from (^{3}He,t) charge-exchange experiments [42].

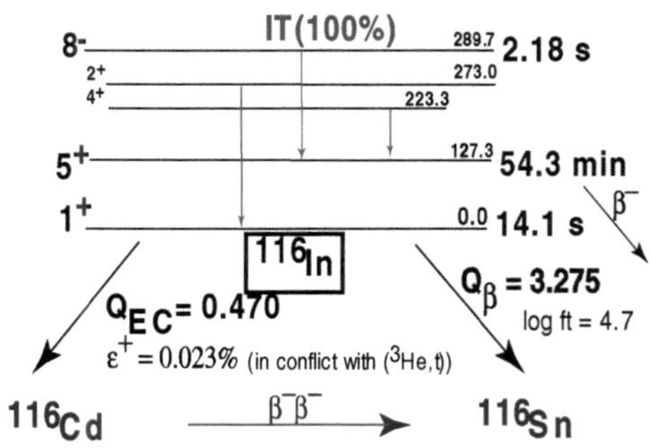

Fig. 4 Partial decay scheme of ^{116}In including the lowest-lying isomeric states

3 First in-trap EC measurements

The modifications and extensions to TITAN-EC described above were implemented and tested in October/November 2011 during a dedicated technical run. Two earlier measurement campaigns [37] with radioactive ^{107}In and ^{124,126}Cs have been performed with one installed HPGe detector, in order to prove the feasibility of the spectroscopy Penning trap technique. A spectrum showing the decay of ^{124}Cs is shown in Fig. 5. Cs X-rays from the 6.3 s $J^{\pi} = 7^+$ isomer decay and Xe X-rays from the electron-capture are well identified. The $J^{\pi} = 7^+$ ^{124}Cs isomer decays via a sequence of transitions, some of them highly converted [43]. The measured spectrum marks the first ever detected EC decay in an ion trap.

New updates of the experimental setup (see Fig. 2) included 4 HPGe detectors (10 mm crystal thickness, 2000 mm^2 surface area), 3 Si(Li) detectors (5 mm thickness, 2000 mm^2 surface area), a custom-made amplifier for the analogue signal processing, a new digitizer (SIS3302 ADC) and a modified data acquisition system, as well as supporting infrastructure including high-voltage and liquid nitrogen supply systems. The objective of the aforementioned technical run was to test the new hardware and to detect X-rays from the decay of ^{116}In, i.e., possible ^{116}Cd X-rays from the EC or ^{116}In X-rays from the excited isomeric states, in order to provide guidance to further beam and hardware developments.

A spectrum taken during the technical run with ^{116}In stored successively for 500 ms in the trap is shown in Fig. 6. Several independent runs have been added equalling a measurement time of 17.8 h. The ^{116}In X-ray transitions at 24.2 keV and 27.3 keV are clearly visible in the spectrum. These transitions are caused by the de-excitation of the $J^{\pi}= 8^-$ isomer into the $J^{\pi}= 5^+$ isomer, because the ^{116}In beam delivered by ISAC is composed of the ground state as well as the two isomers ($J^{\pi}= 5^+$ at 127.3 keV, $J^{\pi}= 8^-$ at 289.7 keV [44]). The electron conversion coefficient for the de-excitation of the $J^{\pi}= 8^-$ isomer is known to be $\alpha_K = 1.15 \pm 0.09$ [45], which also means that the γ de-excitation with a 162.4 keV γ-ray is equally probable (see inset of Fig. 6). The $J^{\pi}= 5^+$ isomer with a half-life of 54.3 min decays exclusively through β^- emission populating many states in ^{116}Sn. Because of the long half-life compared to the storage time, the rather strong $4_2^+ \longrightarrow 2_2^+$ 417 keV γ-ray transition in ^{116}Sn only appeared as a weak line in the present spectrum. Note, the spectrum in Fig. 6 also shows the germanium escape peak from the 24.2 keV line appearing at 14.2 keV.

The electron-capture branching ratio measurements necessitate a normalization method because the number of ions in the trap is not readily known with sufficient

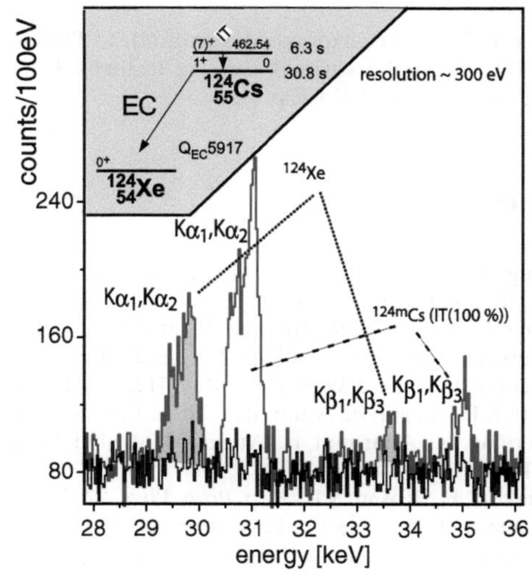

Fig. 5 Decay of ^{124}Cs and its spectrum in the first EC-branching ratio measurement using the spectroscopy Penning trap technique [37]. Shown are the K_α and K_β X-rays of ^{124}Xe and ^{124}Cs

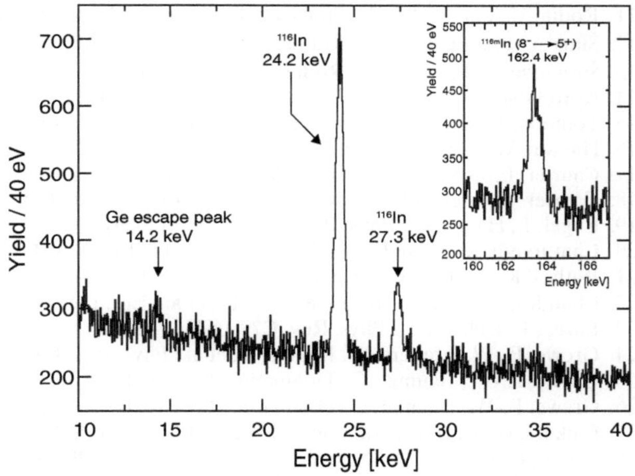

Fig. 6 Raw X-ray spectrum of the ^{116}In in-trap decay measurement taken with an HPGe detector. Shown is the sum of several ^{116}In runs (40 eV binning, 500 ms trapping time) equalling a measuring time of 17.8 h. The occurrence of the background is presently under investigation

accuracy. For the upcoming ECBR experiments at TITAN it is planned to measure the number of β^- decays by directing the β^- particles onto the SSB detector, which was not operational during the present technical run. Another method, which is currently being discussed, is the detection of the γ-rays from excited states in the daughter nuclei, which are fed by the β^- decay using a dedicated HPGe detector mounted close to the trap.

TITAN-EC is set to finish installation of the remaining hardware and to measure the first ECBRs in 2012. The relatively low background environment of ions stored in a Penning trap has the potential of performing measurements of extremely small ECBRs, including those from 1st forbidden decays, which are inaccessible to traditional X-ray spectroscopy techniques.

Acknowledgements This work is supported by National Science and Engineering Research Council of Canada and the Deutsche Forschungsgemeinschaft (DFG) under grant-number no. FR 601/3-1.

One of the authors, D.F., expresses his gratitude to TRIUMF for the fine support during his extended stay in 2011. The authors would also like to thank J. Urquhart (Simon Fraser University) for his contribution to the TITAN-EC project.

References

1. Dilling, J., Bricault, P., Smith, M., Kluge, H.-J., TITAN collaboration: Nucl. Instrum. Methods Phys. Res. **B204**, 492 (2003)
2. Dilling, J., Baartman, R., Bricault, P., et al.: Int. J. Mass Spectrom. **251**, 198 (2006)
3. Frekers, D., Dilling, J., Tanihata, I.: Can. J. Phys. **85**, 57 (2007)
4. Schechter, J., Valle, J.W.F.: Phys. Rev. **D25**, 2951 (1982)
5. Elliott, S.R., Vogel, P.: Annu. Rev. Nucl. Part. Sci. **52**, 115 (2002)
6. Ackerman, N., Aharmim, B., Auger, M., et al., EXO collaboration: Phys. Rev. Lett. **107**, 212501 (2011)
7. Primakoff, H., Rosen, S.P.: Rep. Prog. Phys. **22**, 121 (1959)
8. Suhonen, J., Civitarese, O.: Phys. Rep. **300**, 123 (1998)
9. Escuderos, A., Faessler, A., Rodin, V., Šimkovic, F.: J. Phys. **G37**, 125108 (2010)
10. Rodin, V.A., Faessler, A., Šimkovic, F., Vogel, P.: Nucl. Phys. **A766**, 107 (2006); and Erratum **A793**, 213 (2007)
11. Rodin, V.A., Faessler, A., Šimkovic, F., Vogel, P.: Nucl. Phys. **A793**, 213 (2007)
12. Šimkovic, F., Faessler, A., Müther, H., Rodin, V., Stauf, M.: Phys. Rev. **C79**, 055501 (2009)
13. Šimkovic, F., Faessler, A., Rodin, V.A., Vogel, P., Engel, J.: Phys. Rev. **C77**, 045503 (2008)
14. Kortelainen, M., Suhonen, J.: Phys. Rev. **C75**, 051303 (2007)
15. Tomoda, T., Faessler, A., Schmid, K.W., Grümmer, F.: Nucl. Phys. **A452**, 591 (1986)
16. Haxton, W.C., Stephenson, G.J. Jr.: Prog. Part. Nucl. Phys. **12**, 409 (1984)
17. Caurier, E., Nowacki, F., Poves, A., Retamosa, J.: Phys. Rev. Lett. **77**, 1954 (1996)
18. Caurier, E., Menendez, J., Nowacki, F., Poves, A.: Phys. Rev. Lett. **100**, 052503 (2008)
19. Engel, J., Hagen, G.: Phys. Rev. **C79**, 064317 (2009)
20. Chaturvedi, K., Chandra, R., Rath, P.K., et al.: Phys. Rev. **C78**, 054302 (2008)
21. Rath, P.K., Chandra, R., Chaturvedi, K., et al.: Phys. Rev. **C80**, 044303 (2009)
22. Chandra, R., Chaturvedi, K., Rath, P.K., et al.: Eur. Lett. **86**, 32001 (2009)
23. Barea, J., Iachello, F.: Phys. Rev. **C79**, 044301 (2009)
24. Grewe, E.-W., Frekers, D., Rakers, S., et al.: Phys. Rev. **C76**, 054307 (2007)
25. Grewe, E.-W., Bäumer, C., Dohmann, H., et al.: Phys. Rev. **C78**, 044301 (2008)
26. Grewe, E.-W., Bäumer, C., Dohmann, H., et al.: Phys. Rev. **C77**, 064303 (2008)
27. Frekers, D., Ejiri, H., Akimune, H., et al.: Phys. Lett. **B706**, 134 (2011)
28. Puppe, P., Frekers, D., Adachi, T., et al.: Phys. Rev. **C84**, 051305 (2011)
29. Rakers, S., Bäumer, C., van den Berg, A.M., et al.: Phys. Rev. **C71**, 054313 (2005)
30. Rakers, S., Bäumer, C., van den Berg, A.M., et al.: Phys. Rev. **C70**, 054302 (2004)
31. Ejiri, H., Fushimi, K., Hazama, R., et al.: J. Phys. Soc. Jpn. **64**, 339 (1995)
32. Ejiri, H., Engel, J., Fushimi, K., et al.: Nucl. Phys. **B110**, 375 (2002)
33. Bhattacharya, M., García, A., Hindi, M.M., et al.: Phys. Rev. **C58**, 1247 (1998)
34. Sikler, G., Crespo López-Urrutia, J.R., Dilling, J., et al.: Eur. Phys. J. **A25**, 63 (2005)
35. Brunner, T., Brodeur, M., Ettenauer, S., et al.: J. Phys. Conf. Ser. **312**, 072006 (2011)
36. Brunner, T., Brodeur, M., Champagne, C., et al.: Nucl. Instrum. Methods Phys. Res. **B266**, 4643 (2008)
37. Brunner, T.: Ph.D. thesis, Technische Universität München, Physik Department E12 (2011)
38. Lapierre, A., Brodeur, M., Brunner, T., et al.: Nucl. Instrum. Methods Phys. Res. **A624**, 54 (2010)
39. Brodeur, M., Brunner, T., Champagne, C., et al.: Phys. Rev. Lett. **108**, 052504 (2012)
40. Ringle, R., Brodeur, M., Brunner, T., et al.: Phys. Lett. **B675**, 170 (2009)
41. Ettenauer, S., Simon, M.C., Gallant, A., et al.: Phys. Rev. Lett. **107**, 27251 (2011)
42. Akimune, H., Ejiri, H., Fujiwara, M., et al.: Phys. Lett. **B394**, 23 (1997)
43. Weiss, B., Liang, C.F., Paris, P., Peghaire, A., Gizon, A.: Z. Phys. A **313**, 173 (1983)
44. Firestone, R., Shirley, V.S.: Table of Isotopes CD-rom. John Wiley & Sons (1996)
45. Heckman, P.H., Gubenator, K., Poyhonen, J., Flammersfeld, A.: Z. Phys. **163**, 451 (1961)

DOI 10.1007/978-94-007-7963-1_18
Reprinted from *Hyperfine Interactions* Journal, DOI 10.1007/s10751-013-0896-4

Laser polarization facility

C. D. P. Levy · M. R. Pearson · R. F. Kiefl · E. Mané · G. D. Morris · A. Voss

Published online: 23 October 2013

Abstract An overview of the laser polarizer facility at ISAC is provided. The charge-exchange cell is described in some detail.

Keywords Polarized radioactive beam · Laser spectroscopy · BNMR

1 Polarizer overview

The laser polarization facility at ISAC consists of a section of low-energy beam line (the "polarizer") and associated laser facilities. Figure 1 shows the layout of the beam line components. The beam is nuclear-spin polarized in-flight via collinear laser optical pumping. The facility was originally built to produce polarized $^{8}Li^{+}$ as a magnetic probe for condensed matter experimenters. They require intense beams (10^{7} atoms s^{-1}) of highly polarized $^{8}Li^{+}$ for studying fine details of beta-detected

ISAC and ARIEL: The TRIUMF Radioactive Beam Facilities and the Scientific Program.

C. D. P. Levy (✉) · M. R. Pearson · R. F. Kiefl · E. Mané · G. D. Morris · A. Voss
TRIUMF, 4004 Wesbrook Mall, Vancouver, BC V6T 2A3, Canada
e-mail: levy@triumf.ca

R. F. Kiefl
Department of Physics and Astronomy, University of British Columbia,
6224 Agricultural Road, Vancouver, BC V6T 1Z1, Canada

Present Address:
E. Mané
Ministério da Ciência e Tecnologia e Inovação, Esplanada dos Ministérios,
Bloco E, CEP: 70067-900, Brasília, DF, Brazil

Present Address:
A. Voss
Department of Physics, University of Jyväskyla, P.O.Box 35, 40014 Jyväskyla, Finland

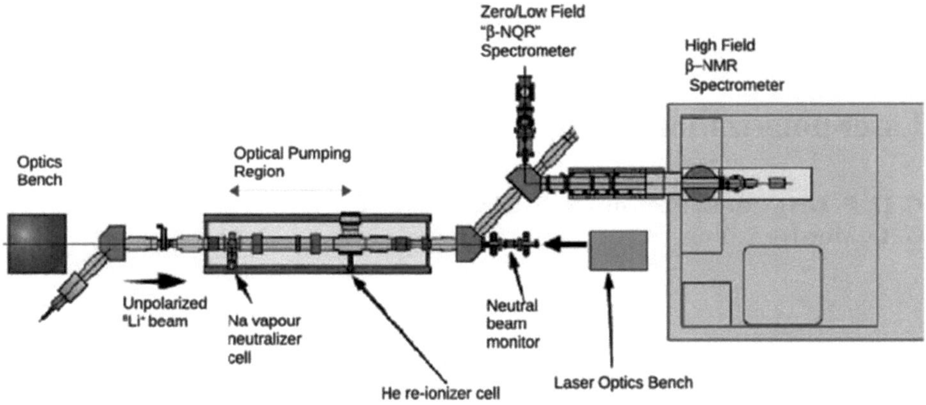

Fig. 1 Layout of the polarizer and downstream beam lines. The ion and laser beams are counter-propagating

nuclear magnetic resonance (BNMR) and beta-detected nuclear quadrupole resonance (BNQR) in thin films and interfaces, and they remain the most frequent users [see contribution to this issue of Hyperfine Interactions]. The facility is also used in other areas of research, which are described in several other contributions to this issue. For example, polarized beams at ISAC are used to investigate nuclear structure and fundamental symmetries, and to measure nuclear moments using BNMR and BNQR. In addition, the polarizer is a suitable location for carrying out high resolution collinear laser fluorescence spectroscopy with unpolarized beams. The signal-to-noise in that case can be enhanced by a factor of $\sim 10^4$ by bunching the incoming ion beam using the TITAN cooler/buncher, and then gating fluorescence detection in coincidence with the bunches. In that case, average beam fluxes of $\sim$50 ions s^{-1} are enough to make spectroscopic measurements [1].

The polarizer design follows well-known principles [2, 3]. As used for condensed matter studies, an incident 28 keV $^8Li^+$ beam is partially neutralized with $\sim$50 % efficiency in a Na vapour charge-exchange cell (CEC). Unneutralized ions exiting the CEC are removed from the beam by electrostatic deflection plates. The fast atomic ^{8}Li beam thus created has a strong recycling D_1 electronic transition in the visible, which is ideal for polarizing the beam to as high as $\sim$80 % with circularly polarized light. In principle, many elements can be polarized through optical pumping. The alkali metals are the most straightforward, with high polarization achieved by pumping with visible wavelength lasers on the $^2S_{1/2}$–$^2P_{1/2}$ or $-^2P_{3/2}$ atomic ground state transitions. Most sodium and lithium isotopes have been polarized as beams at ISAC. Figure 2 shows a photograph of the polarizer section of beam line.

The polarized beam is re-ionized through impact ionization in a windowless, cooled helium gas cell with over 60 % efficiency [4]. Re-ionizing the beam has several advantages: (1) The beam can be directed by electrostatic elements to different experimental stations. (2) Two experiments can share the same beam, through the use of a kicker. (3) The experimental target is not exposed to laser light. (4) The absolute polarization direction, initially longitudinal with respect to the beam motion, is unchanged by electrostatic elements. Therefore the beam polarization can be transformed from longitudinal to transverse simply by steering the beam through

Fig. 2 Photograph of the polarizer section of beam line, showing the CEC on the left and the helium re-ionizer on the right. The magnetic coils provide the guide field along the axis of the polarizer

a net angle of 90°, with no loss of polarization. (5) The beam energy at the target, and hence implantation depth, can be adjusted by applying a variable potential to the target. All of these features have been used to advantage at ISAC.

The 1.9 m long optical pumping region, between the CEC and the helium cell, is situated within a weak longitudinal magnetic field of ~1 mT which defines the polarization direction. Optical pumping in a weak field polarizes both the electronic and nuclear spins, due to the hyperfine coupling between them. The laser frequency is fixed and is Doppler-tuned into resonance with the electronic transition by varying a positive electric potential of order 100 V applied to deceleration electrodes and the electrically floating CEC. Resonance is found either by fluorescence detection in the polarizer or by detection of beta decay in a downstream neutral beam monitor (NBM—see Fig. 1). The NBM contains a Pt thin foil which partially intercepts the diverging polarized beam that has not been re-ionized in the helium gas cell. Backward and forward beta detectors measure the ensuing beta decay asymmetry, which depends on the longitudinal beam polarization. The foil has a 5 mm diameter central aperture to transmit the laser beam and is placed in a 10 mT holding field, which preserves the polarization of the implanted atoms prior to beta decay. In addition to the NBM, experimenters using the polarizer typically have their own means of monitoring beam polarization. For laser spectroscopy work, the CEC bias potential is usually scanned over a range of several kilovolts, both positive and negative, thus scanning the effective laser frequency over as much as several tens of gigaherz.

Some development of polarized non-alkali-metal beams has been carried out. Motivated by a user request, the feasibility of polarizing a fluorine beam was investigated. We demonstrated that the facility can produce F^+ beams with a polarization of ~20 % [5]. The process relies on a recycling transition in which the lower atomic fluorine energy level is metastable, and could be used for any fluorine isotope having a lifetime long enough to be extracted from a hot target (and non-zero nuclear spin of course). Another study investigated the feasibility of using ^{11}Be (nuclear spin 1/2) as a pure magnetic probe in the condensed matter program [6]. (^{8}Li has nuclear spin 2 and therefore also interacts with electric field gradients, which may add undesired complexity to an experiment). Beryllium is an alkaline-earth metal and as such the ion can be polarized directly, without recourse to neutralization and re-ionization, owing to its electronic structure similar to that of an alkali metal atom. It was found that the two dominant beta decay transitions, having opposite-sign asymmetry parameters, produce a net beta decay asymmetry of ^{11}Be too low to be useful in condensed matter BNMR studies. It may be possible to overcome this by counting only beta particles with energies greater than the endpoint energy of the less energetic transition. As part of the beam development, magnetic coils were added to the transport lines between the polarizer and the BNQR/BNMR areas, so as to extend the guide fields all the way to those experimental stations. This was necessary because Be^+ ions are paramagnetic, and without the guide field would lose polarization during transport.

2 Alkali-metal-vapour charge-exchange cell

Charge-exchange neutralization cross sections in alkali-metal vapours are large, in the range 10^{-15}–10^{-14} cm^2. The optimal vapour thickness (atoms cm^{-2}) for our purpose produces ~50 % neutralization efficiency (thicker targets produce excessive energy broadening of the atomic beam). Such targets have been commonly produced by a CEC consisting of a heated horizontal tube and alkali-metal reservoir. The vapour condenses at the cooled ends of the tube and is usually wicked back to the centre by a stainless steel mesh. A problem with this design is the escape of alkali-metal vapour from the ends of the CEC, which is countered by reducing the tube vacuum conductance with a concomitant reduction in beam acceptance. A different approach was developed by Zelenski [7]. In his design, sodium vapour is emitted from a nozzle to form a jet perpendicular to the beam direction. The jet condenses on a cooled surface and the liquid sodium flows back to the alkali-metal reservoir through a separate channel, with very low vapour leakage rates, even at a vapour thickness and beam aperture diameter larger than required in the ISAC polarizer.

A slightly simpler approach (mechanically) has been taken in the polarizer CEC, as illustrated in Fig. 3. Alkali metal in a lower reservoir is heated to produce a vertical vapour jet that intersects the ion beam in the interaction region. A saturated vapour pressure of about 1 millibar would exist in the reservoir if it were sealed. The jet condenses above the interaction region and flows back down to the reservoir along the walls. The recycling rate is typically hundreds of grams per day. The tube connecting the reservoir and interaction region is cooled at the top, thus condensing vapour atoms that might otherwise scatter off the tube wall and escape. At the same vertical level as two conical beam aperture pieces are two more conical pieces in the perpendicular direction, all four identical except that one is capped. One of them faces a glass viewport, which permits direct observation of the interaction region. The room temperature viewport condenses any vapour contacting it and thus acts as a very sensitive witness of leakage from the apertures. The capped piece helps keep wall surfaces away from the interaction region. The length of the CEC in the beam line direction is 127 mm and the beam apertures are 16 mm in diameter.

A double-layer, stainless-steel-mesh wick lines the tube connecting the reservoir and the interaction region. The wick is necessary for stable operation. Stability is also best when operating with constant reservoir heater power, rather than constant reservoir temperature. This is because the total evaporation rate is almost proportional to the heater power (radiative heat losses are relatively small). Any changes in tube conductance or liquid surface area in the reservoir are to a large degree compensated by changes in the reservoir temperature that keep the total evaporation

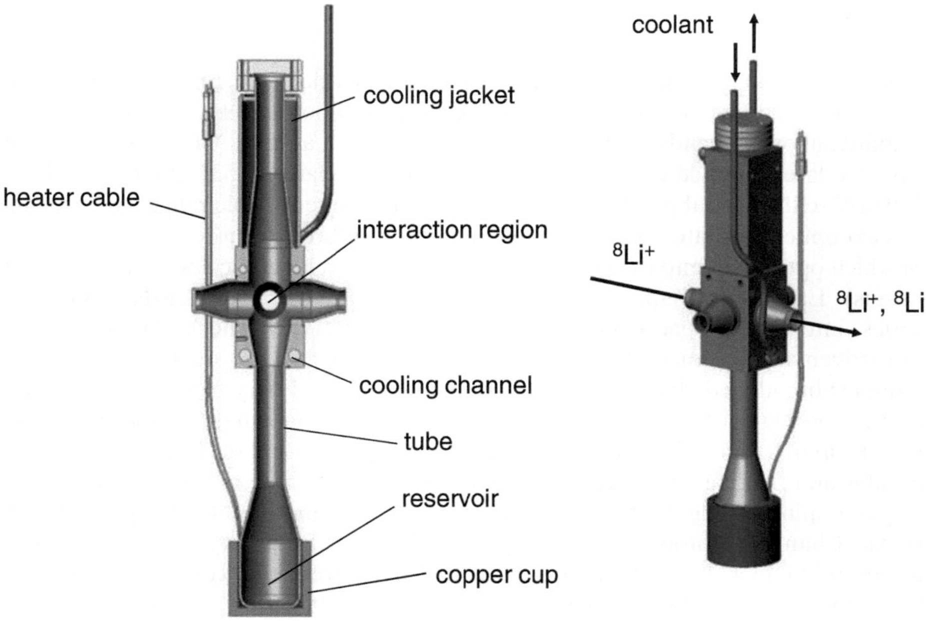

Fig. 3 Views of the alkali-metal vapour cell. Construction is stainless steel except as shown. The stainless steel wick is not shown

rate nearly constant. For example, if the surface area of sodium in the reservoir decreases, then the cooling rate of the reservoir due to evaporation goes down, the reservoir temperature rises, and the evaporation rate per unit area increases. The complexity of the physical processes in the ISAC design, which include transition from viscous to molecular flow, and scattering and condensation on surfaces at different temperatures, probably precludes computer modeling of the jet profile. The net effect, though, should be molecular beaming by emission from a tube [8], narrowed by condensation near the tube exit. This expectation is supported by the observed excellent confinement of rubidium vapour by the CEC when loaded with high purity rubidium. The confinement has not been as good when operating with sodium, with a loss rate of order 1 g per week. It is not known if the loss is in some way caused by oxide contamination present in the sodium, or by the physical properties of sodium, or some other reason. The CEC is loaded in situ. Maintenance is seldom required and the CEC can run for a couple of months accumulated running time between cleanings, when loaded with sodium. A loaded CEC can be sealed in situ and removed to atmosphere for use later. Reference [9] is a detailed report of tests of the CEC loaded with potassium.

A second CEC of very similar design (but having more uniform condensation surface temperatures) has been constructed so that both Na and Rb CECs will be available whenever required. At the time of writing, the second CEC has been operated for a total of ~2 weeks, loaded with 20 g of Na metal. Vapour confinement has been excellent, with only trace amounts deposited on the viewport.

3 Laser system

The laser system for polarizing beams is contained in a clean area one floor below the beam line. The laser room has a 1 m thick concrete floor and ±0.3 °C temperature stability, allowing hands-off laser operation for days at a time. A 1 MHz bandwidth ring dye laser pumped by a diode-pumped solid state laser at 532 nm delivers 100–150 mW to the optical pumping region in the yellow to red visible range. An external electro-optic modulator (EOM) splits the laser into three or more frequencies, two of which optically pump the ground state hyperfine levels (in the case of alkali metal beams). Each laser frequency is split further by 19 MHz and 28 MHz EOMs in series, into closely spaced lines at 9 or 10 MHz intervals. The latter two EOMs are driven hard enough to produce multiple sidebands, in order to match the Doppler-broadened absorption profile of the beam [10]. Every beam atom is within interaction range of at least one laser line due to the atomic natural linewidth. The light is circularly polarized with a fixed quarter-waveplate, and the helicity is reversed at rates up to once every 10 s by moving a half-waveplate in or out of the laser beam. Regular spin flipping cancels many systematic errors and doubles the polarization signal. Changing isotope requires adjusting the laser frequency (mainly due to a different Doppler shift), and changing the first EOM frequency (due to the different hyperfine splitting). Both adjustments can be carried out within a few minutes.

Laser light is sent by mirrors ~20 m to the polarizer where it travels in the counter-propagating direction to the ion beam. A telescope between the polarizer and laser table allows the laser beam size and divergence to be adjusted to overlap the atomic beam in the optical pumping region. The telescope will also focus laser light in front of a fluorescence monitor for laser fluorescence spectroscopy. In that case, two 3 mm diameter alignment apertures can be moved in or out of the laser beam for accurate beam alignment. A laser exit window permits laser power transmitted through the polarizer to be continuously monitored and is required for initial alignment. The exit window is at Brewster's angle to eliminate back reflections, which can be a significant source of noise in sensitive laser spectroscopy. (The fluorescence monitor is designed to minimize detection of scattered laser light from beams coming from one direction only). Narrow-band transmission interference filters further reduce detection of scattered laser light by up to several orders of magnitude.

The laser frequency is prevented from drifting by a system based on measurement of interference fringes. A dye laser sample beam and the beam from a frequency-stabilized helium-neon reference laser are overlapped in a hermetically-sealed, temperature-stabilized 300 MHz free-spectral-range scanning interferometer. The oscilloscope trace showing two fringes is analyzed by computer, and any change in the fringe separation is corrected by feedback to the dye laser, about once per second. The long-term system stability is limited by drift of the HeNe laser to about ± 3 MHz, sufficient for polarizing beams or performing high resolution spectroscopy. A Fizeau wavemeter measures the dye laser frequency with a reproducibility of 30 MHz, allowing the laser frequency to be set within the original interferometer order if the frequency is lost, even after several days. (Different orders present different fringe separations, so it is essential to be in the same order when reacquiring the laser lock). The ability to retain the original laser frequency is particularly useful when running experiments with low intensity beams, since it avoids time spent on lengthy calibration scans.

The dye laser frequency was doubled using a crystal within an external resonant build up cavity to produce 313 nm light for polarizing $^{11}Be^{+}$ ions directly. Typical second harmonic power was 20 mW. The dye laser was locked at the fundamental frequency in the usual way. EOMs are not readily available at ultra-violet wavelengths and were not used for broadening the laser spectrum to match the ion beam Doppler profile. Instead, a series arrangement of seven tubes at slightly different potentials stepped the ion beam energy through a total span of $\sim$2 eV as the ion beam passed through the optical pumping region, thus stepping ions through seven regions having slightly different Doppler-shifted laser frequencies. Only one hyperfine level could be pumped this way. One can imagine a more elaborate electrode system partially overcoming that limitation. Three factors help mitigate the loss of laser power when using second harmonic light for pumping ions. (1) The Doppler profile of the ion beam is not broadened by the charge exchange process, (2) the optical pumping region is longer thanks to the removal of the CEC and helium cell, and (3) the transition probability is higher for ultraviolet transitions. A separate set of UV mirrors and telescope transmitted the light to the polarizer, alongside the visible range mirrors. Other laser facilities belonging to the laser spectroscopy and TRINAT groups, including Ti:sapphire ring lasers and diode lasers, are available for use in the polarizer. Light from these is transported to the polarizer through fibre optic cables.

4 Summary

The laser polarization facility at ISAC is in regular use as a reliable source of polarized alkali-metal beams. The possibility exists to generate other types of polarized beams. The polarizer is also a good location for carrying out collinear laser spectroscopy. The TITAN cooler/buncher and highly developed BNMR and BNQR facilities further extend its capabilities.

References

1. Nieminen, A., Campbell, P., Billowes, J., Forest, D.H., Griffith, J.A.R., Huikari, J., Jokinen, A., Moore, I.D., Moore, R., Tungate, G., Äystö, J.: Phys. Rev. Lett. **88**, 94801 (2002)
2. Anton, K.-R., Kaufman, S.L., Klempt, W., Moruzzi, G., Neugart, R., Otten, E.W., Schinzler, B.: Phys. Rev. Lett. **40**, 642–645 (1978)
3. Otten, E.W.: Treatise Heavy-Ion Sci. **8**, 517–638 (1989)
4. Levy, C.D.P., Baartman, R., Jayamanna, K., Kiefl, R., Kuo, T., Olivo, M., Wight, G.W., Yuan, D., Zelenski, A.N.: Nucl. Phys. A **701**, 253c–258c (2002)
5. Levy, C.D.P., Cocolios, T.E., Behr, J.A., Jayamanna, K., Minamisono, K., Pearson, M.R.: Nucl. Instrum. Meth. A **580**, 1571–1577 (2007)
6. Levy, C.D.P., Pearson, M.R., Morris, G.D., Chow, K.H., Hossain, M.D., Kiefl, R.F., Labbé, R., Lassen, J., MacFarlane, W.A., Parolin, T.J., Saadaoui, H., Smadella, M., Song, Q., Wang, D.: Hyperfine Interact. **196**, 287–294 (2010)

7. Zelenski, A., Alessi, J., Briscoe, B., Dutto, G., Huang, H., Kponou, A., Kokhanovski, S., Klenov, V., Lehrach, A., Levy, P., LoDestro, V., Mori, Y., Okamura, M., Raparia, D., Ritter, J., Takeuchi, T., Wight, G., Zoubets, V.: Rev. Sci. Instrum. **73**, 888 (2002)
8. Steckelmacher, W.: Rep. Prog. Phys. **49**, 1083–1107 (1986)
9. Klose, A., Minamisono, K., Geppert, Ch., Frömmgen, N., Hammen, M., Krämer, J., Krieger, A., Levy, C.D.P., Mantica, P.F., Nörtershäuser, W., Vinnikova, S.: Nucl. Instrum. Meth. A **678**, 114 (2012)
10. Hatakeyama, A., Hirayama, H., Behr, J.A., Izumi, H., Levy, C.D.P., Melconian, D., Shimoda, T.: In: Derenchuk, V.P., von Przewoski, B. (eds.) Proc. Ninth Int. Workshop on Polarized Sources and Targets, Nashville, Indiana, 30 September–4 October 2001. World Scientific, pp. 339–343 (2002)

DOI 10.1007/978-94-007-7963-1_19
Reprinted from *Hyperfine Interactions* Journal, DOI 10.1007/s10751-013-0894-6

β-NMR

Gerald D. Morris

Published online: 23 October 2013

Abstract The *β*-NMR facility at ISAC is constructed specifically for experiments in condensed matter physics with radioactive ion beams. Using co-linear optical pumping, a $^{8}Li^{+}$ ion beam having a large nuclear spin polarisation and low energy (nominally 30 keV) can be generated. When implanted into materials these ions penetrate to shallow depths comparable to length scales of interest in the physics of surfaces and interfaces between materials. Such low-energy ions can be decelerated with simple electrostatic optics to enable depth-resolved studies of near-surface phenomena over the range of about 2–200 nm. Since the *β*-NMR signal is extracted from the asymmetry intrinsic to beta-decay and therefore monitors the polarisation of the radioactive probe nuclear magnetic moments, this technique is fundamentally a probe of local magnetism. More generally though, any phenomena which affects the polarisation of the implanted spins by, for example, a change in resonance frequency, line width or relaxation rate can be studied. The *β*-NMR program at ISAC currently supports a number of experiments in magnetism and superconductivity as well as novel ultra-thin heterostructures exhibiting properties that cannot occur in bulk materials. The general purpose zero/low field and high field spectrometers are configured to perform CW and pulsed RF nuclear magnetic resonance and spin relaxation experiments over a range of temperatures (3–300 K) and magnetic fields (0–9 T).

Keywords *β*-NMR · Nuclear magnetic resonance · Radioactive beam

---

ISAC and ARIEL: The TRIUMF Radioactive Beam Facilities and the Scientific Program.

G. D. Morris
Centre for Molecular and Materials Science, TRIUMF, 4004 Wesbrook Mall,
Vancouver, BC V6T2A3, Canada
e-mail: gmorris@triumf.ca

1 Introduction

Nuclear magnetic resonance (NMR) is a powerful experimental techinque applied in numerous branches of chemistry, bio-sciences, medical diagnosis and physics. The basic working principle is that an initially prepared polarisation of a set of nuclei can be monitored as it interacts with the local magnetic and electronic environment. The signal obtained contains information about the local field and its variation among sites within a molecule or crystal, and its fluctuations in time [1]. In solids, studies as a function of temperature, applied pressure and magnetic fields provide insight into the interactions at play - particularly the electronic structure - that determine the ground state properties of a given material. Two fundamental differences distinguish conventional NMR from beta-detected NMR (β-NMR): the origin of the initial polarisation and the method of detecting the NMR signal. In conventional NMR the initial polarisation of nuclei is obtained by the application of a magnetic field B to induce a very small polarisation of order $\mu B/k_B T \approx 10^{-6}$ to 10^{-4}, where μ is the nuclear magnetic moment and k_B is Boltzman's constant, in thermal equilibrium at temperature T. The application of a radio frequency (RF) magnetic field on resonance perturbs the nuclear spins, whose polarisation will relax back to equilibrum as they precess about the applied field. The signal is picked up as a small voltage induced in a coil by the precessing magnetic moments. At the ISAC β-NMR facility the initial nuclear polarisation of a beam of radioactive nuclei is prepared in-flight by optical pumping just upstream from the spectrometer, acheiving a nuclear polarisation of about 0.7. This beam is implanted into a sample and the subsequent evolution of the polarisation is monitored via the asymmetry intrinsic to the parity-violating weak beta-decay of the probe nuclei. A pair of detectors composed of fast plastic scintillators and phototubes situated on opposite sides (to the left and right or back and front) of the sample detect each energetic beta particle exiting the sample. Because the direction of emission of each beta particle is correlated with the nuclear spin orientation at the moment of decay, the polarisation of the nuclear spins can be determined from a large number of decay events. Due to the relatively large initial polarisation and signal extraction via nuclear beta decay, in terms of the number of nuclei that contribute to the signal, β-NMR is about ten orders of magnitude more sensitive than conventional NMR. In conjunction with the low beam energy and consequent shallow implantation depth into materials, the β-NMR technique can be used to perform novel NMR experiments on films as thin as a few nm and within the near surface region of bulk materials where there are too few nuclei to produce a useful signal by conventional techniques.

The β-NMR facility at ISAC has been developed to provide a complementary probe to the conventional muon spin rotation/relaxation (μSR) facilities which employ energetic muons as probes of bulk materials. Low energy μSR, which also probes thin samples, is perhaps the most closely related technique, although at present with much lower beam intensities [2].

2 Beam properties

Only those radioactive nuclei having a large beta decay asymmetry, sufficient beta energy, a means of production, extraction and polarisation, and lifetime neither too long or too short are suitable for use as probes in materials science. ^{8}Li (spin $I = 2$, lifetime $\tau = 1.21$ s, electric quadrupole moment $Q = +31.4$ mb, gyromagnetic ratio $\gamma = 630.15$ Hz/G) has been, to date, the only species employed at the ISAC β-NMR facility for experiments on materials. ISAC surface ionization targets make ^{8}Li in abundance (about 5×10^7 s^{-1} delivered to the experiment) and extracted beams have very small energy spread of about 1–2 eV as well as small divergence. The decay of $^8\mathrm{Li} \rightarrow {}^8\mathrm{Be} + \bar{\nu}_e + e^-$ proceeds via a single beta decay channel producing electrons with a 13 MeV end point energy, sufficient to reach the beta detectors after passing through thin parts of the cryostat and vacuum windows.

For many experiments the splitting of resonance lines due to quadrupole interactions is an undesirable complication. Without electric quadrupole moments, spin 1/2 nuclei would often be preferable as probes of matter since they are not sensitive to electric field gradients usually present at crystallograpic sites lacking cubic symmetry. Trial runs with polarizing ^{11}Be measured a small beta-decay asymmetry of 0.02 although it was concluded that the nuclear polarisation was saturated. The observed low asymmetry was attributed to the two beta-decay channels, which the beta detectors are unable to distinguish, having nearly equal but opposing decay asymmetries [3]. The 13.8 s lifetime of ^{11}Be is also several times longer than typical T_1 spin relaxation times. This has the effect of reducing the steady state equilibrium polarisation and signal amplitude in measurements performed with continuous beam by a factor of $1/(1 + \tau/T_1)$, or in measurements performed with pulsed beam, forces a long wait of many lifetimes to allow unpolarized ^{11}Be to decay between beam pulses. Consequently, ^{11}Be has not been further developed as a general purpose probe for materials science.

3 Facility spectrometers

Layout of the polarizer and two spectrometers is illustrated in Fig. 1. Nuclear polarisation is generated by optical pumping of neutral Li atoms in flight at a beam energy of about 30 keV, in a section of the beam line about 6 m upstream of the spectrometers. The neutralisation, optical pumping and re-ionisation scheme is described in detail in the accompanying article by Levy et al. [4]. Nuclear polarisation of the beam remains (anti)parallel to the polarizer axis, even as the beam is deflected through electrostatic dipoles and delivered into either of the two spectrometers.

Two spectrometers have been constructed to cover different ranges of applied magnetic fields. Both are equiped with helium flow cryostats which provide stabilized sample temperatures in the range of 3–325 K. In order to minimize the accumulation of residual vacuum gases on clean sample surfaces, the spectrometers operate under ultra high vacuum (UHV) conditions in the 10^{-10} Torr range. Samples are mounted to and removed from each of the cryostats via a vacuum load-lock chamber, a small separate vacuum chamber connected to the main vacuum envelope via an isolation gate valve, which can be accessed, vented and pumped independently pf the main system, thereby preserving the spectrometer UHV. Sections of the beam line upstream of each spectrometer employ small beam apertures and large vacuum pumps to provide a factor of 1,000 reduction in pressure transitioning from the high vacuum at the polarizer exit to UHV at the spectrometers' chambers. When both spectrometers are running data aquistion modes compatible with pulsed beam,

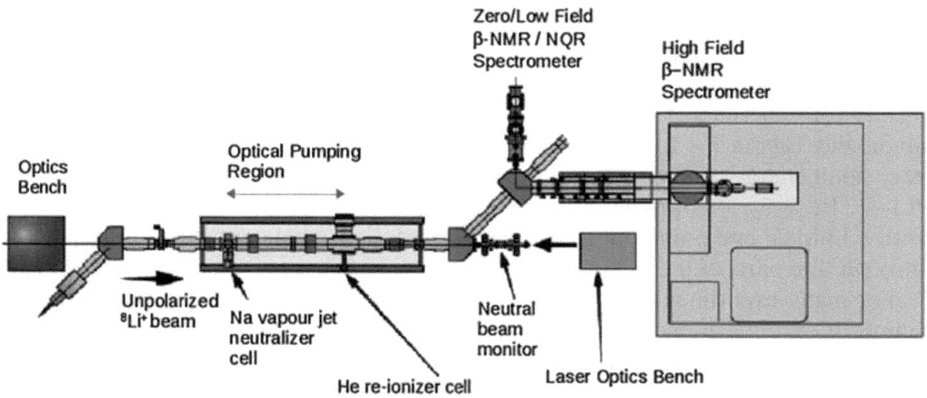

Fig. 1 General layout of the co-linear optical pumping beamline and low- and high-field β-NMR spectrometers in the ISAC low energy hall

the beam can be kicked to both spectrometers in an alternating fashion so both spectrometers run simultaneously. In this way ^{8}Li beam pulses are 0.1–10 s long every 10–20 s.

Each spectrometer is mounted on an electrically isolated platform that may be biased to a high electrostatic retarding potential. Simple electrostatic optics are used to decelerate the ions several cm before impacting the sample surface. Decelleration close to the sample minimizes the lateral spread of the beam on arrival at the sample. Due to the very small energy spread of the beam and its small transverse momentum, well focused beam spots with energies down to a few hundred eV and stopping range of about 5 nm are possible. Typically the entire beam lands within a 2 mm diameter spot on the sample without the use of beam collimation at the spectrometer.

3.1 High Magnetic field spectrometer

The high field spectrometer (see Fig. 2) is constructed around a 9 T superconducting solenoid with field oriented along the incoming beam momentum. The implanted beam has nuclear polarisation normal to the sample surface. Two detectors are situated along the initial polarisation—one inside the cryostat 10 cm in the forward direction with respect to the sample position and one annular detector upstream of the magnet 75 cm in the backward direction. Although they are very different in physical dimensions, focusing of the decay electrons in high magnetic field results in nearly equivalent effective solid angles. The RF magnetic field is generated with a non-resonant helical transmission line having a flat 50 Ω impedance and frequency response up to 45 MHz. The RF system can generate CW or complex-modulated pulsed RF for various types of resonance experiments.

A CCD camera viewing a sapphire scintillator through a vacuum window is used as a beam tuning diagnostic by directly imaging light generated from tracks of β and α particles (the latter from ^{8}Be decay) in the scintillators. For samples that are grown on or mounted on sapphire it is convenient to monitor the beam spot position throughout a set of measurements.

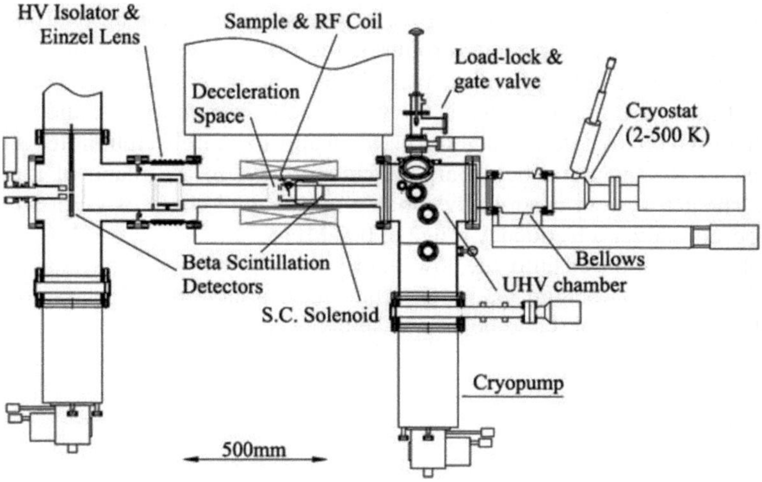

Fig. 2 Schematic side view of the high-field spectrometer. Beam momentum, initial polarisation, and applied DC magnetic field are all normal to the face of the sample

Fig. 3 Schematic view from above of the zero/low-field spectrometer

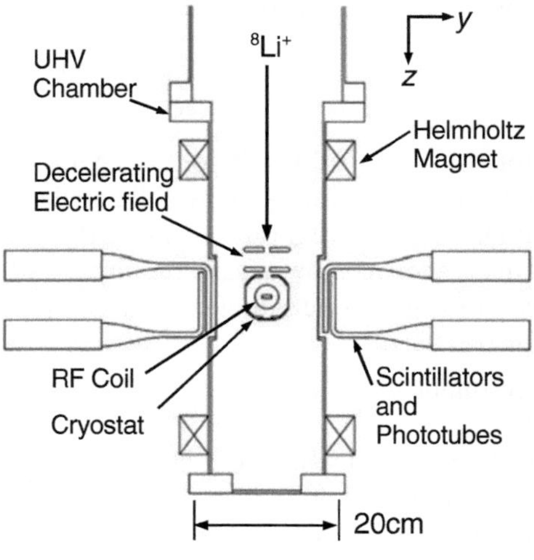

3.2 Zero- and low-field spectrometer

Being optimized for low magnetic fields (0–24 mT), the low field spectrometer (Fig. 3) differs from the high field spectrometer in several fundamental ways. Due to the use of electrostatic dipoles in the low-energy beamline, the orientation of spins of the $^8Li^+$ ions exiting the polarizer is preserved while the beam momentum is bent through 90 degrees. Polarized $^8Li^+$ ions therefore arrive at the low field spectrometer with spin polarisation transverse to their momentum (parallel to y, see Fig. 3) and therefore in the plane of the face of the sample. The external field H_0, if applied, is oriented (anti)parallel the initial polarisation, and so is also parallel to the sample face, an arrangement suitable for measuring the depth dependence of magnetic field in the Meissner state of superconductors. Trim coils generate weak (< 1 mT) fields in the other directions to permit zeroing of the other components of the ambient magnetic field to within 1 μT. Two pairs of fast plastic scintillation detectors and photomultipliers are situated outside the vacuum chanber on the left and right sides, along the initial polarisation. A small coil (in an approximately Helmholtz geometry) generates an RF magnetic field perpendicular to H_0, in the plane of the sample surface. The RF system is similarly capable of CW and complex-modulated pulse mode operation over a frequency range of 0–2 MHz.

Samples are mounted into the low field spectrometer on a four-stage multple sample ladder via a vacuum load-lock. The use of a ladder enables rapid changes among the loaded samples and beam spot imaging scintillator.

4 Measurement modes

Two basic types of experiment can be performed with either specrometer - frequency scans of resonances and time-differential spin relaxation measurements with pulsed beam. As a function of either time or frequency, the experimental signal is extracted as an asymmetry in the number of events n_i accumulated in a pair of detectors situated on opposing sides of the sample ($i =$ left (backward) and right (forward) in the low (high) field spectrometer.)

$$A_0 P = \frac{n_1 - n_2}{n_1 + n_2}, \tag{1}$$

where for ^{8}Li $A_0 \simeq -0.2$ is a factor taking into account the intrinsic decay asymmetry averaged over the detectors' finite solid angles.

At single-counter event rates of about 10^6 s^{-1}, resonance or spin relaxation spectra such as those shown below usually require 20–40 min per run to accumulate. A typical temperature scan on one sample can be completed in a day.

4.1 RF resonances

In fixed magnetic field H_0, the β-NMR resonance is obtained by measuring the beta decay asymmetry spectrum as a function of the frequency of a small transverse RF magnetic field $H_1 \ll H_0$. On resonance, transitions between spin states results in loss of polarisation and hence a dip in beta decay asymmetry. Figure 4 shows the NMR spectrum of ^{8}Li stopped within a 50 nm film of Ag at 145 K. In 3 T two lines are resolved, with positive shifts with respect to the Larmor frequency $\nu_0 = \gamma H_0$. The absence of quadrupole splitting implies the Li occupies a site of cubic symmetry and isolated from defects. In paramagnetic metals the Knight shift $K = (\nu - \nu_0)/\nu_0 = A\chi(T)$ measures the product of Pauli spin suseptibility χ and coupling A between the conduction electrons of the host metal and the Li, which is site dependent. The full temperature dependence of the resonance amplitudes and shifts reveal that at low temperatures the octahedral site dominates; thermally activated diffusion above about 110 K tends to migrate the Li to the substitutional site [5]. Similar experiments have also been performed on bulk and thin film Au [6] and Pd [7].

In type-II superconductors the resonance lineshape provides a direct measurement of the magnetic field distribution resulting from the penetration of magnetic

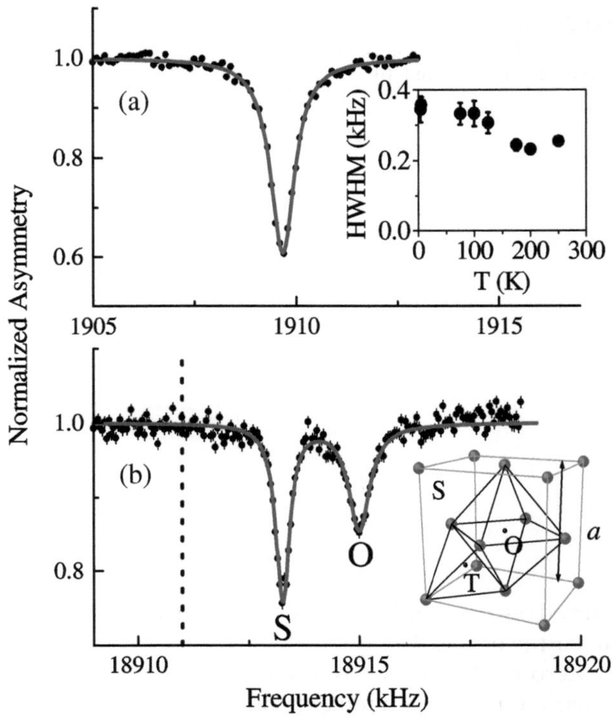

Fig. 4 ^{8}Li NMR resonance in 50 nm Ag film at **a** 0.3 and **b** 3 T. Two lines resolved at high field are assigned to Li occupying the octahedral (O) and substitutional (S) sites with different Knight shifts with respect to the Larmor frequency in MgO indicated by the dashed line. Figure reproduced from [5]

field through the material. β-NMR resonance experiments on bulk superconducting $NbSe_2$ ($T_c = 7.0$ K) resolved the asymmetric line shape due to a triangular array flux-line lattice. The normal core radius extracted from the lineshape at modest 10 mT fields was found to be strongly magnetic field dependent and suggestive of changing vortex-vortex interactions within 20 nm of the surface [8].

The first β-detected nuclear quadrupole resonance (NQR) in zero magnetic field was performed with ^{8}Li stopped in a single crystal of $SrTiO_3$ [9]. The nuclear quadrupole resonance in zero applied magnetic field occurs at resonance frequencies matching transitions among nuclear spin levels determined by the electric field gradient at the probe site, described by the Hamiltonian:

$$\mathcal{H}_q = \frac{e^2qQ}{4I(2I-1)}(3m^2 - I(I+1)) \\ = \nu_q(m^2 - 2) \tag{2}$$

where, for I=2, $\nu_q = e^2qQ/8$ is the NQR frequency, and $eq = V_{zz}$ is the axially symmetric electric field gradient characteristic of the ^{8}Li site. The first zero field β-NQR—measured in $SrTiO_3$—is shown in Fig. 5. The peak position at $3\nu_q$ = 228.81(2) kHz corresponds to the $|m = \pm 2\rangle \rightarrow |m = \pm 1\rangle$ transition, but the amplitude is much larger than would be obtained from simply driving the spin population from $m = \pm 2$ to $m = \pm 1$. The large resonance amplitude is attributed to the nearly complete depolarisation of the ^{8}Li via mixing of spin states due to small non-axial magnetic and/or electric field components. ^{8}Li βNQR experiments have also been

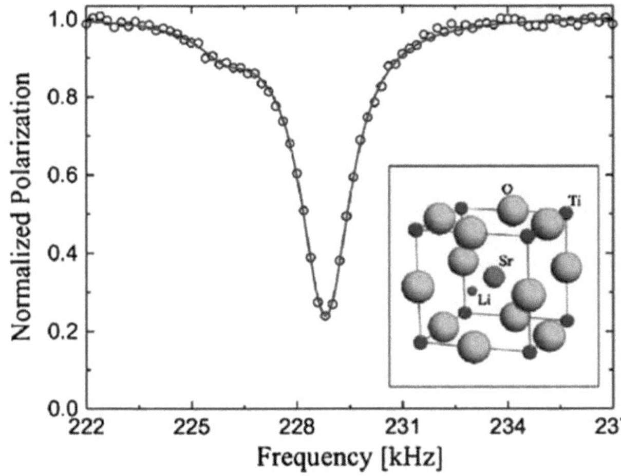

Fig. 5 Zero field ^{8}Li β-detected NQR resonance in <100> $SrTiO_3$. From [10]

carried out on samples of Al_2O_3 and Sr_2RuO_7 in which it is found that the Li occupies at least two sites [10].

Zero field β-NQR has also been applied to making improved measurements of electric quadrupole moments of $^{9,11}Li$, which are needed to constrain theoretical models of these neutron-rich nuclei. The ratio $Q_{11}/Q_9 = 1.077(1)$ was deduced from zero field NQR frequencies under the assumption that Li ions of differing masses occupy the same site in $SrTiO_3$ and therefore experience the same electric field gradient. The measurement improved the uncertainty by a factor of 15 over prior results [11].

4.2 Spin relaxation

In the absence of any applied RF magnetic field, fluctuating magnetic fields of nuclear dipolar or electronic origin drive depolarisation of the Li nuclei at a rate $1/T_1 = (\gamma\delta)^2\tau_c/(1 + (\gamma B)^2\tau_c^2)$ where δ is the magnitude of the fluctuating field, τ_c is the characteristic time of flucutations and γB is the Larmor frequency in field B. In the case of $NbSe_2$ the relaxation of Li nuclear spin polarisation is driven by slowly fluctuating Nb nuclear moments. Application of DC fields of a few mT decouples the Li nuclei, suppressing the Li relaxation (see inset, Fig. 6). This relation between magnetic field at the Li site and Li spin relaxation rate has been applied in making a direct measurement of the magnetic penetration depth of superconducting $NbSe_2$ in the Meissner state [12]. Below the superconducting transition temperature the weak applied magnetic field (oriented parallel to the sample face) is expelled. Within the sample the magnetic field drops away with depth into the surface z according to $B(z) = B_0e^{(-z/\lambda)}$; the relaxation rate therefore becomes a function of depth. The Li stopping range distribution $n(z)$ into the sample can be estimated reliably by computer codes [13]. Given this it is then possible to extract $\lambda(T)$ by a deconvolution. The resulting temperature dependence of λ was found to be consistent with a multiply-gapped superconducting order parameter.

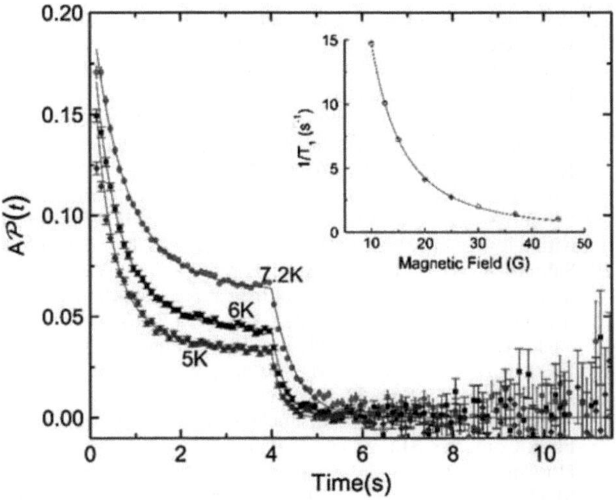

Fig. 6 Decay asymmetry obtained with pulsed beam in a sample of freshly cleaved bulk $NbSe_2$. The left-right decay asymmetry is recorded during the entire beam on/off cycle. While the beam is on the polarisation approaches a constant with amplitude $A_0/(1+\tau/T_1)$ after which the polarisation relaxes to zero. Both beam on and off segments of the data are sensitive to the the relaxation rate $1/T_1$; all the data is fitted with a common value of $1/T_1(T)$. The inset shows the field dependence of the $1/T_1(B)$ in the normal state at 8 K

5 Summary

The large initial polarisation and signal detection via nuclear decay both independent of applied magnetic field make β-NMR about 10 orders of magnitude more sensitive than conventional NMR. Combined with a low and variable implantation energy, the technique becomes a depth-resolved probe of magnetism and local electronic environment on length scales from 2–200 nm. Two spectrometers have been constructed to carry out NMR and zero field NQR RF experiments as well as spin relaxation experiments, generally as a function of magnetic field, temperature and depth. The technique has applications in diverse areas of materials science such as properties of impurities in metals and semiconductors, magnetism and superconductivity in thin films, heterostructures and near-surface interior of bulk samples.

References

1. Abragam, A.: Principles of Nuclear Magnetism. Clarendon, Oxford (1961)
2. Morenzoni, E., Khasanov, R., Luetkens, H., Prokscha, T., Suter, A., Garifianov, N., Glckler, H., Birke, M., Forgan, E., Keller, H., Litterst, J., Ch. Niedermayer, Nieuwenhuys, G.: Low energy muons as probes of thin films and near surface regions. Physica **B326**, 196–204 (2003)
3. Levy, C.D.P., Pearson, M.R., Morris, G.D., Chow, K.H., Hossain, M.D., Kiefl, R.F., Labb, R., Lassen, J., MacFarlane, W.A., Parolin, T.J., Saadaoui, H., Smadella, M., Song, Q., Wang, D.: Development of the collinear laser beam line at TRIUMF. Hyperfine Interact. **196**, 287–294 (2010)
4. Levy, C.D.P., Pearson, M.R., Kiefl, R.F., Mané, E., Morris, G.D., Voss, A.: Laser polarization facility. Hyperfine Interact (this issue). doi:10.1007/s10751-013-0896-4
5. Morris, G.D., MacFarlane, W.A., Chow, K.H., Salman, Z., Arseneau, D.J., Daviel, S., Hatakeyama, A., Kreitzman, S.R., Levy, C.D.P., Poutissou, R., Heffner, R.H., Elenewski, J.E., Greene, L.H., Kiefl, R.F.: Depth-controlled β-NMR of ^{8}Li in a thin silver film. Phys. Rev. Lett. **93**, 157601 (2004)
6. Parolin, T.J.,1, Salman, Z., Chow, K.H., Song, Q., Valiani, J.,1, Saadaoui, H., OH́alloran, A., Hossain, M.D., Keeler, T.A., Kiefl, R.F., Kreitzman, S.R., Levy, C.D.P., Miller, R.I., Morris, G.D., Pearson, M.R., Smadella, M., Wang, D., Xu, M., MacFarlane, W.A.: High resolution β-NMR study of ^{8}Li+ implanted in gold. Phys. Rev. **B77**, 214107 (2008)

7. Parolin, T.J., Salman, Z., Chakhalian, J., Q Song, K Chow, H., Hossain, M.D., Keeler, T.A., Kiefl, R.F., Kreitzman, S.R., Levy, C.D.P., Miller, R.I., Morris, G.D., Pearson, M.R., Saadaoui, H., Wang, D., MacFarlane, W.A.: Beta-NMR of isolated lithium in nearly ferromagnetic palladium. Phys. Rev. Lett. **98**, 047601 (2007)
8. Salman, Z., Wang, D., Chow, K.H., Hossain, M.D., Kreitzman, S.R., Keeler, T.A., Levy, C.D.P., MacFarlane, W.A., R.I. Miller, Morris, G.D., Parolin, T.J., Saadaoui, H., Smadella, M., Kiefl, R.F.: Phys. Rev. Lett. **98**, 167001 (2007)
9. Salman, Z., Reynard, E.P., MacFarlane, W.A., Chow, K.H., Chakhalian, J., Kreitzman, S.R., Daviel, S., Levy, C.D.P., Poutissou, R., Kiefl, R.F.: β-detected nuclear quadrupole resonance with a low-energy beam of $^8Li^+$. Phys. Rev. **B70**, 104404 (2004)
10. Salman, Z., Kiefl, R.F., Chow, K.H., MacFarlane, W.A., Kreitzman, S.R., Arseneau, D.J., Daviel, S., Levy, C.D.P. Maeno, Y., Poutissou, R.: Beta-detected NQR in zero field with a low energy beam of $^8Li^+$. Physica **B374–375**, 468–471 (2006)
11. Voss, A., Pearson, M.R., Billowes, J., Buchinger, F., Chow, K.H., Crawford, J.E., Hossein, M.D., Kiefl, R.F., C.D.P. Levy, MacFarlane, W.A., Mane, E., Morris, G.D., Parolin, T.J., Saadaoui, H., Salman, Z., Smadella, M., Song, Q., Wang, D.: The development of pure -NQR techniques for measurements of nuclear ground state quadrupole moments in lithium isotopes. J. Phys. Conf. Ser. **312**, 092063 (2011)
12. Hossain, M.D., Salman, Z., Wang, D., Chow, K.H., Kreitzman, S., Keeler, T.A., Levy, C.D.P., MacFarlane, W.A., Miller, R.I., Morris, G.D., Parolin, T.J., Pearson, M., Saadaoui, H., Kiefl, R.F.: Low-field cross spin relaxation of 8Li in superconducting $NbSe_@$. Phys. Rev. **B79**, 144518 (2009)
13. Ziegler, J.F., Ziegler, M.D., Biersack, J.P.: Nucl. Instrum. Methods Phys. Res. **B268**, 1818 (2010)

DOI 10.1007/978-94-007-7963-1_20
Reprinted from *Hyperfine Interactions* Journal, DOI 10.1007/s10751-013-0895-5

Nuclear structure explored by β-delayed decay spectroscopy of *spin-polarized* radioactive nuclei at TRIUMF ISAC-1

Intruder configurations in ^{29}Mg and ^{30}Mg, the nuclei in the region of island of inversion

T. Shimoda · K. Tajiri · K. Kura · A. Odahara · M. Suga · Y. Hirayama · N. Imai · H. Miyatake · M. Pearson · C. D. P. Levy · K. P. Jackson · R. Legillon · C. Petrache · T. Fukuchi · N. Hamatani · T. Hori · M. Kazato · Y. Kenmoku · T. Masue · H. Nishibata · T. Suzuki · A. Takashima · R. Yokoyama

Published online: 23 October 2013

Abstract Spin-polarized radioactive nuclear beams at TRIUMF enable a new spectroscopic method which efficiently assigns spins and parities of the daughter levels by taking advantage of the asymmetric β-decay of the polarized parent nucleus. This method was successfully applied to structure studies of ^{29}Mg and ^{30}Mg in connection with the physics of the "island of inversion". In ^{29}Mg, two low-lying levels with intruder configuration were assigned. In ^{30}Mg, coexistence of spherical shape, prolate shape and γ-collectivity was strongly suggested.

Keywords Polarized RI beam · β-delayed decay spectroscopy · ^{29}Mg and ^{30}Mg · Island of inversion · Intruder configurations · Shape coexistence

ISAC and ARIEL: The TRIUMF Radioactive Beam Facilities and the Scientific Program.

T. Shimoda (✉) · K. Tajiri · K. Kura · A. Odahara · M. Suga · T. Fukuchi · N. Hamatani · T. Hori · M. Kazato · Y. Kenmoku · T. Masue · H. Nishibata · T. Suzuki · A. Takashima · R. Yokoyama
Department of Physics, Osaka University, Toyonaka, Osaka 560-0043, Japan
e-mail: shimoda@phys.sci.osaka-u.ac.jp

Y. Hirayama · N. Imai · H. Miyatake
Institute of Particle and Nuclear Studies, KEK, Ibakaki 305-0801, Japan

M. Pearson · C. D. P. Levy · K. P. Jackson
TRIUMF, 4004 Wesbrook Mall, Vancouver, BC V6T 2A3, Canada

R. Legillon · C. Petrache
Université Paris-Sud and Institut de Physique Nucléaire, 91045 Orsay, France

1 Introduction

One of the recent highlights in nuclear physics is the shell evolution in very neutron-rich nuclei with variation of neutron number. In particular, the exotic structure of nuclei around the "island of inversion" [1]. This very neutron-rich region of the nuclear chart with neutron numbers close to the magic number $N = 20$, has been attracting much attention for two decades. The significant fractions of intruder configurations in the ground and/or excited states should cause very different features from those of the magic-number nuclei close to the stability line. There have been many experimental and theoretical works to clarify which nuclei lie within the island of inversion and which level has structure with intruder configurations [2, 3].

However, the experimental information on these nuclei is rather limited. For many of the levels, spin and parity, which are the key quantities to understand the nuclear structure, have not been assigned. For example, in the case of ^{29}Mg whose structure will be discussed in the present work, only the ground-state spin-parity has been assigned [4, 5]. The levels in ^{30}Mg were placed with no spin-parity assignment by the β-delayed γ-spectroscopy of ^{30}Na [6]. A lifetime measurement proposed a 0_2^+ level at 1.788 MeV [7] and a conversion electron measurement confirmed this assignment [8]. The possible shape coexistence in this level is being discussed.

We have developed a powerful tool to assign spin-parity to the levels of the daughter nuclei in β-delayed decay spectroscopy at ISAC-1 at TRIUMF, where the world-highest polarization is available for radioactive alkali beams [9, 10]. This method takes advantage of asymmetric β-decay of the spin-polarized nucleus. The first successful application was performed in 2002 and 2004 on the β-delayed neutron-decay $^{11}\mathrm{Li}_{\mathrm{g.s.}} \xrightarrow{\beta} {}^{11}\mathrm{Be}^* \xrightarrow{\mathrm{n}} {}^{10}\mathrm{Be}^* \xrightarrow{\gamma} {}^{10}\mathrm{Be}_{\mathrm{g.s.}}$. We could firmly assign spins and parities of seven levels in ^{11}Be for the first time and reveal various types of cluster structure in ^{11}Be [11]. In the present work, the second also successful application in 2007 and 2010 to systematic studies of neutron-rich Mg isotopes is briefly discussed. More details are found in [12, 13].

2 Experimental

The principle of the method is based on an asymmetric β-decay shown by the following angular dependence, if we assume an allowed transition:

$$W(\theta) \simeq 1 + AP\cos\theta, \tag{1}$$

where P, θ and A are the polarization of the parent nucleus, the emission angle of the β-ray with respect to the polarization direction and the asymmetry parameter characteristic of the β-transition, respectively. The essential point is that the asymmetry parameter A is a constant depending on the initial and final state spins, I_i and I_f, respectively, as shown in (2) and Table 1. Here, we ignored Fermi transition probabilities for the case of $I_f = I_i$, because in a very neutron-rich nucleus Fermi transitions result in highly excited states, whereas we are discussing a rather

Table 1 Asymmetry parameters for allowed β-decay of ^{29}Na and ^{30}Na

β-decay	I_i^π	I_f^π	A
$^{29}Na_{g.s.} \rightarrow ^{29}Mg$		1/2+	−1.0
	3/2+	3/2+	−0.4
		5/2+	+0.6
$^{30}Na_{g.s.} \rightarrow ^{30}Mg$		1+	−1.0
	2+	2+	−0.33
		3+	+0.67

low-excitation energy region. The large difference in A enables us to unambiguously assign I_f.

$$A \begin{cases} = -1 & (I_f = I_i - 1), \\ \backsimeq \dfrac{-1}{I_i + 1} & (I_f = I_i), \\ = \dfrac{I_i}{I_i + 1} & (I_f = I_i + 1). \end{cases} \tag{2}$$

The AP value for each β-decay is obtained by comparing β-γ coincidence counts in the detectors at $\theta = 0°$ and 180°. In order to cancel out the spurious asymmetry due to instrumental asymmetry, the direction of polarization is reversed by flipping the laser helicity periodically. Note that the polarization P is common for all β-transitions. Therefore, once P is known, for example from the AP value of the β-transition to a spin-known level, the asymmetry parameters for other levels are obtained, i.e. the spin of the level is immediately assigned. For a daughter level which is also populated by γ-transitions from higher levels, it is easily shown that the effects on the asymmetry parameter are estimated and the asymmetry parameter for the specific β-transition is obtained, if the γ- and β-transition intensities are all known.

For the $^{29}Na^+$ and $^{30}Na^+$ beams, polarization of ~50 % was achieved by using the collinear laser optical pumping technique at the polarized beam facility at ISAC-1 [9, 10]. The experiments were performed at the very end of a beamline named "OSAKA" which is located downstream of the polarized beam facility. The direction of polarization was perpendicular to the beam direction and in the horizontal plane.

Figure 1 shows the layout of detectors. The 30.4 keV polarized $^{29}Na^+$ or $^{30}Na^+$ was stopped on a thin Pt foil (10 μm thick) in vacuum. A static magnetic field of ~530 mT was applied to preserve the polarization. The β-rays and the successive γ-rays were detected by 9 detectors surrounding the Pt foil. Each detector consisted of two thin plastic scintillators and a coaxial Ge detector to form a ΔE-ΔE-E detector telescope. This configuration enabled β-ray and γ-ray identification and their energy measurement. Two detectors ("Left" and "Right") placed on the polarization axis measured the asymmetry of β-decay.

The average beam intensities of ^{29}Na and ^{30}Na were 82 and 14 pps, respectively, at the Pt foil. A total of 1.7×10^7 ^{29}Na and 7.1×10^6 ^{30}Na were implanted into the foil during data-acquisition times of 56 and 142 hours, respectively. The polarization was determined from the measured AP-values to be $P(^{29}\text{Na}) = 35 \pm 2$ % and $P(^{30}\text{Na}) = 32 \pm 3$ %. The somewhat smaller polarization than the estimated beam polarization of ~50 % may be due to spin-relaxation in the Pt foil.

Fig. 1 Schematic detector arrangement

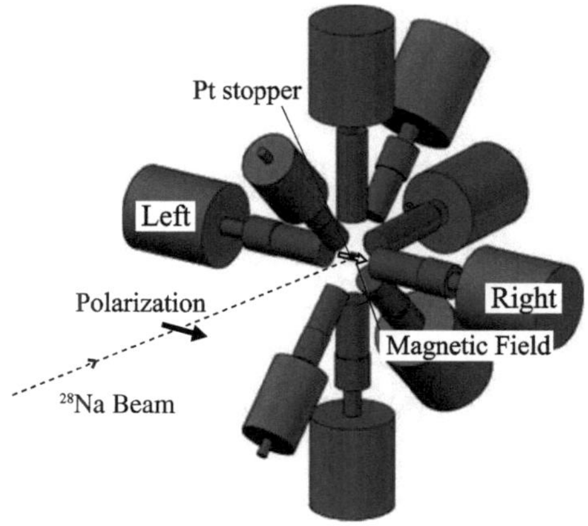

3 Results

3.1 Structure of ^{29}Mg

Figure 2 shows the decay scheme of ^{29}Na revised by the present work. The decay intensities were determined by normalizing the well-established intensity of the 2224-keV γ-ray in the granddaughter ^{29}Al [14] and the neutron decay probability $P(1n) = 21.5(30)$ % [15]. Asterisks in the figure indicate new findings. It is seen that two new γ-transitions have been found and spins and parities (I^{π}) of seven levels have been assigned in ^{29}Mg for the first time. The I^{π} assignments for the two levels at 2.614 (1/2$^+$) and 3.223 (3/2$^+$) MeV, characterized by large β-decay branching ratios, were firmly assigned. Other I^{π}s in parentheses were determined based on the above-mentioned two assignments and detailed γ-transition paths and intensities. Note that the spins and parities of most of the ^{29}Mg states except for the two levels at 1.094 and 1.430 MeV were assigned. These levels are characterized by large log ft-values. It is interesting to note that the 3.985-MeV level which is located above the neutron threshold (3.672 MeV) decays to the ground state by γ-emission. This fact supports the $5/2^+$ assignment.

The experimental results were compared with the shell-model calculations with the universal *sd*-shell interaction B (USDB) [16]. It was found that the level energy, spin-parity and β-transition probability were well reproduced assuming all nucleons are in the *sd*-shell for all the positive-parity levels. However, the two spin-unassigned levels could not be explained by such calculations. This fact strongly suggests that these levels at 1.094 and 1.430 MeV are negative-parity levels with an intruder configuration. This result is consistent with the previous speculation [17]. In fact, the Monte Carlo Shell Model calculation taking into account the pf-shell configurations predicts $7/2^-$ and $3/2^-$ levels at 0.68 MeV and 1.01 MeV, respectively (Utsuno, private communication).

It becomes possible to compare the excitation energies of the negative-parity levels in neutron-rich Mg isotopes. Figure 3 shows the negative-parity levels in odd-mass Mg isotopes, assuming 7/2$^-$ and 3/2$^-$ assignments for the 1.430- and 1.094-MeV ^{29}Mg levels, respectively. It is clearly seen that the excitation energies of the negative-parity levels rapidly decrease at ^{29}Mg with the increase in neutron number. This is the experimental evidence for shell evolution in neutron-rich Mg isotopes.

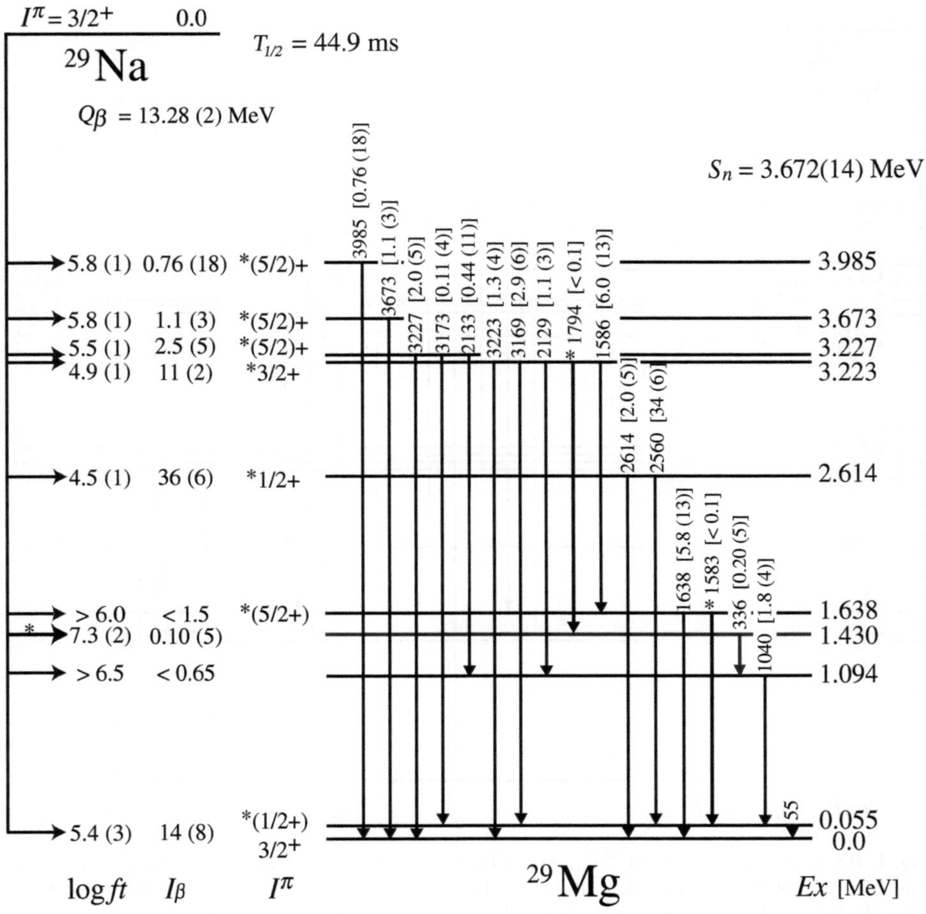

Fig. 2 Revised decay scheme of ^{29}Na. The physical quantities and decay paths marked by asterisks are ones newly found in the present work. The level at 1.430 MeV was observed in the β-decay of ^{29}Na for the first time (emphasized in *blue*). The γ-transition intensities per ^{29}Na β-decay are shown in units of percent, together with the γ-ray energies in keV. The β-decay to the first excited state at 55 keV was observed, but since its intensity could not be evaluated separately from that of the ground state, the summed intensity is shown

Fig. 3 Low-lying negative-parity levels in odd-mass Mg isotopes. The spin-parity assignments of the ^{31}Mg levels and ^{33}Mg$_{g.s.}$ are after [17] and [18, 19], respectively

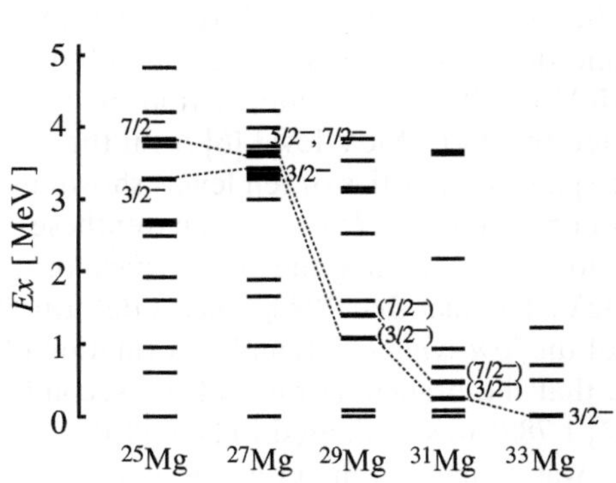

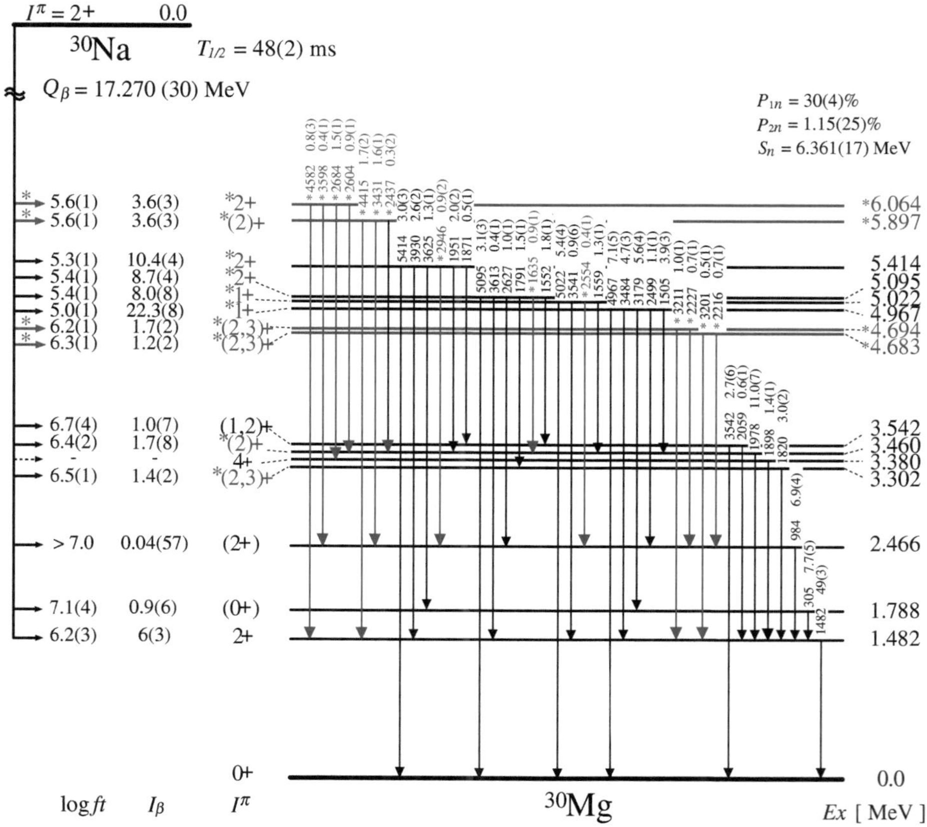

Fig. 4 Revised decay scheme of ^{30}Na. Asterisks denote new results achieved in the present work. The level at 3.380 MeV and the deexciting 1898-keV γ-ray (shown in *blue*), which were known in the ^{14}C(^{18}O,2p) reaction [22] , were observed in the β-decay of ^{30}Na for the first time

3.2 Structure of ^{30}Mg

Figure 4 shows the revised decay scheme of ^{30}Na established by the present work. The decay intensities were determined by referring to the known total β-decay probability P(0n) = 69.2 % [6]. It is seen that fourteen γ-transitions were newly found and detailed γ-γ analyses claimed four new levels at 4.683, 4.694, 5.897 and 6.064 MeV in ^{30}Mg. The confirmation of the new decay path 5.897→1.482 →g.s. eliminated the 4.415-MeV level [6] from the level scheme. It should be emphasized that the spins and parities of ten levels shown in *red* have been assigned for the first time. Among them five I^{π}s without parentheses were firmly assigned.

It is to be noted that the major part of β-decay goes to levels higher than 4.967 MeV. The fact that ^{30}Na$_{\mathrm{g.s.}}$ has a deformed shape [20, 21] suggests a spherical nature of the low-lying levels and deformation of the high-lying levels in ^{30}Mg.

Note that the branching ratio of the second forbidden transition ^{30}Na$_{\mathrm{g.s.}}(2^+) \rightarrow$ ^{30}Mg(0_2^+, 1.788) was evaluated to be 0.9(6) % (log ft = 7.1(4)), whereas no β-decay intensity was claimed in the previous work [7]. This fact suggests deformation of the

Table 2 The γ-transition intensities deexciting the 4.967-MeV level (1^+)

	$E_x^f(I_f^\pi)$	0.0 (0_1^+)	1.482 (2_1^+)	1.788 (0_2^+)	2.466 (2_2^+)	3.460 (2^+)
	$\sigma\lambda$	M1	M1	M1	M1	M1
I_γ	Experiment	1.0	0.66(7)	0.878(8)	0.15(2)	0.55(5)
	Weisskopf est.	1.0	0.35	0.26	0.13	0.03

Both the experimental and Weisskopf estimates are normalized to those of the ground state transition

Table 3 The γ-transition intensities deexciting the 5.414-MeV level (2^+)

	$E_x^f(I_f^\pi)$	0.0 (0_1^+)	1.788 (0_2^+)	1.482 (2_1^+)	2.466 (2_2^+)	3.460 (2^+)
	$\sigma\lambda$	E2	E2	M1	M1	M1
I_γ	Experiment	1.0	0.43(6)	1.0	0.35(8)	0.79(9)
	Weisskopf est.	1.0	0.13	1.0	0.42	0.12

Both the experimental and Weisskopf estimates are normalized to those of the ground state transition

0_2^+ level. This result is consistent with the small transition rate of the 0_2^+ level to the ground state [8], which is known to be spherical [17]. It is also worth noting that the 0_2^+ level is populated by γ-transitions from two higher levels 5 times more than the β-transition, and that the two higher levels are associated with large β-transition probabilities. Namely, (i) the largest β-transition (22.3 %) goes to the 4.967-MeV level (1^+), then leads to the 0_2^+ level with 5.6 % intensity, and (ii) the next largest one (10.4 %) goes to the 5.414-MeV level (2^+) and finally to the 0_2^+ level (0.9 %). The fact that $^{30}Na_{g.s.}$ is well deformed suggests that these two levels also have intruder configurations to a large extent. However, we could not find the deexciting γ-ray from the possible 2^+ state of the rotational band on the 0_2^+ state.

Tables 2 and 3 compare the observed γ-transition probabilities from the 4.967-MeV and 5.414-MeV levels, respectively, with the Weisskopf estimates. It is found that the transitions 4.967 (1^+)→ 3.460 (2^+) and 5.414 (2^+)→ 3.460 (2^+) show a very large enhancement from the Weisskopf estimates. This fact suggests that the 3.460-MeV level also has a deformed shape with an intruder configuration.

Note that the β-decay intensity to the second 2^+ level at 2.466 MeV was found to be less than the experimental sensitivity in the present work, whereas the previous work reported 3.7 % [6]. On the contrary, the β-decay to the first 2^+ level at 1.482 MeV was observed with 6(3) % intensity. Since the 2_1^+ level is interpreted to be spherical [17], the unobserved β-decay intensity to the 2_2^+ level suggests its non-spherical nature, but different deformation from $^{30}Na_{g.s.}$. This level is likely the 2^+ band head of the γ-band predicted by a calculation of large-amplitude collective motion, based on the constrained Hartree-Fock-Bogoliubov plus local quasiparticle random-phase approximation (CHFB+LQRPA) method [23].

4 Summary

We have investigated the structures of ^{29}Mg and ^{30}Mg by taking advantage of TRIUMF's spin-polarized radioactive nuclear beams of ^{29}Na and ^{30}Na, respectively. Because of high polarization of the beams and detailed decay intensities, we could assign spins and parities to seven levels in ^{29}Mg and ten levels in ^{30}Mg for the first time. Since all positive-parity levels in ^{29}Mg were consistent with the shell model calculations with the USDB Hamiltonian, the two unassigned levels at 1.430 and 1.094 MeV associated with large log ft-values were assigned negative-parity with an intruder configuration. In ^{30}Mg, five levels at 1.788 (0_2^+), 2.466 (2_2^+), 3.460 (2^+), 4.967 (1^+), and 5.414 (2^+) MeV exhibited a non-spherical nature. From the β- and γ-branching ratios, we propose that the 1.788-, 3.460-, 4.967- and 5.414-MeV levels have a deformed shape and the 2.466-MeV level is the 2^+ band head of the γ-band.

Acknowledgements The authors gratefully acknowledge TRIUMF staff for their excellent support, encouragement and hospitality. They are also indebted to Drs. Y. Utsuno, K. Matsuyanagi, N. Hinohara and H. Sagawa for valuable discussions. This work was supported in part by Grant-in-Aid for Scientific Research (B) (19340061 and 22340058) in Japan.

References

1. Warburton, E.K., et al.: Mass systematics for $A=29$–44 nuclei: The deformed $A\sim32$ region. Phys. Rev. **C41**, 1147 (1990)
2. Brown, B.A.: Islands of insight in nuclear chart. Physics **3**, 104 (2010)
3. Sorlin, O, Porquet, M.-G.: Nuclear magic numbers: New features far from stability. Prog. Part. Nucl. Phys. **61**, 602 (2008)
4. Bauman, P., et al.: Gamow-Teller beta decay of ^{29}Na and comparison with shell-model predictions. Phys. Rev. **C36**, 765 (1987)
5. National Nuclear Data Center, Brookhaven National Laboratory (http://www.nndc.bnl.gov/), the ^{29}Mg data last update in 2012
6. Bauman, P., et al.: Beta decay of ^{30}Na: Experiment and theory. Phys. Rev. **C39**, 626 (1989)
7. Mach, H., et al.: New structure information on ^{30}Mg, ^{31}Mg and ^{32}Mg. Eur. Phys. J. **A25**, s01 (2005)
8. Schwerdtfeger, W., et al.: Shape coexistence near neutron number N=20: first identification of the $E0$ decay from the deformed first excited $J^\pi=0^+$ state in ^{30}Mg. Phys. Rev. Lett. **103**, 012501 (2009)
9. Levy, C.D.P. et al.: Polarized radioactive beam at ISAC. Nucl. Inst. Methods **B204**, 689 (2003)
10. Levy, C.D.P., et al.: The collinear laser beam line at ISAC. Nucl. Phys. **A746**, 206c (2004)
11. Hirayama, Y., Shimoda, T., et al.: Study of ^{11}Be structure through β-delayed decays from polarized ^{11}Li. Phys. Lett. B **611**, 239 (2005)
12. Kura, K. et al.: Structure of ^{28}Mg studied by β-decay spectroscopy of spin-polarized ^{28}Na: The first step of systematic studies on neutron-rich Mg isotopes. Phys. Rev. **C85**, 034310 (2012)
13. Tajiri, K.: Ph.D. thesis, Osaka University (2012)
14. Détraz, C., et al.: Beta decay of $^{27\text{-}32}$Na and their descendants. Phys. Rev. **C19**, 164 (1979)
15. Langevin, M., et al.: β-delayed neutrons from very neutron-rich sodium and magnesium isotopes. Nucl. Phys. **A414**, 151 (1984)
16. Brown, B.A., Richter, W.A.: New "USD" Hamiltonian for the *sd* shell. Phys. Rev. **C74**, 034315 (2006)
17. Terry, J.R.: Single-neutron knockout from intermediate energy beams of ^{30,32}Mg: Mapping the transition into the "island of inversion". Phys. Rev. **C77**, 014316 (2008)
18. Yordanov, D.T., et al.: Spin and magnetic moment of ^{33}Mg: evidence for a negative-parity intruder ground state. Phys. Rev. Lett. **99**, 212501 (2007)
19. Yordanov, D.T., et al.: Laser spectroscopy in the island of inversion. Hyperfine Interact **196**, 53 (2010)
20. Keim, M.: Recent measurements of nuclear moments far from stability. AIP Conf. Proc. **455**, 50 (1998)

21. Tripathi, V., et al.: Competition between normal and intruder states inside the "island of inversion". Phys. Rev. **C76**, 021301(R) (2007)
22. Deacon, A.N., et al.: Cross shell excitations near the "island of inversion": Structure of ^{30}Mg. Phys. Rev. **C82**, 034305 (2010)
23. Hinohara, N., et al.: Shape fluctuations in the ground and excited 0^+ states of ^{30,32,34}Mg. Phys. Rev. **C84**, 061302 (2011)

DOI 10.1007/978-94-007-7963-1_21
Reprinted from *Hyperfine Interactions* Journal, DOI 10.1007/s10751-013-0897-3

The MTV experiment: a test of time reversal symmetry using polarized ^{8}Li

J. Murata · H. Baba · J. A. Behr · Y. Hirayama · T. Iguri · M. Ikeda · T. Kato · H. Kawamura · R. Kishi · C. D. P. Levy · Y. Nakaya · K. Ninomiya · N. Ogawa · J. Onishi · R. Openshaw · M. Pearson · E. Seitaibashi · S. Tanaka · R. Tanuma · Y. Totsuka · T. Toyoda

Published online: 23 October 2013

Abstract The MTV (Mott Polarimetry for T-Violation Experiment) experiment at TRIUMF-ISAC (Isotope Separator and ACcelerator), which aims to achieve the highest precision test of time reversal symmetry in polarized nuclear beta decay by measuring a triple correlation (*R*-correlation), is motivated by the search for a new physics beyond the Standard Model. In this experiment, the existence of non-zero transverse electron polarization is examined utilizing the analyzing power of Mott scattering from a thin metal foil. Backward scattering electron tracks are measured using a multi-wire drift chamber for the first time. The MTV experiment was commissioned at ISAC in 2009 using an 80 % polarized ^{8}Li beam at 10^7 pps, resulting in 0.1 % statistical precision on the *R*-parameter in the first physics run performed in 2010. Next generation cylindrical drift chamber (CDC) is now being installed for the future run.

Keywords Time reversal symmetry · Test of standard model · Electron polarimeter

ISAC and ARIEL: The TRIUMF Radioactive Beam Facilities and the Scientific Program.

J. Murata (✉) · T. Iguri · M. Ikeda · R. Kishi · Y. Nakaya · K. Ninomiya · N. Ogawa · J. Onishi · E. Seitaibashi · S. Tanaka · R. Tanuma · Y. Totsuka · T. Toyoda
Department of Physics, Rikkyo University, Tokyo 171-8501, Japan
e-mail: jiro@rikkyo.ac.jp

H. Baba
Nishina Center, RIKEN, Saitama 351-0198, Japan

J. A. Behr · C. D. P. Levy · R. Openshaw · M. Pearson
TRIUMF, Vancouver, BC V6T 2A3, Canada

Y. Hirayama
KEK, Tsukuba, Ibaraki 305-0801, Japan

T. Kato · H. Kawamura
Cyclotron and Radioisotope Center, Tohoku University, Miyagi 980-8578, Japan

1 Introduction

Time reversal symmetry is one of the most fundamental symmetries in nature, and its violation is considered equivalent to CP-violation under the CPT-theorem. The present study is one of the experiments searching for a large T-violation independent of the Cabbibo–Kobayashi–Maskawa (CKM) mechanism in nuclear beta decay. Definition of the R-correlation can be found in the following beta decay rate function [1–3].

$$\omega dEd\Omega \propto \left(\cdots + R\sigma \cdot \frac{<\mathbf{J}>}{J} \times \frac{\mathbf{p_e}}{E_e}\right) \quad (1)$$

The existence of a non-zero value of the R parameter implies non-zero T-violating transverse electron polarization perpendicular to the parent nuclear polarization direction. The finest precision was obtained by a radioactive nuclear beta decay experiment performed at PSI [4] using polarized ^{8}Li, with the result $R = (-0.9 \pm 2.2) \times 10^{-3}$. The present experiment is going to achieve one order of magnitude improvement in the precision by applying tracking measurement in order to remove the dominant systematic errors in the PSI experiment, and by utilizing the statistical advantage of the high intensity polarized beam at TRIUMF.

2 Experiment

The present study is testing the R correlation using the high beam intensities and polarization of ^{8}Li at TRIUMF, combined with electron polarimetry using the analyzing power of Mott scattering [5]. The electron transverse polarization in the vertical direction can be extracted from the measured value of the backward scattering left-right asymmetry using the known analyzing power.

In the MTV experiment, a planar drift-chamber is used to track both initial electron tracks and backward scattering tracks (V-tracks). A schematic view of the experimental setup is shown in Fig. 1. At TRIUMF-ISAC, an optically pumped horizontally polarized ^{8}Li beam, at 28 keV with 10^7 pps and 80 % polarization, is implanted on the surface of a 10 μm thick aluminum stopper, which is placed between permanent spin holding magnets producing about 500 Gauss. The beam polarization is measured by the beta-asymmetry counter set around the stopper. A fraction of order 10^{-4} of the decay electrons, emitted from the stopper, are backwardly scattered by a 100 μm lead analyzer foil after penetrating the drift chamber. The incident tracks and backward scattered tracks are measured by the drift chamber. The Mott scattering angle is determined by this tracking information in event by event. Electron energy is measured using stopping plastic scintillation counters, in order to estimate the contribution of electromagnetic final state interaction which does not violate the T-symmetry but contribute on producing the electron transverse polarization.

The MTV project at TRIUMF was proposed in 2008. After confirming the detector performances in commissioning experiments at the KEK-TRIAC using low intensity beam [6], we moved the whole experimental setup to the low energy polarized beam line (Osaka) at TRIUMF [7] in order to achieve the highest statistical precision measurement [8]. The first commissioning experiment (Run-I) was performed in 2009, and the physics run (Run-II) was performed in 2010.

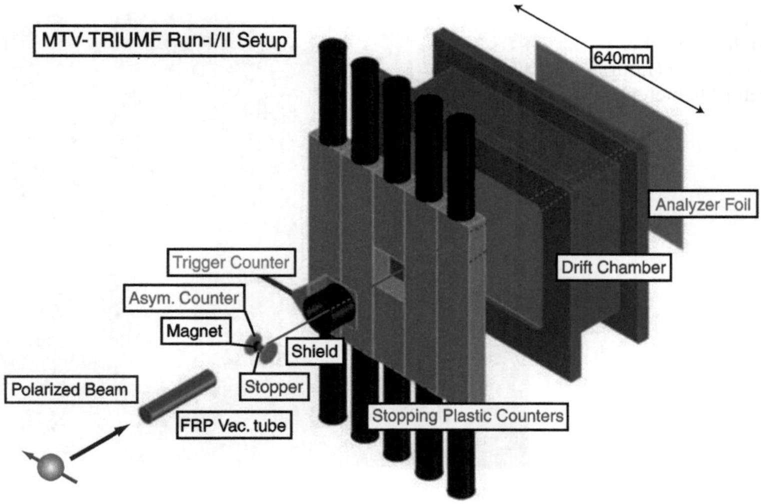

Fig. 1 Experimental setup of the MTV Run-I (2009) and II (2010)

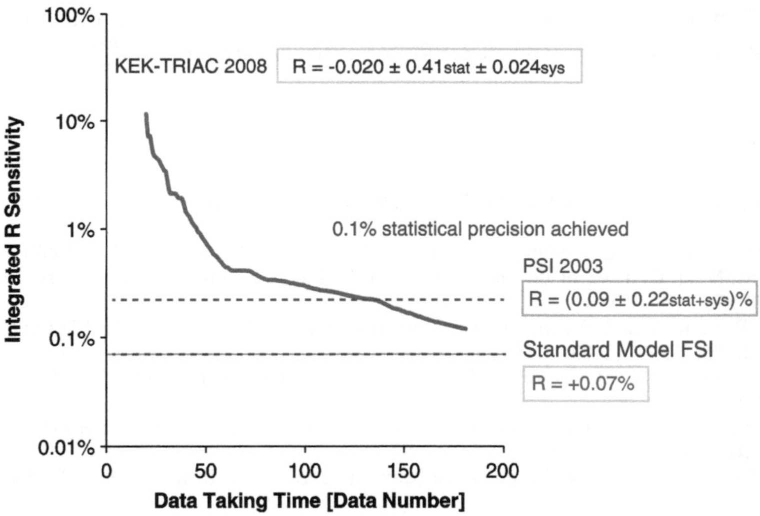

Fig. 2 Obtained statistical precision for Run-II. Integrated statistics from the beginning are used

As shown in Fig. 2, a preliminary result from the offline analysis was obtained with statistical precision of about 0.1 %, which will provide the highest statistical precision data in T-violating beta decay search. The largest systematics comes from a combination of parity violating beta emission asymmetry and asymmetric geometry of the detector, which cannot be cancelled by beam spin flipping. In order to remove all of the systematic effects arising from the asymmetric geometrical configuration of the planar drift chamber, a next generation detector, CDC which is shown in Fig. 3, has been built in Japan. The CDC has 400 anode wires with 4 mm cell size, with 4 sense layers. The small cell size enables us to perform high count rate measurement

Fig. 3 Next generation cylindrical drift chamber (CDC). Size: 70 cm (Long), 20 cm (Outer Diameter)

by reducing a space charge effect. The CDC was commissioned in 2011 at ISAC, showing vital rate stability to be used in the very high beam intensity environment. The MTV project will perform the highest precision test of time reversal symmetry using the CDC, sensitive to fewer systematic effects, beginning in 2012.

References

1. Jackson, J.D. et al.: Coulomb corrections in allowed beta transitions. Nucl. Phys. A **4**, 206–212 (1957)
2. Possible tests of time reversal invariance in beta decay. Phys. Rev. **106**, 517–521 (1957)
3. Severijns, N. et al.: Tests of the standard electroweak model in nuclear beta decay. Rev. Mod. Phys. **78**, 991–1040 (2006)
4. Huber, R. et al.: Search for time-reversal violation in the β decay of polarized ^{8}Li nuclei. Phys. Rev. Lett. **90**, 202301 (2003)
5. Murata, J. et al.: Test of time reversal symmetry using polarized ^{8}Li at TRIUMF-ISAC. J. Phys. CS **312**, 102011 (2011)
6. Kawamura, H.: Study of time reversal symmetry in nuclear β-decay using tracking detector. D.Sc Thesis, Rikkyo University (2010)
7. Levy, C.D.P. et al.: Polarized radioactive beam at ISAC. Nucl. Instrum. Meth. B **204**, 689–693 (2003)
8. Kluge, H.J.: Atomic physics techniques for studying nuclear ground state properties, fundamental interactions and symmetries: status and perspectives. Hyperfine Interact. **196**, 295–337 (2010)

DOI 10.1007/978-94-007-7963-1_22
Reprinted from *Hyperfine Interactions* Journal, DOI 10.1007/s10751-013-0898-2

The radon EDM apparatus

E. R. Tardiff · E. T. Rand · G. C. Ball · T. E. Chupp · A. B. Garnsworthy · P. Garrett · M. E. Hayden · C. A. Kierans · W. Lorenzon · M. R. Pearson · C. Schaub · C. E. Svensson

Published online: 3 December 2013

Abstract The observation of a permanent electric dipole moment (EDM) at current experimentally accessible levels would provide clear evidence of physics beyond the Standard Model. EDMs violate CP symmetry, making them a possible route to explaining the size of the observed baryon asymmetry in the universe. The Radon EDM Experiment aims to search for an EDM in radon isotopes whose sensitivity to CP-odd interactions is enhanced by octupole-deformed nuclei. A prototype apparatus currently installed in the ISAC hall at TRIUMF includes a gas handling system to move radon from a collection foil to a measurement cell and auxiliary equipment for polarization diagnostics and validation. The features and capabilities of the apparatus are described and an overview of the experimental design for a gamma-ray-anisotropy based EDM measurement is provided.

Keywords Electric dipole moment · Octupole enhancement · Radon · Polarization measurement

---

ISAC and ARIEL: The TRIUMF Radioactive Beam Facilities and the Scientific Program

E. R. Tardiff (✉) · G. C. Ball · A. B. Garnsworthy · M. R. Pearson · C. Schaub
Science Division, TRIUMF, 4004 Wesbrook Mall, Vancouver, BC,
Canada V6T 2A3
e-mail: tardiff@gmail.com

E. T. Rand · P. Garrett · C. E. Svensson
Physics Department, University of Guelph, Guelph, ON,
Canada N1G 2W1

T. E. Chupp · W. Lorenzon
Physics Department, University of Michigan, 450 Church St.,
Ann Arbor, MI 48109-1040, USA

M. E. Hayden · C. A. Kierans
Physics Department, Simon Fraser University, 8888 University Drive,
Burnaby, BC, Canada V5A 1S6

1 Introduction

A proposal to investigate the possibility of permanent electric dipole moments was first made by Purcell and Ramsey [1], who noted that, since an EDM is a polar vector that must align with the axial angular momentum vector, it would violate parity (P) conservation. The same is true of time reversal (T) symmetry. Thus, given the CPT theorem an EDM must be symmetric under charge conjugation (C) and violate CP symmetry. Under the Sakharov model of baryogenesis [2], CP violation is a necessary condition for generating a matter-antimatter asymmetry from an initial condition consisting of equal parts of both constituents following the big bang. While the Standard Model contains experimentally observed CP violation through the phase in the CKM matrix, this is not enough to explain the magnitude of the baryon asymmetry. Many proposed Beyond-the-Standard-Model (BTSM) theories contain potentially CP-violating phases which, depending on their values, could generate EDMs at levels above current experimental limits. Figure 1 shows the range of predictions of a few of these theories for the electron, neutron, and mercury EDM along with experimental limits. While the fact that a nonzero EDM has not been observed has already eliminated much of the parameter space of some of these theories, current experimental limits are still orders of magnitude larger than the Standard Model EDMs generated by the CKM phase. Thus, an EDM discovery by any of the currently planned experiments would be evidence of BTSM physics.

The experimental community is currently pursuing EDM measurements in a variety of systems, from particles to atoms to molecules to paramagnetic insulating solids. The most precise EDM measurement to date was performed using the ^{199}Hg atom, setting a limit of $|d\left(^{199}\mathrm{Hg}\right)| < 3.1 \times 10^{-29} e \cdot \mathrm{cm}$ at the 95 % C.L. [11]. Octupole-deformed radon nuclei could be up to 600 times more sensitive than ^{199}Hg to fundamental CP-violating interactions [12, 13], making radon a good candidate for an EDM search. Standard EDM measurements involve observing minor differences in the behavior of the system of interest under the influence of an electric field, often by looking for a change in spin precession frequency when a large electric field is applied parallel or antiparallel to a magnetic field. In the first phase of the Radon EDM measurement, the anisotropy of radiation emitted from a polarized radon sample will provide the precession signal. An apparatus designed for development studies of the gas handling and polarization techniques necessary for the EDM measurement is installed in TRIUMF's ISAC hall [14], and a simulation of the planned EDM apparatus design supports the potential of the Radon EDM experiment to discover an EDM.

2 The development apparatus

While TRIUMF develops the capacity for neutron-rich radon beams, the Radon EDM Experiment has been developing various capabilities necessary for an eventual EDM measurement, generally using isotopes of xenon as convenient analogs for radon. Tests undertaken include verification of a method to transfer noble gas atoms implanted in a thin collector foil to a glass measurement cell [15] and studies of radioactive xenon polarization. A suite of diagnostic equipment allows for spin-exchange optical pumping efficiency to be characterized using adiabatic fast passage

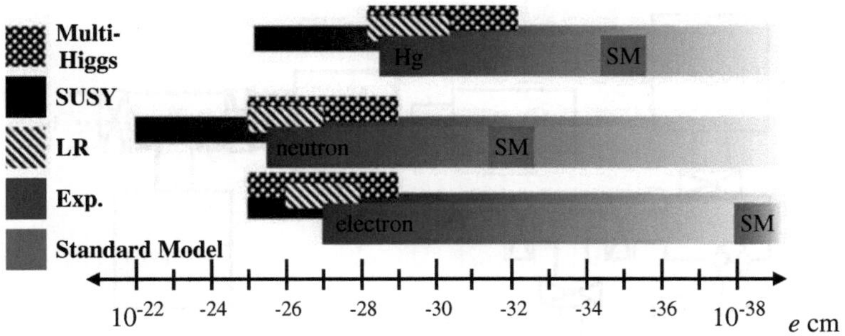

Fig. 1 A comparison of current experimental limits (*gray*) on the EDM of the electron ($d < 1.05 \times 10^{-27}$ [3]), neutron ($d < 2.9 \times 10^{-26}$ [4]), and mercury ($d < 3.1 \times 10^{-29}$ [11]) with the corresponding predictions that can be generated in models with *Left-Right* Symmetry (LR), Multiple-Higgs, and Supersymmetry (SUSY). Also included are the EDMs generated by the CP-odd phase in the Standard Model's CKM matrix [5–10]

(AFP) measurements of natural xenon and electron paramagnetic resonance (EPR) measurements of natural rubidium.

2.1 Gas transfer

The original tests of the Radon EDM Experiment's gas transfer system [15] were performed by monitoring the activity of ^{120}Xe as it moved through the vacuum system. A beam of ^{120}Cs at 30 keV was first implanted in a thin zirconium foil, illustrated in Fig. 2, where it decayed in situ to ^{120}Xe. Zirconium was chosen for this purpose because of the rapid diffusion of xenon in Zr at high temperatures [16]. Resistively heating the foil to around 1,000°C while simultaneously cooling the coldfinger (Fig. 3) to temperatures around 80 K caused gas released from the foil to adsorb to the coldfinger. From there, with the appropriate sequence of valve openings and closures, the gas was pushed into the measurement cell by warming the coldfinger and releasing a burst of purified nitrogen loaded from the gas manifold (Fig. 4). The gas manifold is also used in offline polarization tests to supply the nitrogen and xenon needed for EPR and AFP measurements. Detector stands placed around the apparatus allowed high-purity germanium (HPGe) detectors to be positioned as shown in Fig. 2. This enabled monitoring of the xenon activity at the foil, coldfinger, and cell over the course of the transfer process. The signals generated by these detectors during transfer tests were processed by the 8π DAQ [17]. Transfer efficiencies of around 80 % of the activity from the foil to the cell were observed.

2.2 Polarization measurements

The primary purpose of the Radon EDM development apparatus (Fig. 2) is to optimize the polarization system and to validate the gamma-ray anisotropy precession-detection method. To obtain a precession frequency from the radiation emitted from a sample of radioactive noble gas, the gas must be polarized. In the current design, the measurement cell is pre-loaded with a small amount of natural rubidium. Heating the cell after it has been loaded with noble gas and sealed off from the rest of the vacuum system causes it to fill with rubidium vapor. Optical pumping [18] polarizes

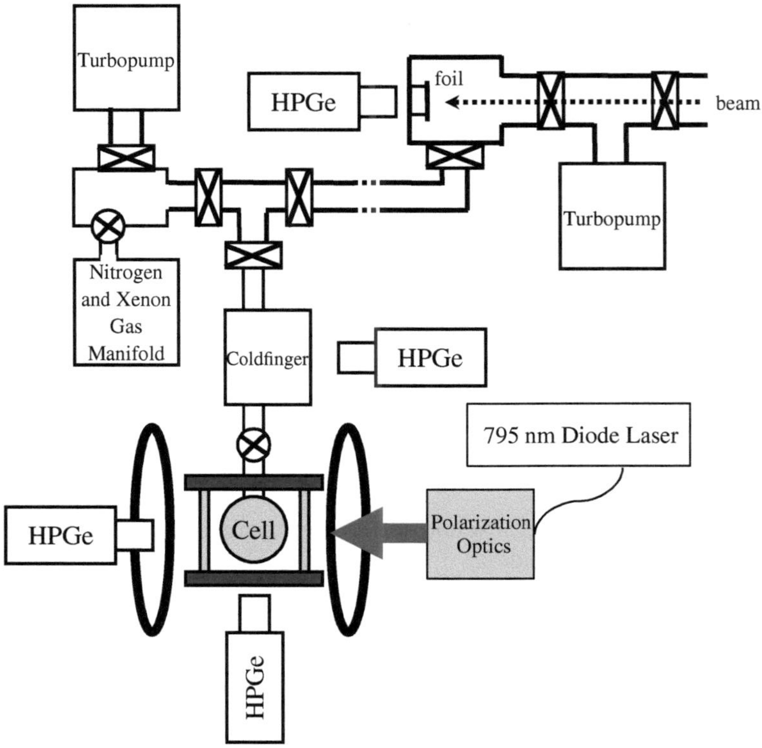

Fig. 2 A schematic overview of the development apparatus installed in the ISAC hall. Key elements of the gas transfer system are shown, along with the arrangement of HPGe detectors that allow the radioactivity to be tracked as it is moved from the collection foil to the measurement cell. EPR and AFP measurements require additional equipment illustrated in Figs. 5 and 7

the rubidium, and spin-exchange collisions transfer that polarization to the noble gas [19].

The measurement cell consists of a glass sphere located in an oven composed of a glass cylinder with plastic top and bottom plates. A pair of Helmholtz coils generates the static magnetic field **B** that defines the polarization axis. Laser light centered at 794.8 nm emitted by a diode laser (Coherent FAP System) travels through a fiber optic cable to a set of polarization optics located within a laser safety enclosure. This light passes through a collimating lens, a polarizing beamsplitter cube, and a $\lambda/4$ plate, resulting in circularly polarized light used for spin-exchange optical pumping of the noble gas nuclei [19].

When a polarized radioactive nucleus with spin ≥ 1 decays to an excited state of its daughter with spin ≥ 1, the directional distribution of the gamma-ray subsequently emitted as the excited state decays is anisotropic. The angular distribution of these gamma-rays is calculable and is a function of the initial and final spin states, the multipolarity of the transition, and the sublevel population distribution of the initial state [20]. In most cases, the maximum difference in emission probability is between 0° and 90° measured with respect to the polarization direction. With this in mind, HPGe detector stands located parallel and perpendicular to the magnetic field direction at the cell allow for optimal anisotropy detection capability.

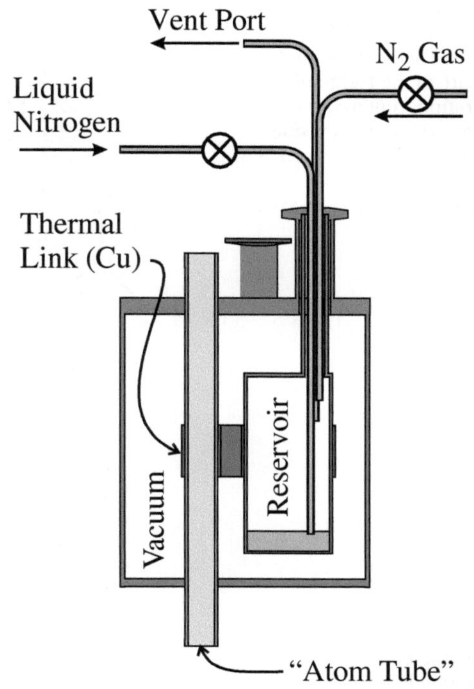

Fig. 3 A schematic of the current coldfinger design. A copper thermal link connects the "atom tube" to a liquid nitrogen reservoir that can be rapidly filled and evacuated. This in turn allows for rapid cryopumping and release of noble gas atoms

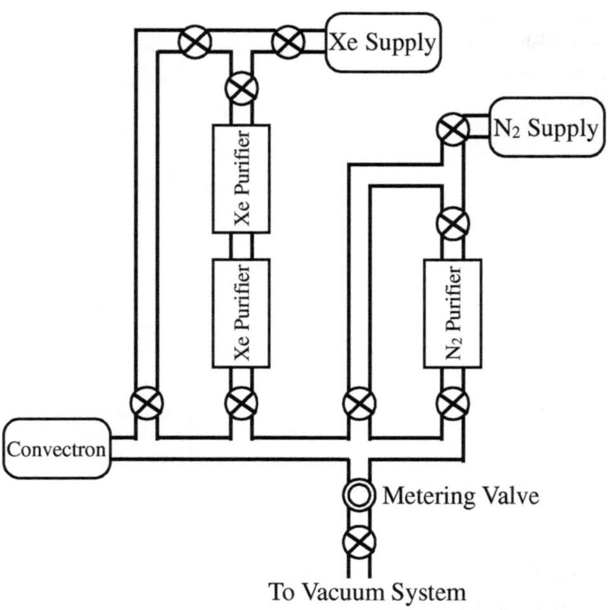

Fig. 4 A schematic of the manifold used to deliver purified nitrogen, xenon, or both to the vacuum system. The supply of natural xenon is low purity, necessitating the use of two Entegris 35KF inert gas purifiers in series. The valve opening to the vacuum system is an air-actuated Swagelok diaphragm valve; all others are hand-operated. The hand-operated metering valve allows for control of the flow rate

Rubidium polarization can be measured using the EPR apparatus illustrated in Fig. 5. Under current operating parameters, a 10 MHz RF magnetic field, $\mathbf{B}_{RF}$, is applied to the optical pumping cell to drive transitions between neighboring rubidium

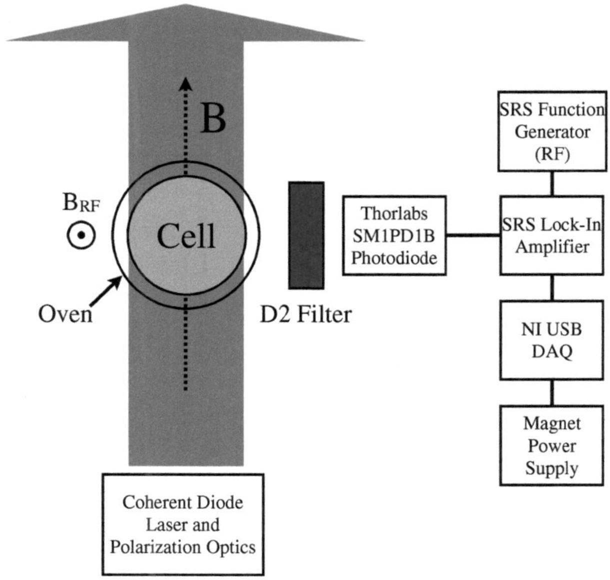

Fig. 5 A diagram of the components of the EPR apparatus. The RF coils (not shown) are in the Helmholtz configuration, centered on the cell. They are located outside the oven and generate an RF magnetic field $\mathbf{B}_{RF}$ perpendicular to the laser axis, as shown

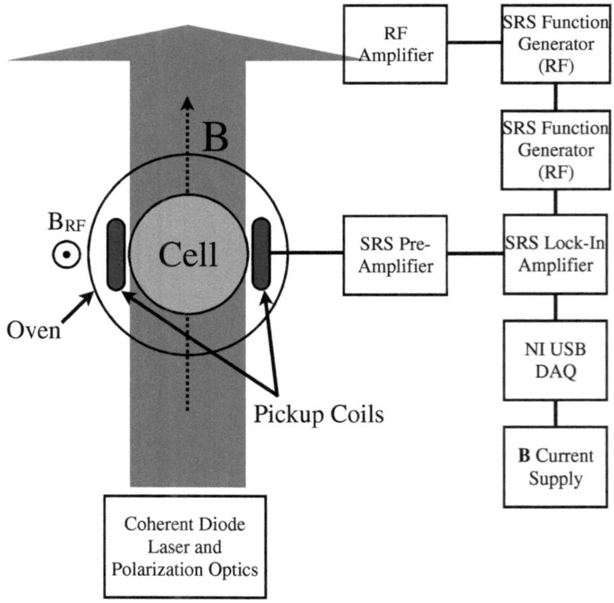

Fig. 6 A diagram of the AFP setup components. As in the EPR setup in Fig. 5, for $\mathbf{B}_{RF}$ only the field axis is shown. The pickup coils are located inside the oven, in a Helmholtz-like configuration perpendicular to both **B** and $\mathbf{B}_{RF}$

hyperfine states when the static magnetic field **B** in the polarization direction is tuned to provide the appropriate energy splitting. The RF is 100 % amplitude modulated at a rate of 20 Hz to facilitate the monitoring of fluorescence light from the cell. This light is detected with a photodiode (Thorlabs SM1PD1B), connected to the input of a Stanford Research Systems (SRS) Lock-In (LI) amplifier (SR560). A National Instruments USB DAQ sweeps **B** through all the rubidium hyperfine resonances (both ^{85}Rb and ^{87}Rb are accessible) while reading the output of the SR560, which is

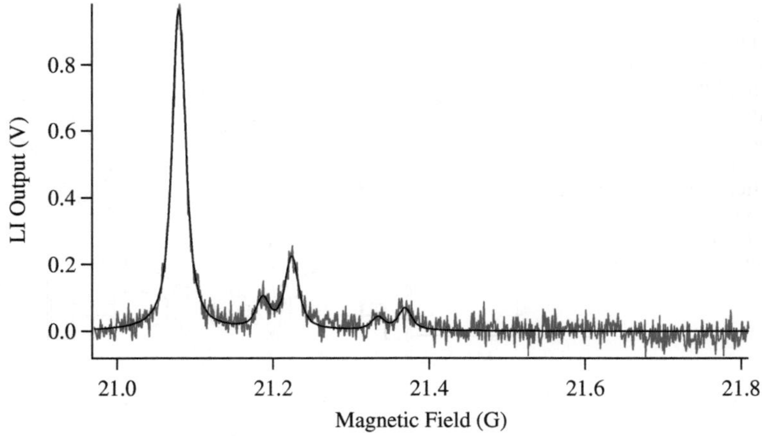

Fig. 7 An example Rb EPR spectrum fit to a Voigt lineshape. These data indicate a ^{85}Rb polarization of 73(3) % at this particular RF power and nitrogen pressure

locked to the RF modulation frequency. Figure 6 shows an example EPR spectrum for a cell containing rubidium and 500 Torr of purified nitrogen. The applied RF signal destroys some of the rubidium polarization, so multiple EPR spectra must be obtained to allow for an extrapolation to zero RF power.

For xenon polarization diagnostic tests, a supply of natural xenon allows for AFP measurements of both ^{129}Xe ($I = 1/2$) and ^{131}Xe ($I = 3/2$). This enables the comparison of dipole and quadrupole relaxation mechanisms in the noble gas polarization system. Two SRS DS345 function generators are used in the setup shown in Fig. 7. One of these provides an 8.8 kHz signal that is amplified and delivered to the RF coils (the frequency corresponds to the lowest self-resonance of the pickup coils used to monitor NMR signals). The other provides a cancellation signal for the lock-in amplifier, which is used to null the RF signal detected by the pickup coils. Similar to the procedure used for EPR measurements, **B** is swept through the xenon NMR resonances, located at 7.4 G and 25.0 G for ^{129}Xe and ^{131}Xe, respectively, in an 8.8 kHz driving field.

3 EDM experimental design and simulations

The final Radon EDM experimental apparatus has not been fully designed. A few basic features of the EDM apparatus design are outlined here. In addition, a Geant4 simulation framework has been written to enable studies of the impact of design modifications on the expected EDM experimental sensitivity.

3.1 An EDM measurement apparatus

Figure 8 shows the detector arrangement for an EDM experiment based on gamma-ray anisotropy detection of radon precession. Not shown are the electrodes that must be incorporated into the measurement cell to generate a large electric field parallel or antiparallel to the magnetic field direction. In the presence of such a field, a nonzero

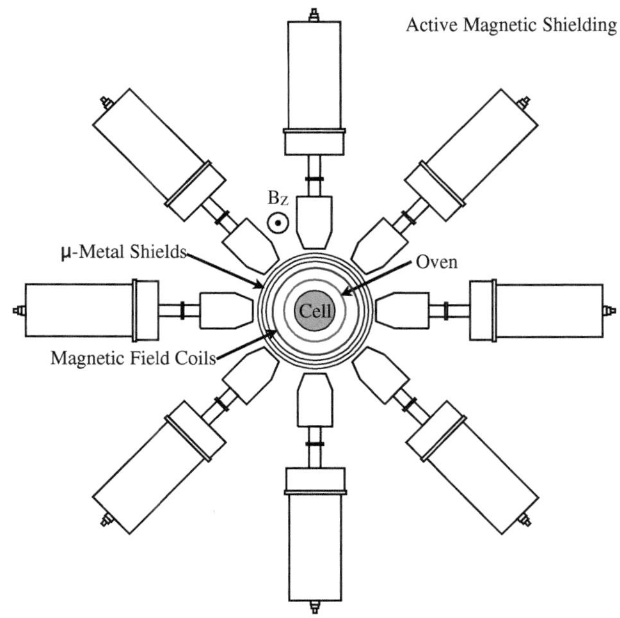

Fig. 8 A sketch of an EDM detection layout. The cell is located in a weak, uniform magnetic field $\mathbf{B}_Z$ and incorporates electrodes to generate a strong electric field parallel or antiparallel to the magnetic field direction. The ring of HPGe detectors is located in the plane perpendicular to the field axis

EDM would modify the precession frequency by an amount $\Delta\omega = 4dE$, where d is the magnitude of the EDM and E is the electric field strength. The collection and transfer apparatus discussed in Section 2.1 is designed to be consistent with operation of the experiment at an optimum duty cycle, in which measurements would be performed on one radon sample while simultaneously collecting the next sample in the Zr foil.

3.2 Simulation framework and EDM sensitivity

Our Radon EDM sensitivity estimations are based on a total of 100 days of data collection at 100 % duty factor. In order to verify that it is reasonable to expect the results of the gamma-ray anisotropy technique to be competitive with the ^{199}Hg result and to allow for the effect of various experimental design choices on the expected EDM sensitivity to be studied, a Geant4 simulation framework has been developed [21]. Figure 9 illustrates the simulation geometry, which incorporates a simplified cell and oven design within a user-selectable quantity of μ-Metal magnetic shielding and monitored by a ring of HPGe detectors developed for the GRIFFIN spectrometer under construction at ISAC [22]. Various detector configurations can be studied to determine their influence on the total gamma-ray efficiency (as in Fig. 10) as well as other physical and geometric aspects of the measurement. Under conditions in which an electric field strength of 5 kV/cm is applied and a spin coherence time $T_2 = 15$ seconds is obtained, the simulation predicts an EDM sensitivity of $4.64 \times 10^{-26} e \cdot \text{cm}$ [21]. Given a linear improvement in sensitivity with both electric field strength and coherence time, this suggests that it ought to be feasible to reach the $1 \times 10^{-26} e \cdot \text{cm}$ level at which an octupole enhancement factor makes the Radon EDM experiment competitive with the ^{199}Hg limit of $3.1 \times 10^{-29} e \cdot \text{cm}$ [11].

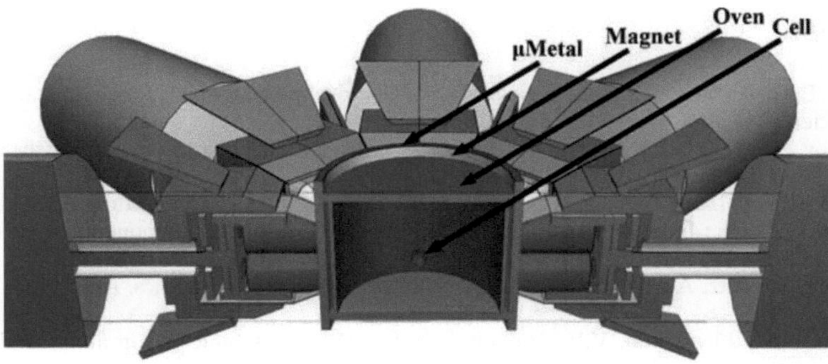

Fig. 9 An example of a Geant4 simulation geometry, including layers of glass, copper, and μ-metal between the radiation source and the ring of HPGe detectors

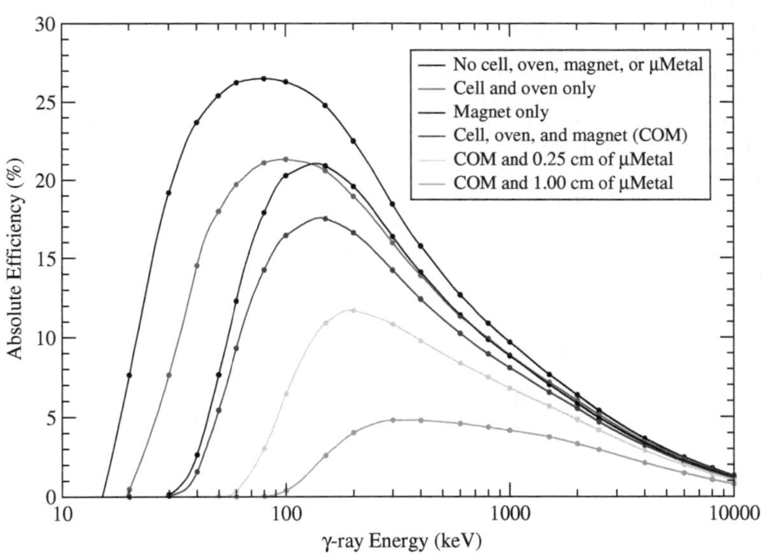

Fig. 10 Simulated gamma-ray efficiencies for a ring of eight GRIFFIN detectors [22] in multiple configurations of the experimental components

Acknowledgements This work would not have been possible without TRIUMF's Gamma Ray Spectroscopy at ISAC group, the TRIUMF Machine Shop, ISAC Operations, ISAC target development, the SFU Faculty of Science Machine Shop, the University of Michigan Physics Department Machine Shop, and the University of Michigan Chemistry Department Glass Blower. We gratefully acknowledge support provided by TRIUMF, the Natural Sciences and Engineering Research Council of Canada, the National Research Council of Canada, and the US Department of Energy (Grant DE-FG02-04ER41331).

References

1. Purcell, E.M., Ramsey, N.F.: On the possibility of electric dipole moments for elementary particles and nuclei. Phys. Rev. **78**, 807 (1950)
2. Sakharov, A.D.: Violation of CP invariance, C asymmetry, and Baryon asymmetry of the universe. JETP Lett. **5**, 24 (1967)
3. Hudson, J.J., et al.: Improved measurement of the shape of the electron. Nature **473**, 493 (2011)
4. Baker, C.A., et al.: Improved experimental limit on the electric dipole moment of the neutron. Phys. Rev. Lett. **97**, 131801 (2006)
5. Pospelov, M., Ritz, A.: Electric dipole moments as probes of new physics. Ann. Phys. **318**, 119 (2005)
6. Ginges, J.S.M., Flambaum, V.V.: Violations of fundamental symmetries in atoms and tests of unification theories of elementary particles. Phys. Rep. **397**, 63 (2004)
7. Ramsey, N.F.: Electric dipole moment of the neutron. Annu. Rev. Nucl. Part. Sci. **40**, 1 (1990)
8. Kizukuri, Y., Oshimo, N.: Neutron and electron electric dipole moments in supersymmetric theories. Phys. Rev. D **46**, 3025 (1992)
9. Pendlebury, J.M., Hinds, E.A.: Particle electric dipole moments. NIM A **440**, 471 (2000)
10. Bernreuther, W., Suzuki, M.: The electric dipole moment of the electron. Rev. Mod. Phys. **63**, 313 (1991)
11. Griffith, W.C., et al.: Improved limit on the permanent electric dipole moment of ^{199}Hg. Phys. Rev. Lett. **102**, 101601 (2009)
12. Spevak, V., Auerbach, N., Flambaum, V.V.: Enhanced T-Odd, P-Odd electromagnetic moments in reflection asymmetric nuclei. Phys. Rev. C **56**, 1357 (1997)
13. Flambaum, V.V., Zelevinsky, V.G.: Enhancement of nuclear schiff moments and time-reversal violation in atoms due to soft nuclear octupole vibrations. Phys. Rev. C **68**, 035502 (2003)
14. Dilling, J., Krücken, R., Ball, G.C.: ISAC overview. Hyperfine Interact. (2013, this issue)
15. Nuss-Warren, S.R., et al.: On-line collection and transfer of radioactive noble gas isotopes. NIM A **533**, 275 (2004)
16. Warner, T., et al.: Diffusion of Xe in Ta, Zr, and Pt. NIM A **538**, 135 (2005)
17. Garnsworthy, A.B., Garrett, P.E.: The 8π spectrometer. Hyperfine Interact. (2013, this issue)
18. Happer, W.: Optical pumping. Rev. Mod. Phys. **44**, 169 (1972)
19. Walker, T.G., Happer, W.: Spin-exchange optical pumping of noble gas nuclei. Rev. Mod. Phys. **69**, 629 (1997)
20. Hartogh, C.D., Tolhoek, H.A., de Groot, S.R.: Some explicit formulae for the angular distribution and polarization of Gamma radiation emitted by oriented nuclei. Physica **20**, 1310 (1954)
21. Rand, E.T.: Geant4 simulations for the radon electric dipole moment search at TRIUMF. M.Sc. thesis, University of Guelph (2011)
22. Svensson, C.E., Garnsworthy, A.B.: The GRIFFIN spectrometer. Hyperfine Interact. (2013, this issue)

DOI 10.1007/978-94-007-7963-1_23
Reprinted from *Hyperfine Interactions* Journal, DOI 10.1007/s10751-013-0899-1

Charged-particle reaction studies for nuclear astrophysics
The TUDA and TACTIC facilities

A. M. Laird · L. Buchmann · T. Davinson · for the TUDA and TACTIC collaborations

Published online: 18 October 2013

Abstract The TUDA and TACTIC facilities at TRIUMF provide versatile experimental capabilities for the study of astrophysically important reactions using both direct and indirect techniques. The present work will outline the research programme to date, describe the experimental facilities and discuss recent developments.

Keywords Charged particle reactions · Silicon detectors · Nuclear astrophysics

1 Introduction

Experimental measurements of reaction cross sections for nuclear astrophysics at TRIUMF fall broadly into two categories. Direct measurements of radiative capture rates are performed at the DRAGON facility (see relevant article in this journal) while studies of the competing charged-particle exit channels, such as (p,α) and (α,p), are undertaken at the TUDA and TACTIC facilities. The measurements at DRAGON are thus complementary to those at TUDA and TACTIC, and these combined capabilities allow a unique experimental perspective on many nuclear astrophysics challenges. Here we describe the experimental capabilities of the TUDA and TACTIC facilities and outline recent developments and future programme.

ISAC and ARIEL: The TRIUMF Radioactive Beam Facilities and the Scientific Program.

A. M. Laird (✉)
Department of Physics, University of York, York YO10 5DD, UK
e-mail: alison.laird@york.ac.uk

T. Davinson
SUPA, School of Physics and Astronomy, The University of Edinburgh,
Edinburgh EH9 3JZ, UK

L. Buchmann
TRIUMF, Vancouver, BC V6T 2A3, Canada

2 TRIUMF UK Detector Array (TUDA)

The principal goal of the TRIUMF-UK Detector Array (TUDA) is the study of nuclear reactions of astrophysical importance, with charged-particle exit channels, using direct and indirect experimental techniques. TUDA was developed and constructed by the University of Edinburgh, UK, with infrastructure provided by TRIUMF and is now used by a collaboration from the University of Edinburgh, University of York, TRIUMF and Simon Fraser University. TUDA is a configurable charged-particle detector array using silicon strip detectors of various geometries with modular instrumentation and data acquisition. It is a powerful and flexible facility which combines excellent spectroscopy performance, optimal angular resolution and large solid angle. TUDA can be operated either in ISAC-I or ISAC-II, depending on the beam energy requirements of the experiment.

The experimental programme at TUDA covers a wide range of astrophysical environments, covering both explosive and non-explosive nucleosynthesis. These include

- reactions influencing the abundance of nova observables: a resonant elastic scattering measurement of ^{21}Na(p,p)^{21}Na, to probe states in ^{22}Mg which influence the production of ^{22}Na (a target of γ-ray astronomy) in novae [1]; a direct measurement of the ^{18}F(p,α)^{15}O reaction inside the Gamow window, to study the destruction of ^{18}F, the radioisotope thought to dominate γ-ray emission from novae [2]; a resonant elastic scattering measurement of ^{18}F(p,p)^{18}F to search for predicted states [3] in ^{19}Ne which may influence the destruction rate of ^{18}F [4],
- breakout reactions from the HCNO cycle leading to the rp-process in X-ray bursts: a time-reversed measurement of the ^{18}Ne(α,p)^{21}Na reaction [5]; a resonant elastic scattering measurement of ^{20}Na(p,p)^{20}Na to investigate the final step of the breakout chain initiated by ^{15}O(α,γ)^{19}Ne [6],
- hydrostatic burning stages: simulataneous measurements of the ^{12}C(^{12}C,p)^{23}Na and ^{12}C(^{12}C,α)^{20}Ne reactions, which are the main carbon burning reactions in intermediate and high mass stars [7],
- isotopic abundances in asymptotic giant branch (AGB) stars: a time-reversed study of the ^{18}F(α,p)^{21}Ne reaction, which may affect the production of ^{19}F in these environments, as well as the ^{21}Ne:^{22}Ne isotopic ratio observed in some meteorites.

2.1 Experimental capabilities

The versatility of TUDA arises from the detector mounting design which allows almost complete freedom in detector position within the 1.5 m long vacuum chamber. The detector configuration can, therefore, be highly optimised for the reaction under investigation. For resonant proton scattering, the highest quality data is obtained at forward laboratory angles. For such measurements, the detectors can be placed at the downstream end of the TUDA chamber to optimise the energy and angular resolution. Conversely, for (d,p) studies, where the protons recoil backwards in the laboratory frame, the detectors can be located at the upstream end, again to optimise energy and angular resolution. For direct measurements, such as (p,α) reactions, where the expected yield is low, efficiency is a higher priority than resolution. Here the detectors are positioned close to the target to maximise solid angle, and if required, configured to optimise efficiency for coincident detection of reaction products. Moreover, as most measurements at TUDA are performed in inverse kinematics, the heavy reaction products are forward-focused in the laboratory frame, and only a limited angular range needs to be covered in order to achieve high detection efficiency.

The detectors commonly used at TUDA include Micron Semiconductor Ltd [8] LEDA (type YY1) and S2 detectors (see Fig. 1). LEDA [9] is a large area, single-sided silicon strip detector which provides 128 independent detector elements. S2 has a smaller diameter but provides 64 detector elements with smaller strip pitch in a double-sided package. Detector signals are read out via RAL 108 (Rutherford Appleton Laboratory) charge sensitive preamplifiers [10], located within the vacuum chamber volume. They are mounted close to the detectors and cooled using a recirculating ethanol-water mix. The next readout stage consists of 8-channel RAL 109 shaping amplifiers cards [10], with a 0.5 μs shaping time and variable gain. The VME-based data acquisition system consists of 32-channel Silena ADC and 128-channel CAEN V1190 TDC modules. The TDCs are run in common-stop mode, with the time reference for the stop signal derived from the ISAC pre-buncher frequency. For each detector element, energy and time-of-flight information is readout in event-by-event mode.

The main beam diagnostics and monitoring elements consist of a Faraday Cup, 4-vane beam monitors and single-element silicon detectors. The Faraday Cup is mounted at the end of the TUDA vacuum chamber and readout into the datastream via a high sensitivity current integrator to provide continuous monitoring of the beam intensity. The 4-vane beam monitors are located at the entrance and exit of the TUDA chamber to facilitate beam tuning. Finally, single-element silicon detectors are positioned on the target ladder and at the exit of TUDA, close to the beam axis, to monitor beam quality and composition. These have proven particularly useful during experiments with ^{18}F beam, when contamination from ^{18}O is unavoidable.

To date, most reaction studies have used solid targets, though TUDA has the infrastucture to utilise a gas cell target. Where a hydrogen target is required, a $(CH_2)_n$ foil is usually used rather than a gas cell. Consequently, the main contribution to the data rate arises from elastic scattering from the carbon in the target and, to a much lesser extent, protons and α-particles from fusion evaporation on the carbon. The carbon elastic scattering events can be used for normalisation, particularly useful when beam intensities are low. These events are also easily identifiable and separated from events of interest by the appropriate software cuts. For example, in Beer et al. [2], direct measurements of the ^{18}F(p,α)^{15}O reaction were performed through coincident detection of the reaction products. Two S2 detectors were located downstream of the target to detect the ^{15}O products and coincident α-particles were detected in a LEDA detector closer to the target. Figure 2 shows the kinematics for raw events at the highest energy measured (see caption for further details). Events of interest were identifed through total energy, co-planarity and polar-angle correlation gates.

2.2 Recent developments

As described above, $(CH_2)_n$ foils are usually utilised when a hydrogen target is needed. In many cases, the elastic scattering rate from the carbon content is not

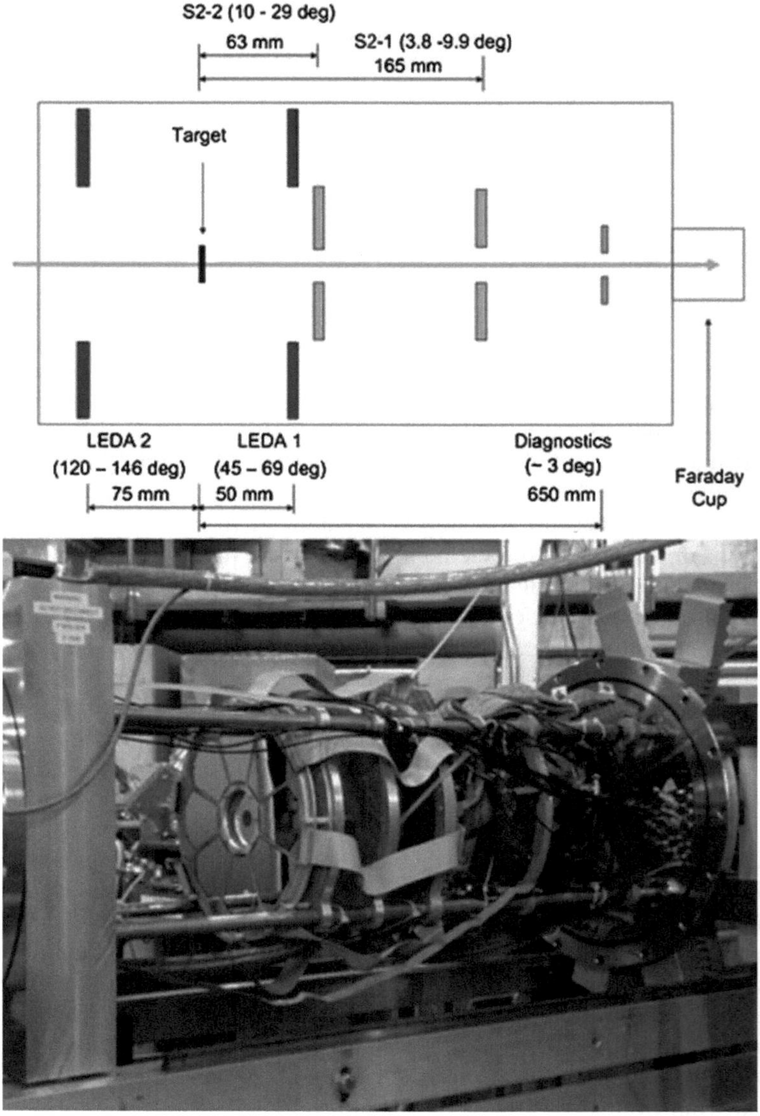

Fig. 1 The TUDA chamber: *top image* outlines one configuration inside TUDA, as used for the experiment by Beer et al. [2]; *bottom image* show photograph of LEDA and S2 detectors installed on TUDA mounting system

a limiting factor in the experiment. However, where higher beam intensities are available, it is this elastic scattering rate that limits the intensity that can be exploited. To address this restriction, developments in the infrastucture of TUDA have been implemented to allow a new technique to be explored. This technique consisted of filling the TUDA vacuum chamber with target gas and running the detectors inside the gas. By detecting the reaction products in coincidence, the centre-of-mass energy of the reaction can be reconstructed and the yield as a function of position determined.

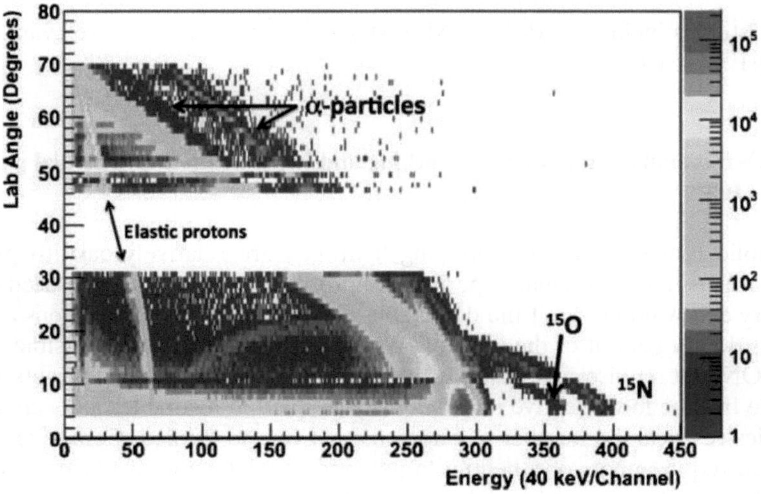

Fig. 2 Kinematics of raw event data for $^{18}F/^{18}O$ on CH_2 at 12.96 MeV lab beam energy. The most intense locus is due to elastic scattering of mass 18 from ^{12}C in the target. The two weaker loci at higher energies are due to α-particles from the $^{18}F(p,\alpha)^{15}O$ and $^{18}O(p,\alpha)^{15}N$ reactions. The loci at small angles, curving from above to below the elastic locus are ^{15}N and ^{15}O from the respective (p,α) reactions

To minimse energy loss in the gas, a re-entrant flange was installed to move the effective entrance to the chamber approximately 70 cm downstream, and a nickel window mounted on the flange. The vacuum system was modified to allow the chamber to be filled with gas, and a bypass between the chamber and the beamline added to allow both regions to be pumped down and vented at the same rate. As the first measurement using this new technique required a hydrogen target, significant additional safety measures were required to minimise the risk of hydrogen ignition. These included amending pump down procedures such that the rear flange of the chamber acted a pressure relief system; automatic shut-offs on the preamp power supplies and detector bias voltage supply, in case of pressure increase inside the chamber; and flushing the vacuum chamber with dry nitrogen before and after every hydrogen fill. A measurement of the $^{21}Ne(p,\alpha)^{18}F$ reaction has already been undertaken and the analysis is in progress. If the required resolutions are achieved, the technique will be exploited to perform futher (p,α) as well as (α,p) studies, particularly at the low, more astrophysically relevant, energies where the carbon elastic scattering rate is prohibitive.

2.3 Future programme

For more than a decade, TUDA has exploited the radioactive beams delivered by the ISAC facility to perform exciting experiments in nuclear astrophysics. As new beam species are developed and higher intensity beam become available, the TUDA collaboration will continue to push the capabilities of the facility to undertake challenging, high impact research. The future programme will include direct measurements of the HCNO breakout reactions, $^{18}Ne(\alpha,p)^{21}Na$, and $^{14}O(\alpha,p)^{17}F$.

Indirect studies include ^{26}Al(d,p)^{27}Al to study the analogue states relevant for the ^{26}Al(p,γ)^{27}Si reaction.

3 TRIUMF Annular Chamber for Tracking and Identification of Charged particles (TACTIC)

While solid reaction targets containing hydrogen are relatively easy to produce, helium targets have particular experimental problems. Gas cells can be used but the necessary exit windows limit the detectable reaction products to light ions as heavy products do not get out of the window. Windowless gas targets are possible, as with DRAGON, but require significant infrastructure. Helium-implanted targets are also available but the low effective target density coupled to available radioactive beam intensities restricts the number of reactions that can be studied. Consequently, the combined requirements of a helium target, and the detection of low energy heavy ions with high efficiency, in a high background environment, makes the study of (α,n) and (α,p) reactions with radioactive beams extremely challenging. To address these challenges, a novel active target specifically for nuclear astrophysics measurements is required. Consequently, the TACTIC detector has been designed, built and tested as a collaborative effort between TRIUMF and the University of York, UK.

3.1 TACTIC design

The TACTIC chamber was designed to study helium-induced reactions, of astrophysical interest, with radioactive beams. The key design criteria were:

1. the target region should be invisible to the active region of the detector to fully exploit low intensity radioactive beams (i.e the detector should not track incident beam particles, only scattered beam and reaction products);
2. should be capable of tracking and identifying low energy heavy ions;
3. high detection efficiency;
4. should be able to handle the background rate associated with running in a radioactive beam environment;
5. should be able to run with a variety of gas pressures and mixtures.

The final design consisted of a cylindrical configuration with a central region, aligned on the beam axis, bounded by two concentric cylindrical wire cathode cages, which confine beam-induced ionisation. Surrounding the cathode cages lies the detection region. Ionisation produced herein is drifted outwards by a radial electrostatic field towards a Gas Electron Multiplier (GEM). The GEM amplifies the electron signals which are then detected in curved anode pads which define the outside of the detector. The whole detector is filled with target gas, usually a helium-CO_2 mix. The cylindrical anode is segmented into 60 pads in z (beam axis) and 8 pads in ϕ providing 480 separate detector elements. The anode signals are read out via shaping preamplifiers and then digitised by the VME-based VF48 flash ADCs. Full pulse shapes are recorded and basic parameters such as pulse height and relative timing derived. The pulse heights allow the energy loss characteristics of the detected particle to be calculated, and from the relative timing the angle of the particle track can be deduced. This information, together with the pad locations, allows the particles to be identified and the tracks reconstructed. For futher details, see Fox et al. [11].

3.2 Future developments and programme

The TACTIC detector has been tested extensively with radioactive sources as well as stable and radioactive beam. These tests have identified several areas of optimisation required before a full experiment can be mounted. Systematic studies of detector response as a function of beam intensity have demonstrated that the limitation due to pile-up in the preamplifiers occurs at much higher beam intensities than that for the maximum data throughput in the data acquisition. Consequently, upgrades to the VF48 firmware are ongoing to increase the readout rate by up to a factor of ten. Concurrently, new calibration and track reconstruction procedures are being developed to improve the energy resolution.

The initial experimental goal of TACTIC is a measurement of the ^{8}Li(α,n)^{11}B reaction, thought to play a role in the production of seed nuclei for the r-process. The future programme will explore further (α,n) and (α,p) reactions and aim to expand to include hydrogen target studies (e.g. ^{7}Be(p,p)).

4 Summary and future outlook

The TUDA facility has already proven to be an extremely powerful tool for nuclear astrophysics studies and, with new beams becoming available, continues to have an exciting and challenging programme for the foreseeable future. TACTIC, once fully commissioned, will be a novel and complementary facility. Together with DRAGON, these detectors offer a suite of capabilities for nuclear astrophysics with radioactive beams unavailable elsewhere.

Acknowledgements UK authors would like to acknowledge the support of the Science and Technology Facilities Council (STFC), the Engineering and Physical Sciences Research Council (EPSRC) and the Royal Society. Canadian collaborators would like to acknowledge the support of the Natural Sciences and Engineering Research Council of Canada (NSERC). The TUDA and TACTIC collaboration would also like to thank the Beam Development and Delivery groups for their continuing support of the experimental programmes.

References

1. Ruiz, C., Davinson, T., Sarazin, F., Roberts, I., Robinson, A., Woods, P.J., Buchmann, L., Shotter, A.C., Walden, P., Clarke, N.M., Chen, A.A., Fulton, B.R., Groombridge, D., Pearson, J., Murphy, A.S.: Multichannel R-matrix analysis of elastic and inelastic resonances in the ^{21}Na + p compound system. Phys. Rev. **C71**, 025802 (2005)
2. Beer, C.E., Laird, A.M., Murphy, A.J. St., Bentley, M.A., Buchman, L., Davids, B., Davinson, T., Diget, C.A., Fox, S.P., Fulton, B.R., Hager, U., Howell, D., Martin, L., Ruiz, C., Ruprecht, G., Salter, P., Vockenhuber, C., Walden, P.: Direct measurement of the ^{18}F(p,α)^{15}O reaction at nova temperatures. Phys. Rev. **C83**, 042801 (2011)
3. Dufour, M., Descouvemont, P.: The ^{18}F(p,α)^{15}O low energy S-factor: a microscopic approach. Nucl. Phys. **A785**, 381 (2007)
4. Murphy, A.J. St., Laird, A.M., Angulo, C., Buchmann, L., Davinson, T., Descouvemont, P., Fox, S.P., José, J., Lewis, R., Ruiz, C., Vaughan, K., Walden, P.: Simultaneous measurement of the

^{18}F(p,p)^{18}F and ^{18}F(p,α)^{15}O reactions: implications for the level structure of ^{19}Ne, and for ^{18}F production in novae. Phys. Rev. **C79**, 058801 (2009)

5. Salter, P.J.C., Aliotta, M., Davinson, T., Al Falou, H., Chen, A., Davids, B., Fulton, B.R., Galinski, N., Howell, D., Lotay, G., Machule, P., Murphy, A.J. St., Ruiz, C., Sjue, S., Taggart, M., Walden, P., Woods, P.J.: Measurement of the ^{18}Ne(α,p$_0$)^{21}Na reaction cross section in the burning energy region for x-ray bursts. Phys. Rev. Lett. **108**, 242701 (2012)
6. Murphy, A.J. St., Aliotta, M., Davinson, T., Ruiz, C., Woods, P.J., D'Auria, J., Buchmann, L., Chen, A.A., Laird, A.M., Sarazin, F., Walden, P., Fulton, B.R., Pearson, J., Brown, B.A.: Level structure of ^{21}Mg: nuclear and astrophysical implications. Phys. Rev. **C73**, 034320 (2006)
7. Tunstall, C.: Cross section measurements for the ^{12}C + ^{12}C fusion reaction at low energies. M.Sc thesis, Department of Physics, University of York, UK (2008)
8. Micron Semiconductor Limited UK: www.micronsemiconductor.co.uk. Accessed 1 Apr 2013
9. Davinson, T., Davinson, T., Bradfield-Smith, W., Cherubini, S., DiPietro, A., Galster, W., Laird, A.M., Leleux, P., Ninane, A., Ostrowski, A.N., Shotter, A.C., Vervier, J., Woods, P.J.: Louvain-Edinburgh Detector Array (LEDA): a silicon detector array for use with radioactive nuclear beams. Nucl. Inst. Methods **A454**, 350 (2000)
10. Thomas, S.L., Davinson, T., Shotter, A.C.: A modular amplifier system for the readout of silicon strip detectors. Nucl. Inst. Meth **A288**, 212 (1990)
11. Fox, S.P., Amaudruz, P.A., Bruskiewich, P., Buchmann, L., Chipps, K.A., Hager, U., Laird, A.M., Martin, L., Ruprecht, G., Shotter, A.C., Walden, P.: TACTIC: a new detector for nuclear astrophysics experiments. J. Phys. Conf. Ser. **312**, 052007 (2011)

DOI 10.1007/978-94-007-7963-1_24
Reprinted from *Hyperfine Interactions* Journal, DOI 10.1007/s10751-013-0900-z

The Doppler shift lifetimes facility

B. Davids

Published online: 23 October 2013

Abstract The Doppler Shift Lifetimes chamber is an experimental facility designed to measure the mean lifetimes of excited nuclear states that decay within femtoseconds via the detection of their Doppler-shifted γ ray transitions.

Keywords Nuclear structure · Nuclear astrophysics · Lifetimes · Widths

1 Introduction

The mean lifetimes of excited nuclear states are of considerable interest in nuclear structure and nuclear astrophysics research. Motivated primarily by the latter, the Doppler Shift Lifetimes (DSL) facility is a scattering chamber designed to provide a clean environment conducive to detecting the γ ray emission from excited nuclear states populated in heavy ion-induced transfer reactions in inverse kinematics.

2 Facility description

The DSL scattering chamber is made from thin Al and contains provisions for mounting cooled ^{3}He-implanted target foils along with a Si surface barrier detector telescope at 0°. A schematic view of the DSL facility is shown in Fig. 1. The scattering chamber was designed with a cold trap to ensure a clean target surface and also to prevent losses of the implanted ^{3}He. This was achieved using a narrow differential pumping aperture followed by a copper cylinder enclosing the path of the beam to the target. The copper cylinder is cooled using liquid nitrogen. To avoid any condensation of impurities on the surface of the target, the copper cylinder is not in direct

ISAC and ARIEL: The TRIUMF Radioactive Beam Facilities and the Scientific Program.

B. Davids (✉)
TRIUMF, 4004 Wesbrook Mall, Vancouver, BC, Canada V6T 2A3
e-mail: davids@triumf.ca

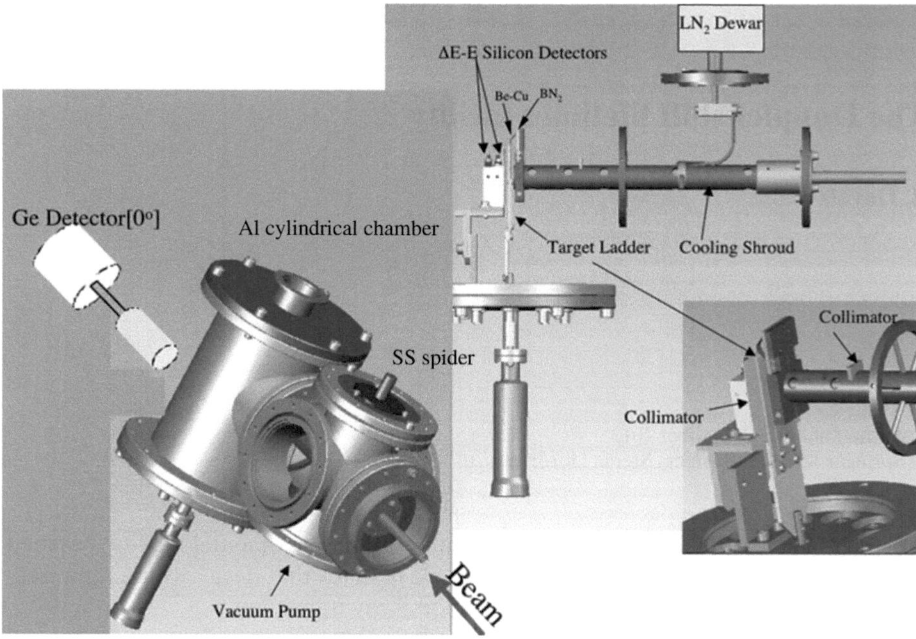

Fig. 1 Schematic view of the DSL facility, showing the liquid nitrogen-cooled shroud along the beam axis, the target ladder, the Si surface barrier detector telescope, and the high purity germanium detector used to measure Doppler-shifted γ rays

contact with the target ladder. Indirect contact of the cold copper cylinder with the copper target ladder is achieved using BeCu fingers mounted on a boron nitride plate, which provides electrical isolation as well. This arrangement maintains a temperature difference between the copper cylinder and the target ladder. In this way the target can be cooled below room temperature to ensure that ^{3}He does not diffuse out when heated by bombardment with a beam power of up to 300 mW. Moreover, the colder surfaces surrounding the target foil and the beam path in front of it reduce the buildup of carbon and other contaminants on the target itself during the experiment.

In all of our measurements target foils were prepared at the Université de Montréal by implanting 30 keV ^{3}He ions into Au and Zr foils, yielding an areal ^{3}He number density of 6×10^{17} cm^{-2}. The concentration of ^{3}He in the foil is monitored via yields of elastically scattered ^{3}He during bombardment.

3 Experimental programme

The ^{15}O$(\alpha, \gamma)^{19}$Ne reaction leads to the initial breakout from the hot CNO cycles that operate in Type I x-ray bursts, which are thermonuclear runaways on accreting neutron stars in binary star systems. Recent calculations [1] suggested that if the rate of this reaction were below a certain threshold, then the periodic x-ray bursts observed from some accreting neutron stars would not occur. Since its completion in 2005, we have used the DSL facility in conjunction with the ISAC-I accelerator to

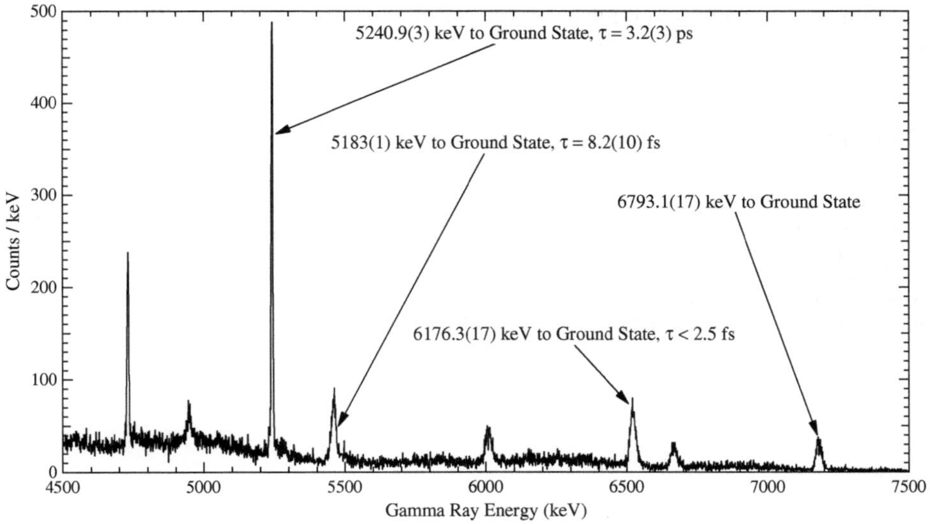

Fig. 2 Spectrum of high energy γ rays detected at 0° in coincidence with α particles emitted during the bombardment of a ^{3}He-implanted Au foil by a 50 MeV ^{16}O beam. Observed transitions from excited states in ^{15}O are indicated along with the mean lifetimes of the states taken from the literature [6]

make two measurements of the mean lifetimes of excited states in ^{19}Ne relevant to the rate of the $^{15}\mathrm{O}(\alpha,\gamma)^{19}\mathrm{Ne}$ reaction on accreting neutron stars.

In our studies of ^{19}Ne lifetimes, a 34 MeV ^{20}Ne beam was incident on a ^{3}He-implanted Au foil, populating the levels of interest via the $^{3}\mathrm{He}(^{20}\mathrm{Ne},\alpha)^{19}\mathrm{Ne}$ reaction. The ^{20}Ne beam and recoiling ^{19}Ne nuclei were stopped in the Au foil. De-excitation γ rays were detected in coincidence with α particles using an 80 % high-purity germanium (HPGe) p-type coaxial detector placed at 0° with respect to the beam axis. Lifetimes were determined from a lineshape analysis of the γ-ray energy spectrum. The α particles were identified by means of energy loss (ΔE) and residual energy measurements in a silicon surface barrier detector telescope consisting of a 25 μm thick detector for ΔE and a 500 μm thick detector for residual energy measurements. The results of these lifetime measurements [2, 3] were used to estimate the $^{15}\mathrm{O}(\alpha,\gamma)^{19}\mathrm{Ne}$ rate and its uncertainty; when the sensitivity of x-ray burst models to the latter was re-examined, a curious insensitivity was found [4], implying the need for further theoretical work.

The lifetime of the 6.79 MeV state in ^{15}O is one of the dominant uncertainties in determining the rate of the $^{14}\mathrm{N}(p,\gamma)^{15}\mathrm{O}$ reaction, on which the estimated ages of the oldest stars in the Milky Way Galaxy strongly depend. In 2008, we moved the DSL facility to ISAC-II, where it has been used with a TIGRESS γ-ray detector [5] to measure the mean lifetime of this state. This measurement was performed using the ISAC-II accelerator in order to reach a bombarding energy at which the state of interest was known to be populated; a TIGRESS n-type γ-ray detector was employed due to its greater resistance to fast neutrons, which are more copiously produced at higher energies. A spectrum of high energy γ rays detected at 0° in coincidence with α particles emitted during the bombardment of a ^{3}He-implanted Au foil by a 50 MeV ^{16}O beam in ISAC-II is shown in Fig. 2. These data are currently under analysis with the aim of determining the mean lifetime of the 6.79 MeV state in ^{15}O and thereby constraining the ^{14}N$(p, \gamma)^{15}$O reaction rate.

References

1. Fisker, J.L., Görres, J., Wiescher, M., Davids, B.: Astrophys. J. **650**, 332 (2006)
2. Kanungo, R., et al.: Phys. Rev. C **74**(4), 045803 (2006)
3. Mythili, S., Davids, B., et al.: Phys. Rev. C **77**(3), 035803 (2008)
4. Davids, B., Cyburt, R.H., José, J., Mythili, S.: Astrophys. J. **735**, 40 (2011)
5. Svensson, C.E., et al.: J. Phys. **G31**, S1663 (2005)
6. Ajzenberg-Selove, F.: Nucl. Phys. A **523**, 1 (1991)

DOI 10.1007/978-94-007-7963-1_25
Reprinted from *Hyperfine Interactions* Journal, DOI 10.1007/s10751-013-0901-y

Nuclear astrophysics measurements with DRAGON

Jennifer Fallis

Published online: 23 October 2013

Abstract The DRAGON facility in ISAC-I at TRIUMF was designed to measure radiative capture reactions of importance for astrophysics in inverse kinematics. This work will provide a short summary of experiments to date, outline the status of the separator and its detectors, as well as discuss recent developments.

Keywords Recoil mass separator · Radiative capture · Nucleosynthesis · Novae · Supernovae

1 Introduction

DRAGON, the Detector of Recoils And Gammas Of Nuclear reactions, located in the medium energy area of the ISAC facility at TRIUMF, has been designed with the specific purpose of measuring radiative capture reactions, predominantly reactions of astrophysical importance. The range of beam energies that can be delivered by ISAC-I, in combination with the availability of both stable and radioactive beams, allows for measurements of reactions of interest in a large variety of different astrophysical environments. Astrophysically motivated studies to date have included measurements of importance to:

- production of isotopic observables in nova, such as the contribution of ^{21}Na$(p,\gamma)^{22}$Mg to the production of ^{22}Na (γ-ray astronomy) [1, 2], ^{26g}Al$(p,\gamma)^{27}$Si to the destruction of ^{26g}Al (γ-ray astronomy) [3], ^{33}S$(p,\gamma)^{34}$Cl

ISAC and ARIEL: The TRIUMF Radioactive Beam Facilities and the Scientific Program.

J. Fallis (✉)
TRIUMF, 4004 Wesbrook Mall, Vancouver, BC V6T 2A3, Canada
e-mail: jfallis@triumf.ca

to the production of ^{34}Cl (γ-ray astronomy) and destruction of ^{33}S (presolar grains) [4, 5], as well as $^{17}O(p, \gamma)^{18}F$ [6] and $^{18}F(p, \gamma)^{19}Ne$ [7] to determine the abundance of ^{18}F (γ-ray astronomy),
- important nucleosynthesis pathways in novae such as a measurement of $^{23}Mg(p, \gamma)^{24}Al$ which is an important link between the Ne-Na and Mg-Al cycles [8],
- stellar He burning through the investigation of both $^{12}C(\alpha, \gamma)^{16}O$ [9], and $^{16}O(\alpha, \gamma)^{20}Ne$ [10, 11], to study the production and destruction of ^{16}O,
- ^{44}Ti production in core-collapse supernova explosions via $^{40}Ca(\alpha, \gamma)^{44}Ti$ [12],
- s-process abundances in massive stars by measuring $^{17}O(\alpha, \gamma)^{21}Ne$ which is potentially a significant neutron poison [13], and
- stellar and big bang nucleosynthesis through the study of the $^{3}He(\alpha, \gamma)^{7}Be$ reaction at higher energies [14].

Additionally, non-astrophysically motivated reactions, such as $^{16}O(^{12}C,\gamma)^{28}Si$ to explore the $^{12}C-^{16}O$ cluster structure in ^{28}Si [15], have been studied.

2 Experimental

Reactions at DRAGON are studied in inverse kinematics to take advantage of the radioactive beams delivered by ISAC, nearly all of which are too short-lived to ever be used as a target for measurements in regular kinematics. As astrophysical reactions of interest typically involve the capture of protons and alpha particles, a windowless gas target is used. Gas targets have certain advantages: they are uniform, free of impurity layers, and their thickness can be changed with little effort. Additionally the stopping power of the target can be measured directly by measuring the energy of the beam using the first magnetic dipole of the separator with and without gas in the target. Thirty BGO (bismuth germanate) γ detectors surround the target, covering 90 % of the 4π solid angle. For narrow, isolated resonances the position of the BGOs in which the γs are detected can be used to determine the position of the reaction in the target. When combined with the stopping power measurement, this allows for a determination of the resonance energy of the reaction, provided sufficient statistics are acquired [3]. Two surface barrier Si detectors are located within the gas target, and positioned at 30° and 57° with respect to the beam axis. The elastic scattering rates measured on these detectors provide a means of monitoring and normalizing the beam current [2].

An additional benefit of measuring reactions in inverse kinematics is that it allows for the recoil nucleus itself to be measured, along with the de-excitation γ rays from the reaction. This, however, requires significant suppression of the unreacted beam as recoil product yields are low, typically $10^{-10} - 10^{-14}$ per incident beam nucleus. This beam suppression is achieved through two stages of mass separation, each consisting of a pair of magnetic and electrostatic dipoles (Fig. 1) [16]. Variable width slits are placed at strategic locations throughout the separator. The first set of slits is located after the first magnetic dipole and selects ions of a known charge state. The second and third sets of slits are located at the focus after each electrostatic dipole and select for a single mass/charge (A/q) ratio. If the beam suppression of the separator itself is not sufficient, applying additional software cuts on the data, such as a measurement of the local time-of-flight after the second stage of separation [17]

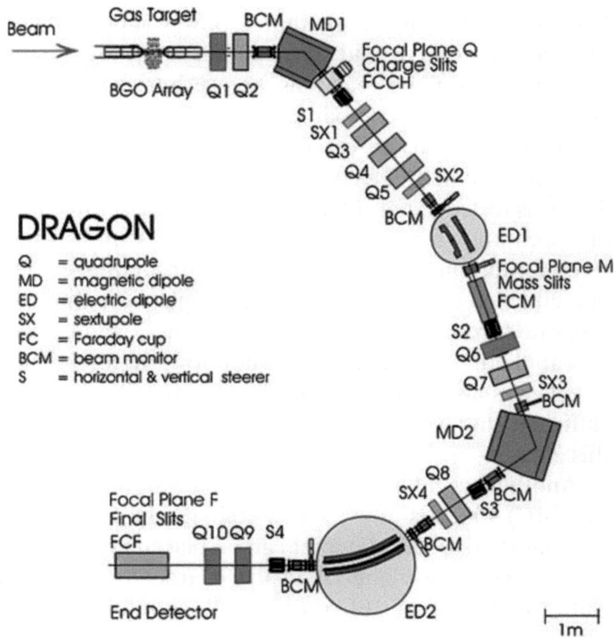

Fig. 1 Schematic representation of the DRAGON recoil mass separator

and/or a coincidence requirement between detected γs and heavy ions arriving at the end detector, can increase the background suppression by several orders of magnitude [16].

The DRAGON end detector can be either a double-sided silicon strip detector, which provides position and energy information [18], or an ionization chamber with a segmented anode, which allows for measurements of both energy loss and total energy deposited [19]. The choice of which end detector to use for a given experiment depends on the masses of the beam and recoil particles and their respective energies. A recent study of ^{17}O$(p, \gamma)^{18}$F at $E_{c.m.} = 300$ keV used both detectors in sequence and confirmed that the results agreed within uncertainty as expected [6].

3 Recent developments

While DRAGON continues to make measurements using beams up to mass $A = 30$ for which is was designed, we have also begun to move beyond the design specifications having published a measurement of ^{40}Ca$(\alpha, \gamma)^{44}$Ti [12] and recently tested the beam suppression for ^{58}Ni$(p, \gamma)^{59}$Cu [20] and ^{84}Kr$(\alpha, \gamma)^{88}$Sr. One of the obstacles to measuring reactions at higher masses was that the charge state distribution of the beam and recoils after the gas target resulted in A/q values too high to bend through the separator. This was solved by including a retractable SiN foil right after the inner gas cell before the differential pumping tubes [19, 21]. This acts as a solid state stripper providing the higher charge states required. The highest mass and energy combination tested to date has been ^{84}Kr at $E_{\text{beam}} = 1.25$ MeV/u. Charge states up to 26^+ were produced by the SiN foil, and it was demonstrated that the separator could bend charge states down to 24^+ before reaching the current limit on the first magnetic dipole power supply. Another concern was whether or not the separator would have sufficient beam suppression at these higher masses. Fortunately, for the reactions we are interested in measuring, the outgoing recoil cone angle is small compared to the acceptance of the separator. GEANT3 simulations indicate that the variable slits can be closed in significantly for such reactions without adversely affecting the transmission of the recoils through the separator. This was tested experimentally in April of 2011 by measuring the resonance strength of a well known resonance in ^{58}Ni$(p,\gamma)^{59}$Cu for a range of slit width settings [20]. Even for the standard slit settings, the unsuppressed 'leaky' beam did not overwhelm our end detector (in this case the ionization chamber), and the recoils could be separated from the 'leaky' beam once the γ and heavy ion coincidence requirement was applied. The beam suppression for ^{84}Kr$(\alpha,\gamma)^{88}$Sr with the separator tuned for recoils was also studied. While no ^{88}Sr recoils were seen, the beam suppression was sufficient that we expect to be able to determine an upper limit for the reaction at this energy.

Another recent addition to the separator diagnostics is a small ionization chamber (mini-IC) which is now located after the first magnetic dipole and just upstream of the charge slits. This detector can be placed in such a way that it intercepts the beam one charge state higher than the charge state tuned through the separator. This will provide a means of continuously monitoring the composition of mixed beams, especially in cases involving long lived radioisotopes that can't be quickly measured using the standard β-decay monitors or NaI detectors. The mini-IC was used for the first time in the recent ^{18}F$(p,\gamma)^{19}$Ne experiment.

4 Conclusion

In the past 10 years, since the first radioactive beam was delivered to DRAGON, a wide range of reactions contributing to many aspects of astrophysical nucleosynthesis have been measured. With the simple addition of a SiN foil, recent tests have demonstrated that the mass range of reactions within DRAGON's reach is even larger than previously expected. Combined with ongoing ISAC beam development, we can expect that DRAGON will continue to explore many more astrophysically important reactions in the years to come.

Acknowledgements The authors gratefully acknowledge funding from the Natural Sciences and Engineering Research Council of Canada. In addition we would like to thank the beam delivery and ISAC operations groups at TRIUMF without whom this work would not be possible.

References

1. Bishop, S., Azuma, R.E., Buchmann, L., Chen, A.A., Chatterjee, M.L., D'Auria, J.M., Engel, S., Gigliotti, D., Greife, U., Hernanz, M., Hunter, D., Hussein, A., Hutcheon, D., Jewett, C., José, J., King, J., Kubono, S., Laird, A.M., Lamey, M., Lewis, R., Liu, W., Michimasa, S., Olin, A., Ottewell, D., Parker, P.D., Rogers, J.G., Strieder, F., Wrede, C.: Phys. Rev. Lett. **90**(16), 162501 (2003)
2. D'Auria, J.M., Azuma, R.E., Bishop, S., Buchmann, L., Chatterjee, M.L., Chen, A.A., Engel, S., Gigliotti, D., Greife, U., Hunter, D., Hussein, A., Hutcheon, D., Jewett, C.C., José, J., King, J.D., Laird, A.M., Lamey, M., Lewis, R., Liu, W., Olin, A., Ottewell, D., Parker, P., Rogers, J., Ruiz, C., Trinczek, M., Wrede, C.: Phys. Rev. C. **69**(6), 065803 (2004)

3. Ruiz, C., Parikh, A., José, J., Buchmann, L., Caggiano, J.A., Chen, A.A., Clark, J.A., Crawford, H., Davids, B., D'Auria, J.M., Davis, C., Deibel, C., Erikson, L., Fogarty, L., Frekers, D., Greife, U., Hussein, A., Hutcheon, D.A., Huyse, M., Jewett, C., Laird, A.M., Lewis, R., Mumby-Croft, P., Olin, A., Ottewell, D.F., Ouellet, C.V., Parker, P., Pearson, J., Ruprecht, G., Trinczek, M., Vockenhuber, C., Wrede, C.: Phys. Rev. Lett. **96**(25), 252501 (2006)
4. Parikh, A., Faestermann, T., Krücken, R., Bildstein, V., Bishop, S., Eppinger, K., Herlitzius, C., Lepyoshkina, O., Maierbeck, P., Seiler, D., Wimmer, K., Hertenberger, R., Wirth, H.F., Fallis, J., Hager, U., Hutcheon, D.A., Ruiz, C., Buchmann, L., Ottewell, D., Freeman, B., Wrede, C., Garcia, A., Delbridge, B., Knecht, A., Sallaska, A., Chen, A., Clark, J.A., Deibel, C., Fulton, B.R., Laird, A., Greife, U., Guo, B., Li, E., Li, Z., Lian, G., Wang, Y., Liu, W., Parker, P., Setoodehnia, K.: In: Proceedings of Science Nuclei in the Cosmos XI, POS(NIC XI)052 (2010)
5. Fallis, J., Parikh, A., Bertone, P.F., Bishop, S., Buchmann, L., Chen, A.A., Christian, G., Clark, J.A., D'Auria, J.M., Davids, B., Deibel, C.M., Fulton, B.R., Greife, U., Guo, B., Hager, U., Herlitzius, C., Hutcheon, D.A., José, J., Laird, A.M., Li, E.T., Li, Z.H., Lian, G., Liu, W.P., Martin, L., Nelson, K., Ottewell, D., Parker, P.D., Reeve, S., Rojas, A., Ruiz, C., Setoodehnia, K., Sjue, S., Vockenhuber, C., Wang, Y.B., Wrede, C.: Phys. Rev. C. (in press)
6. Hager, U., Buchmann, L., Davis, B., Fallis, J., Fulton, B., Galinski, N., Greife, U., Hutcheon, D.A., Ottewell, D., Rojas, A., Ruiz, C., Setoodehnia, K.: Phys. Rev. C. **85**(3), 035803 (2012)
7. Akers, C., Laird, A.M., Fulton, B.R., Ruiz, C., Bardayan, D.W., Buchmann, L., Christian, G., Davids, B., Erikson, L., Fallis, J., Hager, U., Hutcheon, D., Martin, L., Murphy, A.St.J., Nelson, K., Spyrou, A., Stanford, C., Ottewell, D., Rojas, A.: Phys. Rev. Lett. **110**, 262502 (2013)
8. Erikson, L., Ruiz, C., Ames, F., Bricault, P., Buchmann, L., Chen, A.A., Chen, J., Dare, H., Davids, B., Davis, C., Deibel, C.M., Dombsky, M., Foubister, S., Galinski, N., Greife, U., Hager, U., Hussein, A., Hutcheon, D.A., Lassen, J., Martin, L., Ottewell, D.F., Ouellet, C.V., Ruprecht, G., Setoodehnia, K., Shotter, A.C., Teigelhöfer, A., Vockenhuber, C., Wrede, C., Wallner, A.: Phys. Rev. C. **81**(4), 045808 (2010)
9. Matei, C., Buchmann, L., Hannes, W.R., Hutcheon, D.A., Ruiz, C., Brune, C.R., Caggiano, J., Chen, A.A., D'Auria, J., Laird, A., Lamey, M., Li, Z., Liu, W., Olin, A., Ottewell, D., Pearson, J., Ruprecht, G., Trinczek, M., Vockenhuber, C., Wrede, C.: Phys. Rev. Lett. **97**(24), 242503 (2006)
10. Hager, U., Brown, J.R., Buchmann, L., Carmona-Gallardo, M., Erikson, L., Fallis, J., Greife, U., Hutcheon, D., Ottewell, D., Ruiz, C., Sjue, S., Vockenhuber, C.: Phys. Rev. C. **84**, 022801(R) (2011)
11. Hager, U., Buchmann, L., Davids, B., Fallis, J., Greife, U., Hutcheon, D.A., Irvine, D., Ottewell, D., Rojas, A., Reeve, S., Ruiz, C.: Phys. Rev. C. **86**(5), 055802 (2012)
12. Vockenhuber, C., Ouellet, C.O., The, L.S., Buchmann, L., Caggiano, J., Chen, A.A., Crawford, H., D'Auria, J.M., Davids, B., Fogarty, L., Frekers, D., Hussein, A., Hutcheon, D.A., Kutschera, W., Laird, A.M., Lewis, R., O'Connor, E., Ottewell, D., Paul, M., Pavan, M.M., Pearson, J., Ruiz, C., Ruprecht, G., Trinczek, M., Wales, B., Wallner, A.: Phys. Rev. C. **76**(3), 035801 (2007)
13. Taggart, M., Hager, U., Laird, A., Ruiz, C., Hutcheon, D., Ottewell, D.F., Fallis, J., Erikson, L., Bentley, M., Brown, J., Buchmann, L., Chen, A.A., Chen, J., Chipps, K., D'Auria, J., Davids, B., Davis, C., Diget, C.A., Fox, S.P., Fulton, B.R., Galinski, N., Greife, U., Herwig, F., Hirschi, R., Howell, D., Martin, L., Mountford, D., Murphy, A., Pignatari, M., Reeve, S., Ruprecht, G., Sjue, S., Veloce, L.: In: Proceedings of Science Nuclei in the Cosmos XI, POS(NIC XI), vol. 045 (2010)
14. Nara Singh, B.S., Sjue, S.K.L., Davids, B., Hass, M., Adsley, P., Buchmann, L., Carmona-Gallardo, M., Fallis, J., Fox, S.P., Fulton, B., Galinski, N., Hager, U., Hutcheon, D.A., Laird, A.M., Martin, L., Ottewell, D., Reeve, S., Ruiz, C., Ruprecht, G., Shotter, A., Tengblad, O.: In: Nuclear Physics and Astrophysics V. Journal of Physics: Conference Series (2011)
15. Goasduff, A., Courtin, S., Haas, F., Lebhertz, D., Jenkins, D.G., Beck, C., Fallis, J., Ruiz, C., Hutcheon, D.A., Amandruz, P.A., Davis, C., Hager, U., Ottewell, D., Ruprecht, G.: In: Schmitt, C., Navin, A., Rejmund, M., Lacroix, D., Goutte, H. (eds.) FUSION11 EPJ Web of Conferences, vol. 17 (2011)
16. Hutcheon, D., Buchmann, L., Chen, A.A., D'Auria, J.M., Davis, C.A., Greife, U., Hussein, A., Ottewell, D.F., Ouellet, C.V., Parikh, A., Parker, P., Pearson, J., Ruiz, C., Ruprecht, G., Trinczek, M., Vockenhuber, C.: Nucl. Inst. Methods Phys. Res. B. **266**, 4171 (2008)
17. Vockenhuber, C., Erikson, L.E., Buchmann, L., Greife, U., Hager, U., Hutcheon, D.A., Lamey, M., Machule, P., Ottewell, D., Ruiz, C., Ruprecht, G.: Nucl. Inst. Methods Phys. Res. A. **603**, 372 (2009). doi:10.1016/j.nima.2009.02.016
18. Wrede, C., Hussein, A., Rogers, J.G., D'Auria, J.: Nucl. Inst. Methods Phys. Res. B. **204**, 619 (2003)

19. Vockenhuber, C., Buchmann, L., Caggiano, J., Chen, A.A., D'Auria, J.M., Davis, C.A., Greife, U., Hussein, A., Hutcheon, D.A., Ottewell, D., Ouellet, C.O., Parikh, A., Pearson, J., Ruiz, C., Ruprecht, G., Trinczek, M., Zylberberg, J.: Nucl. Inst. Methods Phys. Res. B. **266**, 4167 (2008)
20. Simon, A., Fallis, J., Spyrou, A., Laird, A.M., Ruiz, C., Buchmann, L., Fulton, B.R., Hutcheon, D., Martin, L., Ottewell, D., Rojas, A.: Eur. Phys. J. A. **49**(5), 60 (2013)
21. Vockenhuber, C., Ouellet, C.O., Buchmann, L., Caggiano, J., Chen, A.A., D'Auria, J.M., Frekers, D., Hussein, A., Hutcheon, D.A., Kutschera, W., Jayamanna, K., Ottewell, D., Paul, M., Pearson, J., Ruiz, C., Ruprecht, G., Trinczek, M., Wallner, A.: Nucl. Inst. Methods Phys. Res. B. **259**, 688 (2007)

DOI 10.1007/978-94-007-7963-1_26
Reprinted from *Hyperfine Interactions* Journal, DOI 10.1007/s10751-013-0902-x

The electromagnetic mass analyser EMMA

B. Davids

Published online: 23 October 2013

Abstract EMMA is a recoil mass spectrometer for ISAC-II at TRIUMF that will be used to separate the recoils of nuclear reactions from the beam and to disperse them according to their mass-to-charge ratios. EMMA has been designed for both efficiency and selectivity, possessing large acceptances in angle, mass, and energy without sacrificing the necessary beam suppression and mass resolving power.

Keywords Nuclear structure · Nuclear astrophysics · Recoil mass spectrometer · Electromagnetic separator · Recoil separator

1 Introduction

TRIUMF is constructing an electromagnetic mass analyser, EMMA, for use with the radioactive heavy-ion beams available from its ISAC-II facility. EMMA is a recoil mass spectrometer designed to separate the recoils of nuclear reactions from the beam and to disperse them in a focal plane according to their mass-to-charge ratio (m/q). Measurements of position, energy loss, residual energy, and time-of-flight are expected to uniquely identify the transmitted recoils. In addition to having a large solid angle of 16 msr, the spectrometer will accept recoils within a large range of m/q ($\pm$4 %) and energies ($\pm$20 %) about the central values. These large acceptances result in high detection efficiencies approaching 50 % for the recoils of many fusion-evaporation reactions. The trajectories of monoenergetic ions of a single mass within the spectrometer are calculated to be isochronous within 0.1 %, allowing high resolution time-of-flight measurements and large real-to-random ratios in coincidence experiments. These properties are anticipated to make EMMA a recoil mass

ISAC and ARIEL: The TRIUMF Radioactive Beam Facilities and the Scientific Program.

B. Davids (✉)
TRIUMF, 4004 Wesbrook Mall, Vancouver, BC V6T 2A3, Canada
e-mail: davids@triumf.ca

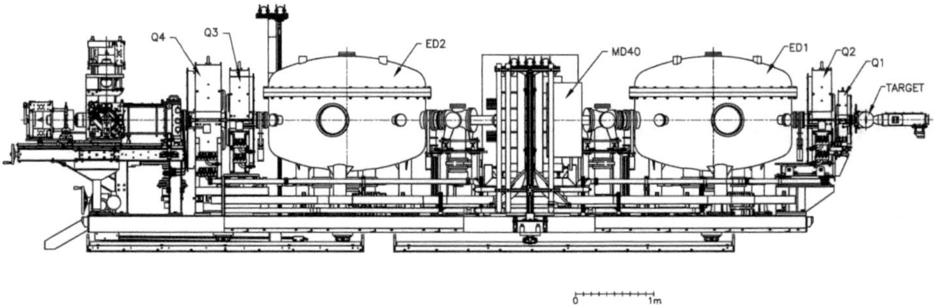

Fig. 1 Conceptual view of EMMA

spectrometer of very high quality that will enable previously impossible experiments with ISAC-II beams. Separation of reaction products from the primary beam at 0° allows the detection of recoils from fusion-evaporation reactions as well as transfer reactions induced by radioactive heavy ions, which emerge from the target in narrow cones centred about the beam direction. The capacity to disperse ions according to m/q combined with multiwire gas detectors in the focal plane will allow high resolution determinations of the atomic masses and atomic numbers of recoils. These capabilities of large acceptance, beam rejection at 0°, and high mass resolution are likely to make EMMA an exceptional instrument for nuclear physics research. When coupled with the unique radioactive ion beams from ISAC-II and the advanced γ-ray spectrometer TIGRESS [1], EMMA will position TRIUMF as a world leader in the field. EMMA is depicted in Fig. 1.

The superconducting ISAC-II linear accelerator has enabled the delivery of intense, high-quality beams of radioactive ions with masses up to 94 u and maximum energies of at least 6.5 A MeV. These beams will allow the study of the single-particle structure of exotic nuclei, the evolution of nuclear structure and shapes far from stability and at high spin, and nuclear astrophysics. Fusion-evaporation and transfer reactions initiated by radioactive ions in inverse kinematics promise to yield invaluable information about these subjects that cannot be obtained by other means. The study of many of these reactions will require detection and identification of the heavy recoil nucleus, in addition to light charged particles, neutrons, and γ rays. EMMA is designed to be ideally suited to study the products of fusion-evaporation reactions induced by the heavy radioactive beams of ISAC-II because its large acceptance implies high transmission efficiency and its high mass resolving power provides high selectivity and excellent beam rejection capability.

One area of research in which EMMA is expected to make an important contribution is the production and study of nuclei far from stability under extreme conditions, such as high excitation energy or angular momentum. This will be accomplished through the study of weak, otherwise inaccessible reaction channels by using EMMA as a mass filter in coincidence experiments. For example, in the study of high-spin states, events in TIGRESS detectors positioned around the target will be gated by signals from a particular nucleus in the focal plane detectors of EMMA. This technique permits the low-background study of weak reaction channels without the concomitant large loss of efficiency normally encountered when these measurements are carried out with small-acceptance detectors. Without the recoil detection and identification, the weakest channels, which are often the most interesting, would be totally obscured by the large number of γ rays from more copiously produced nuclei. The large energy, mass, and angular acceptances of EMMA will be crucial in these experiments, and will provide high sensitivity by allowing triple coincidence measurements in which two γ rays are detected in coincidence with the recoil.

While being well suited to the detection of recoil nuclei from fusion-evaporation reactions, EMMA will also be able to detect the projectile-like recoils of transfer reactions in inverse kinematics with high efficiency and good beam rejection capability. In one- and two-nucleon transfer reactions induced by heavy projectiles on light targets such as (d,p), (p,^{3}He), and (d,t), the recoil nuclei are strongly forward focussed and have relatively small energy spreads. They can therefore be detected with geometric efficiencies near unity. The detection of recoils from these transfer reactions will represent one of the most important uses of the spectrometer. In studies of both fusion-evaporation and transfer reactions, EMMA will be used in conjunction with TIGRESS. We anticipate that between 1/2 and 2/3 of TIGRESS experiments will require EMMA, which has led to detailed considerations of how EMMA can be designed to take full advantage of it.

2 Status and outlook

After completing the ion optical design of EMMA [2], we were awarded a grant in April 2006 by the Natural Sciences and Engineering Research Council of Canada to be expended over 5 years, with the understanding that TRIUMF would furnish the additional capital required to complete the spectrometer. Following the initial award, we began writing detailed specifications for the magnets and electric dipoles. In July 2007, after a tendering process, a contract to build the two electric dipoles, the dipole magnet, and four quadrupole magnets was awarded to Bruker BioSpin GmbH of Germany.

Two of the most critical parts of EMMA are the electric dipoles. The electrodes, field clamps, insulating supports, and vacuum vessels have been produced by Bruker according to our specifications while we have designed and built the high voltage power supplies at TRIUMF. As of now, three positive and three negative high voltage supplies have been constructed and tested at TRIUMF up to electric potentials of at least $\pm$ 325 kV. Figure 2 shows one of the supplies being prepared for testing in our SF_6 pressurized high voltage testing facility. Only two of each type of HV supply are needed for EMMA; the other two are spares. All six supplies were shipped to the Bruker factory in Karlsruhe, Germany for the high voltage testing of both complete electric dipole assemblies in December 2011. The anode and cathode were fabricated from solid titanium and are shown in Fig. 3 during a measurement of the separation between the electrodes, required to be 125.0 ± 0.1 mm.

Fabrication of the magnets is complete and magnetic measurements have shown they will meet our needs. Delivery of the large electromagnetic components took place in 2012 and 2013.

Fig. 2 TRIUMF-built EMMA 350 kV power supply

Fig. 3 Titanium electrodes and ceramic insulating supports for an EMMA electric dipole shown during a measurement of the separation between the anode and cathode

References

1. Svensson, C.E. et al.: J. Phys. **G31**, S1663 (2005)
2. Davids, B., Davids, C.N.: Nucl. Instrum. Methods Phys. Res. **A544**, 565 (2005)

DOI 10.1007/978-94-007-7963-1_27
Reprinted from *Hyperfine Interactions* Journal, DOI 10.1007/s10751-013-0903-9

HERACLES : a multidetector for heavy-ion collisions at TRIUMF

Patrick St-Onge · Jérôme Gauthier · Barton Wallace · René Roy

Published online: 23 October 2013

Abstract HERACLES is a multidetector that has been modified to study heavy-ion collisions, using an ion beam with an energy range between 8 to 15 MeV per nucleon. It has 78 detectors axially distributed around the beam axis in 6 rings allowing detection of multiple charged fragments from nuclear reactions. HERACLES has 4 different types of detectors, BC408/BaF_2 phoswich, Si/CsI(Tl) telescope, BC408/BC444 phoswich and CsI(Tl) detectors. $^{25}Na +^{12} C$, $^{25}Na +^{27} Al$, $^{25}Mg +^{12} C$ and $^{25}Mg +^{27} Al$ reactions have been used to characterize the multidetector. Element identification up to Z = 12 is achieved with the BC408/BaF_2 phoswich detectors, up to Z = 15 with the Si/CsI(Tl) telescopes and up to Z = 12 with the BC408/BC444 phoswich detectors. Isotopic identification is reached with the CsI(Tl) detector up to Z = 2.

Keywords Nuclear fragmentation · Multidetector · Heavy-ion reactions · ISAC-II · TRIUMF

1 Introduction

Properties of the nuclear matter at high density and temperature are not well known. Heavy-ion collisions give the possibility to study nuclear matter at high density and temperature. The recent availability of radioactive ion beams gives new possibilities to study collisions of nuclei with different isospin [1]. With the ISAC II facility at TRIUMF, radioactive ion beams are available with energies up to 15 MeV per nucleon (AMeV). HERACLES is a multidetector that has been used to study heavy-ion collisions at intermediate energy over 25 AMeV [2–5]. To use

ISAC and ARIEL: The TRIUMF Radioactive Beam Facilities and the Scientific Program.

P. St-Onge (✉) · J. Gauthier · B. Wallace · R. Roy
Département de physique, de génie physique et d'optique, Université Laval, Québec, Canada
e-mail: patrick.st-onge.1@ulaval.ca

HERACLES efficiently at ISAC II, we had to modify greatly the multidetector. In fact, every detector in the new version of HERACLES has been modified to lower the energy threshold required for particle identification. In Section 2, an experiment at ISAC II with the new HERACLES is described. Section 3 gives the multidetector's characteristics and first results.

2 Experiment

The main goal of experiment S1067 at TRIUMF is to study the nuclear symmetry energy using radioactive ion beams with different neutron-to-proton values (N/Z). For the July 2011 experiment, three ion beams were used. A radioactive ion beam of ^{25}Na at 9.23 AMeV, a stable ion beam of ^{25}Mg at 9.23 AMeV and a mixed beam with ^{4}He, ^{12}C and ^{16}O at 6.563 AMeV for energy calibration. The reactions used were $^{25}Na +^{12} C$, $^{25}Na +^{27} Al$, $^{25}Mg +^{12} C$ and $^{25}Mg +^{27} Al$.

3 Experimental setup

HERACLES is composed of 6 rings of detectors centered on the beam axis with polar angles (θ) between 4.8° and 46°. Ring 0 ($4.8° < \theta < 6°$) is composed of 6 BC408/BaF_2 scintillators in phoswich mode. Ring 1 ($6° < \theta < 10°$) is composed of 8 telescopes with 50 μm thick silicon detectors and CsI(Tl) scintillators. Ring 2 ($10.5° < \theta < 16°$) and ring 3 ($16° < \theta < 24°$) are each composed of 16 BC408/BC444 scintillators in phoswich mode. Ring 4 ($24° < \theta < 34°$) and ring 5 ($34° < \theta < 46°$) are each composed of 16 CsI(Tl) scintillators (see Table 1).

3.1 BC408/BaF_2 detectors

The BC408/BaF_2 phoswich detector is made of a fast plastic scintillator of 100 μm thick and a BaF_2 crystal scintillator. The thin BC408 100 μm plastic scintillator provides a low energy threshold from 2.2 AMeV for He to 5.5 AMeV for Mg. A pulse shape analysis of the signal from the detectors gives element identification, obtained with a fast-slow representation. The fast component is from the BC408 and the slow component is from the BaF_2. The BC408/BaF_2 phoswich detectors are designed for mass identification from $A = 1$ up to $A = 12$ using time of flight measurements. BC408/BaF_2 detectors are placed at a distance of 1.45 m from the target.

Performance A time resolution of less than 1 ns is necessary to obtain mass identification using time of flight measurements. For the July 2011 experiment at 9.23 AMeV, a beam buncher was not available on the ISAC-II beamline; the time resolution was estimated at 3 ns. Therefore, there was no mass identification using BC408/BaF_2 detectors for that experiment. The analysis of the fast-slow representation provides element identification up to $Z = 12$ (see Fig. 1).

3.2 Si/CsI(Tl) detectors

The Si/CsI(Tl) telescope detector is made of a 50 μm thick silicon detector and a CsI(Tl) crystal scintillator. The thin 50 μm silicon detector has a low threshold from

Table 1 Configuration of detectors in HERACLES

Ring No.	ΔE detector	E detector	θ_{min} (°)	θ_{max} (°)	N	$\Delta\phi$ (°)	ΔE thickness (μm)
0	BC408	BaF_2	4.8	6	6	15	100
1	Si	CsI(Tl)	6	10	8	18	50
2	BC408	BC444	10.5	16	16	22.5	100
3	BC408	BC444	16	24	16	22.5	100
4	–	CsI(Tl)	24	34	16	22.5	–
5	–	CsI(Tl)	34	46	16	22.5	–

θ is the polar angle from the beam axis, N is the number of detectors per ring, ϕ is the azimuthal angle

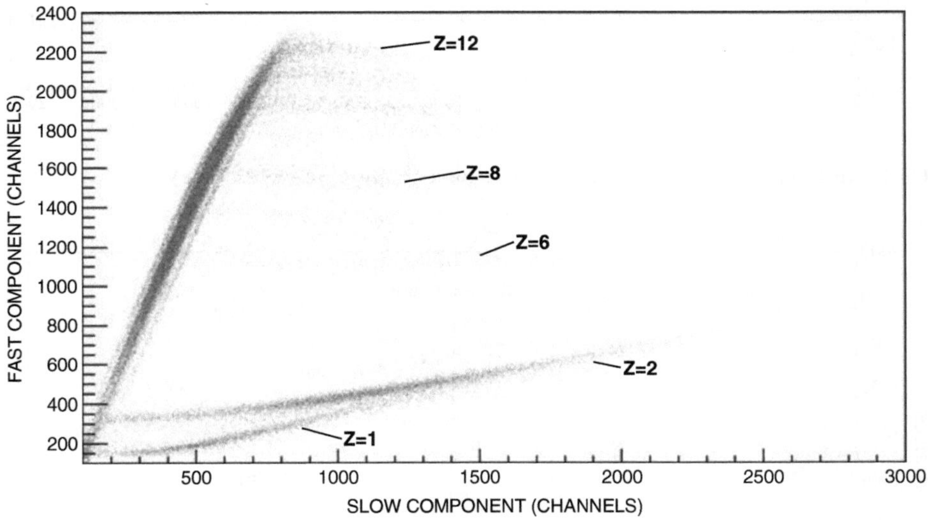

Fig. 1 Fast-slow representation of a BC408/BaF_2 phoswich detector for $^{25}Mg+^{12}C$ at 9.23 AMeV

2 AMeV for He to 5 AMeV for Mg. Fragment identification is done with a ΔE-E representation, the ΔE component coming from the Si and the E component, from the CsI(Tl).

Performance The analysis of the ΔE-E representation provides element identification up to Z = 15. Isotopic identification is reached for ^{1}H, ^{2}H, ^{3}H, ^{4}He, ^{7}Be and ^{9}Be (see Figs. 2 and 3).

3.3 BC408/BC444 detectors

The BC408/BC444 phoswich detector has the same fast plastic scintillator as the BC408/BaF_2 detector, but the second stage of detection is BC444, a slow plastic scintillator. A pulse shape analysis of the signal from the detector gives element identification.

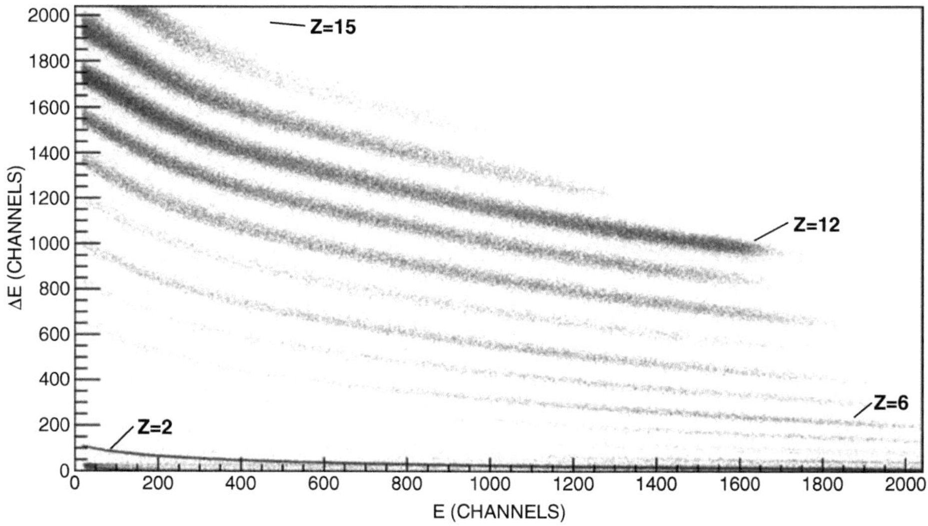

Fig. 2 ΔE-E representation of a Si/CsI(Tl) detector for $^{25}Mg + ^{12}C$ at 9.23 AMeV

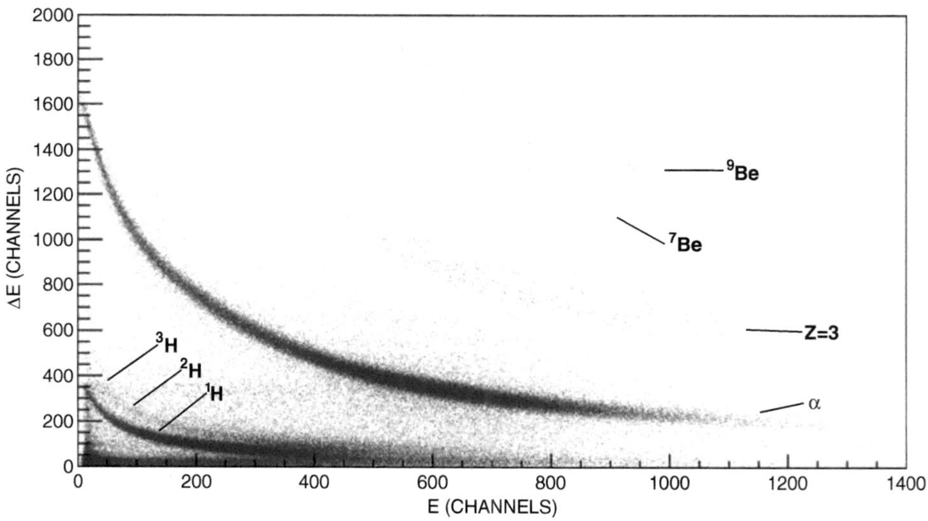

Fig. 3 ΔE-E representation of a high gain Si/CsI(Tl) detector for $^{25}Mg + ^{12}C$ at 9.23 AMeV

Performance The analysis of the fast-slow representation provides element identification up to Z = 12 (see Fig. 4).

3.4 CsI(Tl) detectors

The CsI(Tl) detector has only one stage of detection, the CsI(Tl) scintillator. The threshold is lower than for the other detectors of HERACLES. The decay time of

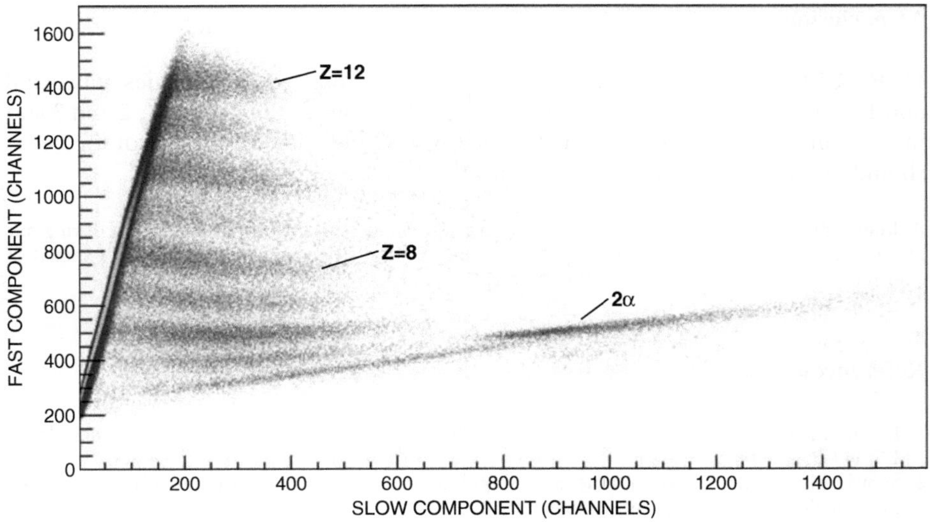

Fig. 4 Fast-slow representation of a BC408/BC444 detector for $^{25}Mg + ^{12}C$ at 9.23 AMeV

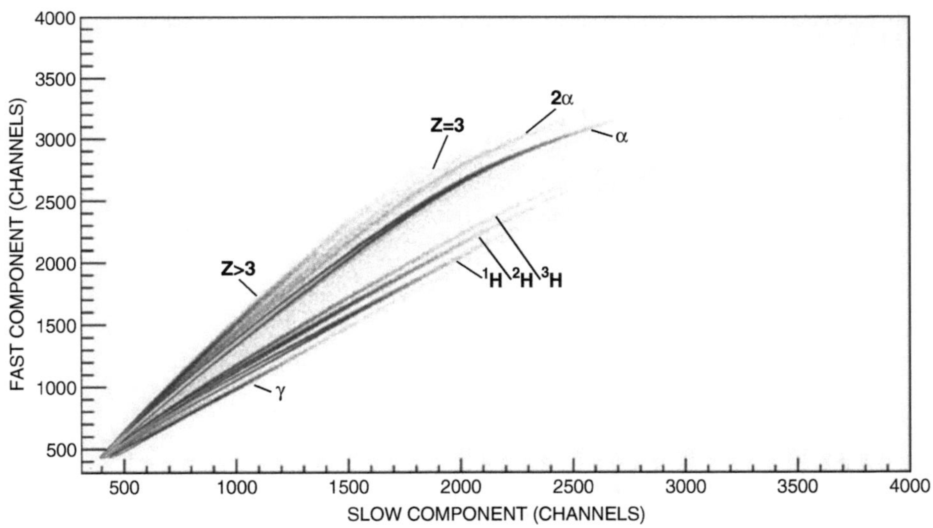

Fig. 5 Fast-slow representation of a CsI(Tl) detector for $^{25}Mg + ^{12}C$ at 9.23 AMeV

CsI(Tl) scintillators has two components, one fast (0.6 μs) and one slow (3.5 μs). A pulse shape analysis of the signal from the fast and slow components gives the identification.

Performance The analysis of the fast-slow representation provides isotopic identification for ^{1}H, ^{2}H, ^{3}H, ^{4}He with energy per nucleon greater than 4 AMeV and element identification up to Z = 3 (see Fig. 5).

4 Conclusion

HERACLES has been modified for heavy-ion collisions for low energies between 8 and 15 AMeV. The first results show element identification in rings 0, 1, 2 and 3 and mass identification for $Z < 3$ in rings 4 and 5. A better time resolution of the beam should provide mass identification in ring 0.

Acknowledgements We would like to thank the staff from TRIUMF for their support during our experiment periods. This work is supported by the Natural Sciences and Engineering Council of Canada.

References

1. Lie, B.A., et al.: Isospin physics in heavy-ion collisions at intermediate energies. Int. J. Mod. Phys. **E7**, 147–230 (1998)
2. Samri, M., et al.: Emission time and sequence in a $^{58}Ni+^{12}C$ fusionlike source at 34.5 MeV/nucleon. Phys. Rev. C **68**, 014609 (2003)
3. Moustabchir, R., et al.: Target proximity effect and dynamical projectile breakup at intermediate energies. Nucl. Phys. A **739**, 15–29 (2004)
4. Gingras, L., et al.: Origins of intermediate velocity particle production in heavy ion reactions. Phys. Rev. C **65**, 061604(R) (2002)
5. Beaulieu, L., et al.: Source size scaling of fragment production in projectile breakup. Phys. Rev. C **54**, 973 (1996)

DOI 10.1007/978-94-007-7963-1_28
Reprinted from *Hyperfine Interactions* Journal, DOI 10.1007/s10751-013-0904-8

IRIS: The ISAC charged particle reaction spectroscopy facility for reaccelerated high-energy ISOL beams

R. Kanungo

Published online: 23 October 2013

Abstract The development of a new charged particle reaction spectroscopy station, IRIS, at ISAC II TRIUMF is described. The facility will be used for studying transfer reactions and inelastic scattering of rare isotopes in inverse kinematics.

Keywords Transfer reactions · Inelastic scattering · Solid hydrogen target · Silicon detector · Ionization chamber · Inorganic scintillator

1 Introduction

The charged particle spectroscopy in direct reactions of rare isotopes is one of the most effective ways to unveil new features in the proton- and neutron-rich nuclei. The fast rare isotope beams from in-flight production allowed the discovery of exotic structures like the neutron halo [1] and skin through direct reactions. The low-energy reaccelerated beams from the ISOL production technique on the other hand allow complementary precision studies on the nature of the halo and structure of exotic nuclei.

These exotic forms of rare isotopes exhibit unusual ordering of nuclear orbitals in regions far away from the line of stability. The arrangement deviates largely from the scheme predicted by the conventional nuclear shell model formulated by Mayer [2] and Haxel et al. [3]. It is found that the higher lying orbitals move lower down in energy intruding into the preceeding shell gap. Thus, the nucleon magic numbers that form the fundamental basis of the nuclear shell structure are modified [4–6]. In order to identify and understand this change in nuclear shell structure it is necessary to know which orbitals the valence nucleons occupy and with what occupation probability.

ISAC and ARIEL: The TRIUMF Radioactive Beam Facilities and the Scientific Program.

R. Kanungo (✉)
Astronomy and Physics Department, Saint Mary's University,
923 Robie Street, Halifax, NS B3H 3C3, Canada
e-mail: ritu@triumf.ca

One-nucleon transfer reactions provide angular momentum selectivity to decisively determine the unknown spin of energy levels in the exotic nuclei [7]. The orbital occupied by the valence nucleons is reflected in the shape of the angular distribution. Reactions of interest in the IRIS facility are those using hydrogen isotopes as target, such as, (p, d), (d, p), $(d,^3\mathrm{He})$. Contrary to reactions using stable isotopes, due to the short lifetime of the unstable nuclei, the targets in these reactions are the light hydrogen isotopes and the projectile is the heavy unstable isotope. This laboratory condition of reversed roles of projectile and target in relation to conventional reactions is termed as inverse kinematics.

The (p, d) and $(d,^3\mathrm{He})$ reactions, where respectively one neutron and one proton from the unstable projectile nucleus, A, is transferred to the target, provide information on the configuration mixing in the ground state of the unstable nucleus of interest with mass number A. They also serve as the tool to determine the spin and excitation energy of the residual nucleus A-1 after the reaction.

The (d, p) and (d, n) reactions are those where the unstable projectile nucleus picks up a nucleon from the target thereby producing a more neutron-rich or more proton-rich nucleus, respectively. It is therefore one of the ways to access the ground and excited states of a more neutron/proton-rich isotope. These reactions can also serve as alternate ways to understand the neutron capture (n, γ) and proton-capture (p, γ) reactions relevant to the nucleosynthesis reaction networks for the rapid-neutron and rapid-proton capture processes.

The two-nucleon transfer reactions of the type (p, t), $(p,^3\mathrm{He})$ and their inverses are important ways to learn about pairing correlation in the exotic isotopes [8]. Due to the neutron-rich surface distribution interesting features not accessible in stable nuclei can emerge in the exotic nuclei. The pair transfer also allows to search for pairing vibration states. The cross section for populating such an excited state in comparison to the ground state population is one of the ways to understand about new shell closures or breakdown of known ones.

2 Description of the facility

The IRIS facility is designed to study the one and two-nucleon transfer reactions and inelastic scattering of exotic nuclei. in the energy range from $\sim$ 3–15A MeV and is currently in its commissioning phase. The layout of the facility and schematic detector arrangement is shown in Fig. 1 The energetic rare isotope beams can be tagged for isobaric contaminants as they lose energy in passing through a low-pressure ionization chamber ((a) in Fig. 1) before interacting with the reaction target. This identification of the incoming beam is particularly important for medium-heavy and heavy nuclei that are difficult to accelerate as pure beams. The challenge in the design of the ionization chamber is to allow charge identification with minimal energy loss such that the beam properties are not greatly degraded before interacting with the target. For this purpose a compact ionization chamber of $\sim$ 16 cm $\times$ 5 cm $\times$ 5 cm with a coplanar anode configuration (Fig. 2) was constructed. The chamber is segmented into 16 anodes which can be combined together in sections to optimize for desired energy loss depending on each case. The signals from individual anodes are processed by Cremat charge sensitive preamplifiers followed by MESYTEC shaping amplifiers and ADCs. The ionization chamber is designed to operate with isobutane at 10–25 mbar. Thin silicon nitride (50 nm) or mylar (900 nm) windows with a dimension of 10 mm $\times$ 10 mm are

Fig. 1 Snapshot of the IRIS beamline setup with the detector layout shown schematically on it. **a** ionization chamber **b** upstream YY1 silicon array **c** upstream S3-type silicon detector **d** downstream YY1 silicon array **e** CsI(Tl) array **f** downstream ΔE S3-type silicon detector **g** S3-type E detector **h** beam counting scintillator

Fig. 2 Model view of the ionization chamber

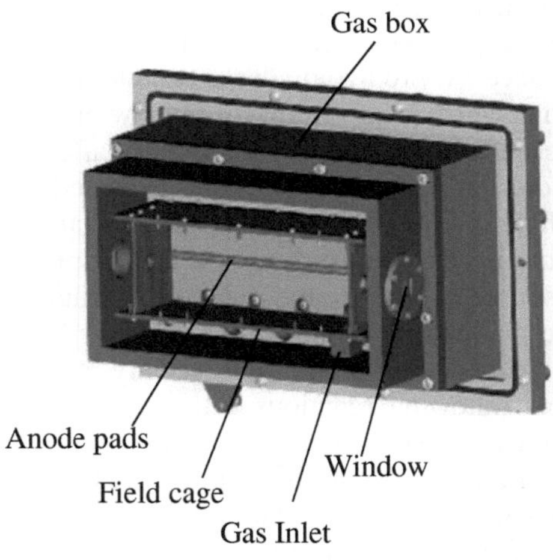

used to separate the gas volume from the vacuum. The characterization study of the ionization chamber with ^{18}O beam is currently in progress.

The novel feature of IRIS is the development of a thin solid hydrogen target to be used for studying the desired reactions. The solid hydrogen target cell (Fig. 3a) will be backed by a thin Ag foil. The target cell with the foil is cooled by a Sumitomo cryocooler with a helium compressor to a temperature of $\sim$ 4K. The hydrogen gas is then sprayed through a diffuser onto the Ag foil forming a solid hydrogen target. By controlling the gas volume the desired target thickness can be achieved. Typical thicknesses will range from $\sim$ 50–150 μm. The target assembly is surrounded by a

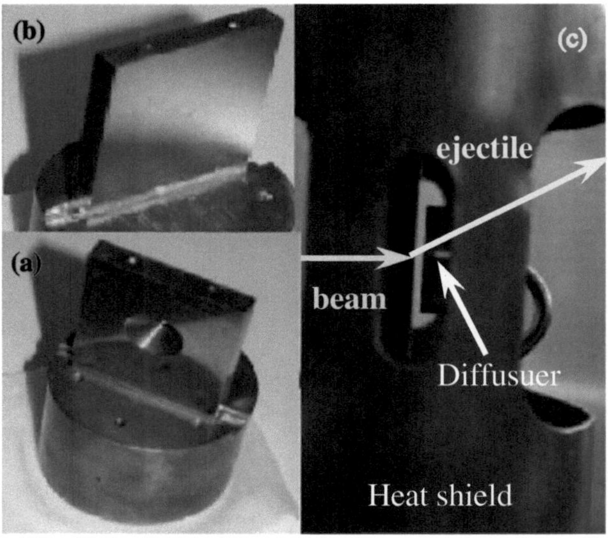

Fig. 3 **a** View of the solid hydrogen target cell **b** View of the backing foil. **c** View of the heat shield

copper cylinder (Fig. 3) connected to the cryocooler assembly whose temperature is around 30K. This acts as a heat shield to restrict heating of the target from the ambient temperature. The reaction products from the target are emitted through the opening in the heat shield. The desired orientation of the target will be such that the light target-like ejectiles after the reaction do not pass through the backing foil. This is necessary to avoid a large broadening of the scattering angle due to multiple scattering which is important for the low-energy ejectiles. For higher energy ejectiles the presence of the foil will not greatly affect the resolution. The determination of the resolution is in progress. IRIS is designed to also have the possibility of using thin polyethylene foils $(CH_2)_n$ and $(CD_2)_n$ as targets. Apart from the restricted reaction yield that is possible from these foils, the presence of carbon in the target acts as a major source of background reactions. Tritium implanted foils for use as triton target are also planned.

The main focus of the IRIS facility is to detect the charged particle reaction products following reactions with isotopes of hydrogen as target. Therefore, the detection system is designed to detect both the light target-like reaction ejectiles and well as the heavy beam-like reaction residues. The light particles that are emitted at backward angles in the laboratory frame usually have rather small energies less than 1 MeV. Therefore, these reaction channels are identified using the energy-angle kinematic correlation. This involves detecting the particles using an annular array of 500 μm thick segmented silicon detectors ((b) in Fig. 1). The forward scattered particles have higher energy, which allows identifying them through a ΔE-E correlation. This is achieved using a detector telescope with a 100 μm thick segmented silicon detector layer ((d) in Fig. 1) followed by a 12 mm thick annular CsI(Tl) array ((e) in Fig. 1) that matches the silicon array in overall configuration. The CsI(Tl) detectors form an array of 16 individual crystals each of which are readout using Hamamtsu silicon photodiodes (Fig. 4a). The annular silicon detector array for both upstream and downstream are of MICRON Semiconductor YY1 type which has 8 independent azimuthal detector sectors. Each sector is segmented into 16 rings which provides the scattering angle information (Fig. 4b).

Fig. 4 Snapshots of **a** CsI(Tl) array and **b** YY1 silicon detector array

In the upstream direction, a smaller silicon array of 500 μm thick MICRON Semiconductor-S3 type detector ((c) in Fig. 1) provides additional smaller scattering angle coverage for particles that pass through the hole of the larger YY1 array. Further downstream to the YY1-CsI(Tl) telescope a detector telescope is placed made up of a 60 μm thick layer of S3-type silicon followed by a 500 μm thick silicon of same type ((f,g) in Fig. 1). This allows a ΔE-E identification of the heavy reaction residue. The detector arrays can be placed at any distance from the target ranging from 8 cm to 75 cm, the choice of which will be optimized for the specific reaction to be studied.

The unreacted beam passes through the hole in all the detectors and is eventually stopped and counted using a radiation hard YAP:Ce inorganic scintillator ((h) in

Fig. 1) readout by a photomultiplier tube. The assembly is placed in the last separate vacuum chamber.

3 Summary

In summary, the transfer reactions and inelastic scattering of exotic nuclei will provide new and valuable information on the evolution of nuclear structure. The ISAC facility at TRIUMF is well suited for such studies. While these reactions at ISAC are currently being performed using polyethylene foil targets, a new development has been made to enhance the scope of studying these reactions with very low intensity beams (few hundred counts per sec.) using a thin solid hydrogen target at the IRIS facility, together with identification of the incident beam species using a low-pressure ionization chamber.

Acknowledgements R. Kanungo is thankful to S. Ishimoto, G. Sheffer, I. Tanihata and A. Miller for valuable discussions. Support from Canada Foundation for Innovation, Nova Scotia Research and Innovation Trust and NSERC, Canada is gratefully acknowledged.

References

1. Tanihata, I., et al.: Measurements of interaction cross sections and nuclear radii in the light p-shell region. Phys. Rev. Lett. **55**, 2676 (1985)
2. Mayer, M.G.: On Closed Shells in Nuclei. Phys. Rev. II **75**, 1969 (1949)
3. Haxel, O., Jensen, J.H.D., Suess, H.E.: On the “Magic Numbers” in nuclear structure. Phys. Rev. **75**, 1766 (1949)
4. Otsuka, T., et al.: Evolution of nuclear shells due to the tensor force. Phys. Rev. Lett. **95**, 232502 (2005)
5. Ozawa, A., et al.: New Magic Number, N = 16, near the neutron drip line. Phys. Rev. Lett. **84**, 5493 (2000)
6. Kanungo, R., et al.: Observation of new neutron and proton magic numbers. Phys. Lett. B **528**, 58 (2002)
7. Satchler, G.R.: Direct Reactions, p. 685–687. Oxford University Press, New York (1983)
8. Yoshida, S.: Note of the two-nucleon stripping reaction. Nucl. Phys. **33**, 685 (1962)

DOI 10.1007/978-94-007-7963-1_29
Reprinted from *Hyperfine Interactions* Journal, DOI 10.1007/s10751-013-0905-7

The TRIUMF-ISAC gamma-ray escape suppressed specrometer, TIGRESS

G. Hackman · C. E. Svensson

Published online: 23 October 2013

Abstract The TRIUMF-ISAC Gamma-Ray Escape Suppressed Spectrometer is a high energy-resolution, high efficiency γ-ray detector array for radioactive beam experiments. It consists of high-purity germanium detectors with scintillator suppressors and waveform digitization. TIGRESS can operate in modes optimized for either maximum efficiency or for high peak-to-total, and can switch between these modes within a day. The digitized waveforms provide 6 mm FWHM position resolution for incident ^{137}Cs γ rays. The array hosts a suite of auxilliary detectors for charged particles and neutrons.

Keywords In-beam γ-ray spectroscopy · Accelerated radioactive beams

1 Background: scientific motivation and design goals

Some of the most exciting recent results in nuclear structure are associated with the evolution of novel nuclear behavior at the extremes of nuclear existence [1, 2], including (but not limited to) halo nuclei [3, 4], dissolution of the near-stability shell gaps [5, 6] and emergence of new magic numbers with special stability [7], and proton-neutron pairing [8]. These behaviors can be explored through the measurement of the excitation properties of exotic nuclei, for example through in-beam γ-ray spectroscopy with accelerated exotic radioactive ion beams (RIBS).

ISAC and ARIEL: The TRIUMF Radioactive Beam Facilities and the Scientific Program.

G. Hackman (✉)
Science Division, TRIUMF, 4004 Wesbrook Mall, Vancouver, B.C., Canada V6T 2A3
e-mail: hackman@triumf.ca

C. E. Svensson
Department of Physics, University of Guelph, Guelph, ON, Canada N1G 2W1
e-mail: sven@uoguelph.ca

Fig. 1 TIGRESS, with 12 detectors in the close-packed configuration

Heavy-ion collisions at or above the Coulomb barrier with RIBs directed upon stationary targets can lead to a wide range of reaction channels and excitation modes. Nuclei excited in the collision process will then emit one to ~40 γ rays, with typical energies from ~50 keV to ~8 MeV. These emitting nuclei will be fast-moving sources ($v/c \simeq 0.03$ to 0.10) so the detected photons will be Doppler shifted in the laboratory frame. These features generally point towards high purity germanium (HPGe) detectors with anti-Compton shields and with high accuracy in γ-ray vector determination. In RIB experiments, beams intensity is limited by RIB production technology and cannot be easily increased, which implies that high total γ-ray detection efficiency cannot be sacrificed. Also, the most sensitive experiments require the use of additional sophisticated radiation detectors, such as heavy-ion recoil spectrometers, light-ion counter arrays, neutron detectors, and conversion electron spectrometers, which requires due care in the mechanical layout of the γ-ray detectors to accomodate these. Finally, to maximize the physics output of a RIB facility, the experimental end-stations must allow for rapid reconfiguration; campaigns on a given production target will drive the beam scheduling moreso than experimental setups [9].

The TRIUMF-ISAC Gamma Ray Escape Suppressed Spectrometer, TIGRESS, is used at the ISAC-II facility [10] primarily for RIB experiments. It consists of up to 16 units of so-called clover HPGe multi-crystal detectors. Each clover is outfitted with a set of scintillator suppressor shields. The scinitllators and clovers are designed so that the HPGe may be inserted to a high-ϵ mode with maximum photopeak efficiency, or inter-clover suppressors may be inserted for a high-PT with maximum escape suppression and hence peak-to-total (or peak-to-background) ratios. The design was specified such that the array can be completely reconfigured from high-ϵ to high-PT and back in under a day. Waveforms from the detectors are sampled by 100 MHz 14-bit flash ADCs, with energy, timing, and pretrigger information derived by digital signal processing on high-density FPGAs. Trigger, clock, and data are transferred from multi-channel ADC/FPGA units to multi-level trigger and data collector cards. At the top level the trigger is validated and read requests are broadcast to the ADCs. TIGRESS has operated with arrays of highly

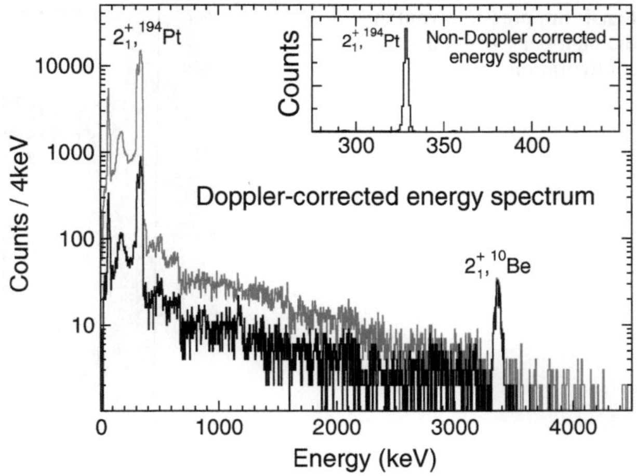

Fig. 2 Gamma-ray energy spectrum from inelastic scattering of ^{10}Be [11]. The *grey* spectrum shows the full γ-ray spectrum in coincidence with a heavy ion, while the *black* spectrum was gated on heavy-ion energies in a range expected for 3.386 MeV projectile excitation

Fig. 3 Schematic showing HPGe outer contact segmentation and cutaway view indicating volume spanned by outer contacts

segmented silicon detectors for charged-particle detection, and will also be used with plunger, CsI(Tl) detector, neutron detector, and recoil separator auxilliary devices. Figures 1 and 2 show the array in its configuration for experiments in 2011 and a representative spectrum from a RIB experiment respectively.

2 Segmented HPGe clover detectors

The high energy-resolution γ ray spectrometers in TIGRESS consist of four HPGe n-type bulleted-coaxial detectors in a single cryostat [12]. The crystals are nominally 60 mm in diameter and 90 mm long before they are machined and tapered (Fig. 3). Each crystal has a photopeak efficiency of $\geq$38 % relative to the ^{60}Co NaI(Tl) standard [13]. The cryostat, whose shape is identical to the EXOGAM [14] clover detectors, is also tapered. This allows for close-packing of the detctors in a rhombicuboctahedron (truncated cube) geometry. The inner-core coaxial contact

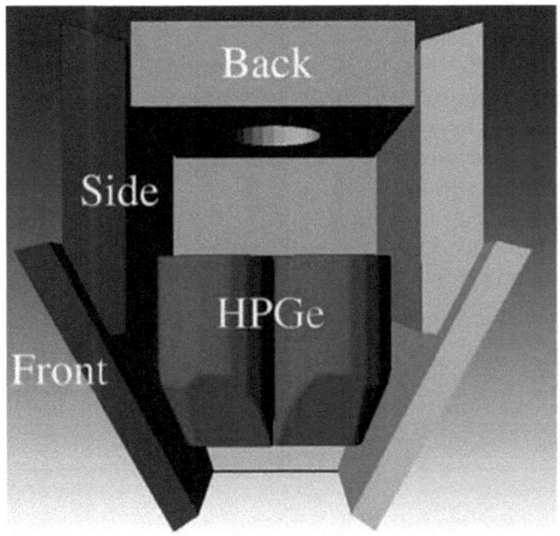

Fig. 4 Suppressor layout in the high-peak-to-total configuration

holds positive bias and collects charge from the full volume of the crystal. The outer surface of the detector has eight electrically isolated contacts, with four quadrants around the axis of the core contact and a lateral segmentation 30 mm from the front of the crystal. The center contact is instrumented with a cold FET and feedback front-end network within the cyrostat volume, while for the outer contacts the network is at room temperature. All contacts are instrumented with charge-sensitive preamplifiers.

In multi-crystal detectors such as the TIGRESS detectors, γ rays that enter one crystal and escape have a high probability of striking another crystal. The incident photon energy, then, can be measured by adding the energy deposition in two (or more) neighboring crystals. This add-back results in a relative efficiency for a full TIGRESS clover detector of between 215 and 220 %, depending on the efficiency of the individual crystals in the cryostat.

3 Escape suppression shields

Each clover is also outfitted with a set of scintillators [15] to detect escaping photon energy as shown in Fig. 4. Four CsI(Tl) crystals in two enclosures, and an eight-crystal BGO side-catcher in a single, hollow square, are bolted to the clover cryostat. These are backplug and sidecatcher suppressors, respectively, and detect small-angle scattering out the back and flat (orthogonal) side of the cryostat. An additional set of eight BGO scintillators arranged in pairs into trapezoidal enclosures fit around the front, tapered part of the clover cryostat for large-angle Compton scattering, especially from the front of the HPGe volume. Light from each scintillator is detected in a pair of phototubes and read with a single charge-sensitive preamplifier [16]. The segmentation of the suppressors allows for application of a wide variety of suppression strategies, depending on the expected energy and multiplicity of γ rays in the specific experiment; this has been investigated in detail [17].

The clover detectors with their fixed suppressors are mounted on a carriage that allows the unit to be inserted and withdrawn through the sixteen square faces of a truncated cube, with the ion beam entering and exiting through the other two square faces. The front suppressors enter the array through the edges of the truncated cube squares. This allows the clovers to be pushed forward so the fronts of the cryostats circumscribe an 11.0 cm radius sphere. This is the high-ϵ configuration. As an alternative, the clovers, together with their fixed suppressors, may be withdrawn to a radius of 14.5 cm, and the front suppressors inserted between them. In this configuration, the front tips of the suppressor collimators are also inscribed by an 11.0 cm radius sphere. In this configuration the front scintillator plates afford the maximum coverage for large-angle Compton scattering, and as such is the high-P/T mode. The original design specified that it should be possible to reconfigure between the high-P/T and high-ϵ modes in under a day.

The mechanical support structure holds eight clover and suppressor units on a central corona perpendicular to the beam axis, and four more at each of 45° and 135° to the beam axis. These parts of the frame supporting the off-perpendicular positions are referred to as front and back lampshades respectively. The lampshades and their clovers may be removed to accomodate large downstream detector or optics devices. This was anticipated primarily for large-acceptance downstream recoil spectrometers but can also be used to accomodate heavy-ion gas counters or counters for (forward-focused) neutrons following heavy-ion reactions.

During the 2010 experimental campaign, TIGRESS was outfitted with ten clover units (including suppressors) on the central corona and backward lampshades, to accomdate the SHARC silicon detector array described below. The campaign began with the array in high-P/T mode. This partial TIGRESS array was measured to have an absolute photopeak efficiency of 4.7 %, including addback. During a break between experiments, the array was reconfigured in one hour to the high-ϵ mode. In this configuration and with SHARC still in place, the absolute photopeak effiency was 7.4 %. This included addback within a clover unit but not between neighbouring cryostats.

4 Data acquisition system

The data acquisition system [18] uses a hierarchal, scalable multi-level triggering system and digital waveform sampling. Signals from the HPGe and suppressor charge-sensitive preamplifiers are digitized by TIG-10 modules. Each of these VXI-C modules consists of ten channels of 100 MHz 14-bit flash ADCs. The waveforms are continuously sampled. Contemporaneously, a large FPGA on each channel evaluates features of the waveform in real time. A clip-delay and leading-edge discriminator are evaluated for two pretriggers; the first validates the signal as a possible valid radiation event (a hit), while the latter is used in the triggering system (a Level 0 trigger). A hit triggers the evaluation of the amplitude of the signal by a moving-window pole-zero-corrected algorithm and a digital implementation of a rise-time-compensated constant fraction discriminator (CFD).

The Level 0 per-channel pretriggers are evaluated on the TIG-10 to generate four possible Level 1 triggers. These are transmitted by time-division multiplexing over an LVDS serial link to the next trigger level in TIG-C collector cards. These VME cards have thirteen ports, twelve of which communicate with lower-level cards such as TIG-10 digitizers. Up to four Level 2 triggers are evaluated based on the Level 1 trigger inputs. These are transmitted by the remaining port up to a

TIG-M Master card, which is physically identical to a TIG-C card but with special master firmware installed. Once a master trigger is evaluated it is sent back down to the lower levels through the TIG-Cs ultimately to the TIG-10 digitizers. At this point the digitized signal trace and evaluated features (energy and time) are validated and queued for transmission. The serial link is also used to transmit the TIG-10 data to the TIG-Cs, where they are stored in a FIFO for readout over the VME backplane into an frontend computer. Parameters for clipping, hit and trigger threshold, energy evaluation, CFD delay and threshold are transmitted through from the VME computers through the TIG-Cs, then to the TIG-10s via the LVDS interface as well. Run control and data storage are managed by MIDAS [19].

5 Position sensitivity

As has been noted previously, TIGRESS is intended for in-beam γ-ray spectroscopy of accelerated RIBs, and as such the emitted γ rays are Doppler shifted. The laboratory-frame γ-ray energy must be transformed into the rest frame of the emitting nucleus of interest. This requires knowing the direction of the γ ray and also the velocity and direction of the emitting ion when the emission occurs. High granularity charged-particle detectors are common and straightforward to use. While the slowing down of a beam in target prior to or after the nuclear reaction prior to γ emission results in an uncertainty in the velocity, this is often small. The largest contribution often comes from γ-ray detection. With an unsegmented detector, the uncertainty in the γ-ray direction is the opening angle of the detector. In the case of TIGRESS detctors in the close-packed configuration, this is approximately 20° and, although it depends in detail on the γ-ray energies and source velocities, the resulting uncertainty in Doppler correction will result in a peak with an order-of-magnitude poorer energy resolution than the intrinsic resolution of HPGe. The eight-fold electrical segmentation of TIGRESS can provide an effective granularity of ~10°. To approach Doppler broadening of the same magnitude as the intrinsic resolution of HPGe over a wide range of experimental conditions, it is necessary to measure the incident γ-ray direction to ~3° or better. This corrsponds to an effective pixel size of ~8 mm in position, in a plane perpendicular to the direction of γ-ray emission. It also requires reconstruction of the γ-ray multiple scattering within the detector, at least to the extent of identifying the most-probable first interaction corresponding to the incident photon vector. This puts a more challenging constraint of measuirng individual interactions with an uncertainty of better than ~5 mm position reconstruction in three dimensions, which is why the TIGRESS outer contacts are segmented longitudinally. Early investigations with a TIGRESS prototype showed that the TIGRESS detector geometry exhibited a sensitivity of 0.44 mm [20] as defined in [21].

The horizontal position resolution of TIGRESS clovers is demonstrated with a collimated ^{137}Cs source using a technique similar to those applied in the large tracking arrays [22]. Electric fields and weighting fields for the TIGRESS geometry are calculated with FEMLAB [23]. Electron and hole trajectories are calculated for various initial positions in the crystal, using the mobilities of [24]. From these, the time- and interaction-position-dependent charge collection waveforms are calculated and then convolved with the PSPICE calculated response of the TIGRESS pream-

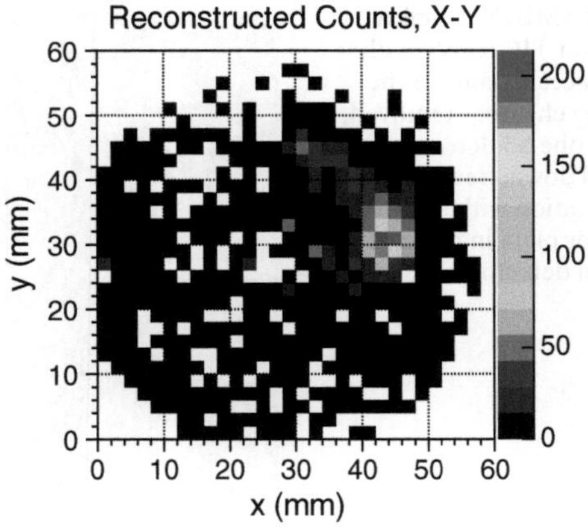

Fig. 5 Reconstructed first-interaction location for a collimated 662 keV γ ray

plifiers. Also, a representative set of single-interaction measurements, employing the concidence measurement technique described in [21], are collected using a small TIG-10-based data acquisition system. Like other highly segmented detectors [25], the TIGRESS detectors exhibit linear and differential crosstalk with a magnitude comparable to the expected induced signals. The differential crosstalk is simulated by an RC-CR transfer function, and both the amplitude of crosstalks (linear and differential) and the time constants of the transfer function are fit to the subset of coincidence measurements. The cross-talk is then included in the signal calculations to generate a set of basis waveforms on a uniform lattice of (1 mm)^3 voxels. The TIGRESS clover was illuminated with the collimated source and waveforms were collected with no coincidence condition. The basis waveforms were searched by minimizing a figure of merit similar to a χ^2 but with a power of 0.7 rather than 2. The minimization assumes one interaction first, but if the figure of merit is poor, a second search is undertaken with two interactions, either two in one segment or one in each of two segments. The search starts by comparing against a coarse, regular subset of the calculated basis waveforms in 8 mm steps over the entire detector volume, then narrowing the search volume and increasing the granularity of the comparison by factors of 2. The highest-energy interaction is taken to represent the most probable position of the incident γ ray. Figure 5 shows the reconstructed position of 662 keV γ rays based on measured waveforms for a collimated source incident at a given position. The full-width at half maximum of this position peak above background is 6 mm, which includes both the uncertainty in reconstruction and the divergence from the 1.5 mm diameter, 7 cm long collimator for the ^{137}Cs source.

6 Auxilliary detectors

Every experiment with TIGRESS to date has used a silicon charged particle detector. The BAMBINO facility, shown in Fig. 6, was used in the first TIGRESS experiments. BAMBINO consists of up to two S2 or S3 CD-style [29] segmented annular detectors and feedthroughs. The two CDs may be arranged as a dE-E telescope or may be placed on either side (upstream and downstream) of the target ladder, which hold

Fig. 6 BAMBINO facility installed at TIGRESS, with the side access panel of the scattering chamber removed, showing one S3 detector mounted downstream of the target location with a blank collimator plate in place of an upstream detector

up to five targets or apertures and which can be biased to suppress delta electrons. The nominal position places an S3 detector 30 mm either upstream or downstream of the target ladder. At those position, two S3 detectors will span laboratory-frame polar scattering angles from 20° to 49.4° at forward angles or 131.6° to 160° at backwards angles, for a total angular coverage of 3.63 srad, or 28.9 % of a full sphere. The innermost rings subtend 1.66° while the outer rings subted 0.82°. Azimuthal angular resolution is 11.25°, given by the 32 sectors of the S3 detector. The target chamber vacuum vessel is made of aluminum and has an outer radius of 102 mm (diameter 8″) and thickness of 1.5 mm (0.06″). The S3 holder is designed so that with appropriate standoffs it may be placed anywhere from 3 mm to 70 mm from the target. The Si signals are processed through charge-sensing amplifiers [16] with nominal sensitivities of 10 or 100 mV per MeV, respectively. The Si signals have <100 ns rise times and a decay constant fall time of ~80 μs , very similar to the signals from the HPGe detectors. This makes it very straightforward to integrate them into the triggering, waveform processing, and data acquisition system; they are read out by TIG-10 modules. Further descriptions of the BAMBINO facility and experements may be found in [4, 6, 26–28].

More recently, TIGRESS has been used with the Silicon High Angular-resolution detector for Reactions and Coulex, SHARC [30]. This is a barrel of up to eight double-sided strip Si arranged in boxes of four detectors each fore and aft of the target location detector, and up to eight quadrants of annular-sector detectors closing the ends of the barrel (see Figs. 7 and 8). The rectangular box detectors have 48 and 24 segments respectively. This provides an angular resolution on average of approximately 1°, depending on the angle. The total geometric coverage of a fully populated SHARC array is 90 %, although contact wires, inter-strip dead layers and such reduce the total effective angular coverage to 2π sr. In the first campaign only one-half of the Si detectors and eight clover detectors were used, so all detectors were read with TIG-10 modules. In a subsequent campaign, the Si were read out with higher-density TIG-64 cards, which include 64 14-bit 50 MHz flash ADCs in a VME package. Like the TIG-10 cards, the TIG-64s generate level-1 triggers and

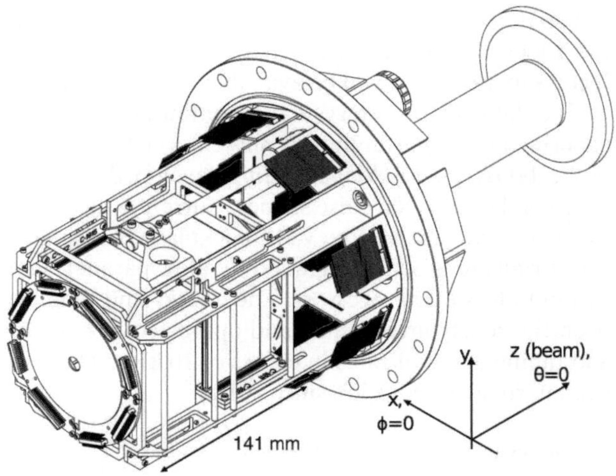

Fig. 7 Line drawing of SHARC, adapted from [30]

Fig. 8 Photograph of SHARC about to be inserted into its scattering chamber at TIGRESS

transfer trigger, global accept, and evaluated energy, time and waveform information to TIG-C cards by the same LVDS protocol.

7 Future prospects

In addition to silicon detectors, other auxilliary detectors are in development and expected to become available over the next few years. The TIGRESS Integrated Plunger [31] will combine radiation-hard scintillators and a plunger apparatus for lifetime measurements in exotic nuclei produced by weak fusion-evaporation channels. Tests of small commercially available PIN diodes and of CsI(Tl) detectors have been undertaken towards this goal. Also, a deuterated liquid scintillator array can replace the downstream lampshades for neutron detection, DESCANT [32]. SPICE, an in-beam electron spectrometer with permanent magnet electron transport and silicon detectors, is under development [33]. TIGRESS will also be compatible with the ElectroMagnetic Mass Analyzer, EMMA, a high-resolution, high-acceptance recoil separator that will be available for experiments in the coming few years [34].

The data acquisition system is continuously being refined and upgraded for new experimental opportunities. A new class of 1 GHz digitizers compatible with the rest of the TIGRESS readout are being developed, to allow for neutron-γ discrimination based on DESCANT pulse shapes. Currently the data bottleneck is that the collector cards are read out over the VME backplane by a single VME processor. The collector card firmware is being upgraded to allow parallel, simultaneous readout of the data by multiple VME processors. The energy and time evaluation algorithms have also been rigorously investigated and optimized to maintain excellent energy resolution and timing at count rates in excess of 10 kHz. These data acquisition upgrades have been in regular operation since 2012.

Acknowledgements Construction and operation of TIGRESS is provided by the Natural Science and Engineering Research Council of Canada (NSERC) and TRIUMF. The authors thank J. N. Orce and M. J. Djongolov for their assistance in preparing this manuscript.

References

1. Krücken, R.: Contemp. Phys. **52**, 101 (2011)
2. Ball, G.C., et al.: J. Phys. G. **38**, 024003 (2011)
3. Tanihata, I., et al.: Phys. Rev. Lett. **100**, 192502 (2008)
4. Kanungo, R., et al.: Phys. Lett. B **682**, 391 (2010)
5. Wimmer, K., et al.: Phys. Rev. Lett. **105**, 252501 (2010)
6. Hurst, A.M., et al.: Phys. Lett. B **674**, 168 (2009)
7. Wigner, J.A., et al.: Phys. Rev. C **81**, 044303 (2005)
8. Cederwall, B., et al.: Nature **469**, 68 (2011)
9. Svensson, C.E., et al.: Nucl. Instrum. Meth. B **204**, 660 (2003)
10. Laxdal, R.E., et al.: Comissioning and Early Experiments with ISAC-II. PAC07 (Albequerque, NM, USA), 2593 (2007)
11. Orce, J.N., et al.: Phys. Rev. C **86**, 041303 (2012)
12. Scraggs, H.C., et al.: Nucl. Instrum. Meth. Phys. Res. A **543**, 431 (2005)
13. IEEE: ANSI/IEEE Standard 325-1986, IEEE Standard Test Procedure for Germanium Gamma-Ray Detectors. IEEE, New York (1986)
14. Azaiez, F.: Nucl. Phys. **A654**, 1003c (1999)
15. Scionix, Bunnik, NL, http://www.scionix.nl (2012). Accessed 3 July 2012
16. Swan Research, Owosso, MI, USA, http://swanresearch.xorgate.com (2012). Accessed 3 July 2012
17. Schumaker, M.A., Svensson, C.E.: Nucl. Instrum. Meth. Phys. Res. A **575**, 421 (2007), and references therein
18. Martin, J.-P., et al.: IEEE Trans. Nucl. Sci. **55**, 84 (2008)
19. http://midas.triumf.ca (2012). Accessed 3 July 2012
20. Svensson, C.E., et al.: Nucl. Instrum. Meth. Phys. Res. A **540**, 348 (2005)
21. Vetter, K., et al.: Nucl. Instrum. Meth. Phys. Res. A **452**, 223 (2000)
22. Lee, I.-Y., Simpson, J.: Nucl. Phys. News **20**, 23 (2010)
23. COMSOL Inc., Burlington, MA, http://www.comsol.com (2012). Accessed 3 July 2012
24. Bruyneel, B., Reiter, P., Pascovici, G.: Nucl. Instrum. Meth. Phys. Res. A **569**, 774 (2006)
25. Bruyneel, B., et al.: Nucl. Instrum. Meth. Phys. Res. A **599**, 196 (2009)
26. Schumaker, M.A., et al.: Phys. Rev. C **78**, 044321 (2008)
27. Schumaker, M.A., et al.: Phys. Rev. C **80**, 044325 (2009)
28. Wilson, G., et al.: Rutherford Centenial Conference on Nuclear Physics, Manchester UK, J. Phys.: Conf. Ser. **381**, 012097 (2012)
29. Micron Semiconductor, Sussex, UK, http://www.micronsemiconductor.co.uk (2012). Accessed 3 July 2012

30. Diget, C.Aa., et al.: J. Instrum. **6**, 02005 (2011)
31. Starosta, K., et al.: Proceedings: CGS14, Guelph, Ontario, Canada (2012)
32. Wong, J.: Design Study of DESCANT: Deuterated Scintillator Array for Neutron Tagging (M.Sc. Thesis) University of Guelph (2008)
33. Ketelhut, S., et al.: submitted to Nucl. Instrum. Meth. Phys. Res. A (2013)
34. Davids, B., Davids, C.N.: Nucl. Instrum. Meth. Phys. Res. A **544**, 348 (2005)

DOI 10.1007/978-94-007-7963-1_30
Reprinted from *Hyperfine Interactions* Journal, DOI 10.1007/s10751-013-0906-6

ARIEL overview

J. Dilling · R. Krücken · L. Merminga

Published online: 24 October 2013

Abstract The Advanced Rare Isotope Laboratory (ARIEL) will expand the scientific capabilities of TRIUMF's rare isotope program by providing more exotic isotope species with very high intensities and by adding two production targets that will provide beams in parallel to the existing ISAC target station. Together, these three stations will fully enable the exploitation of the numerous existing experimental facilities at ISAC, including experimental facilities for medical isotope research, nuclear astrophysics, material studies, fundamental nuclear studies, and searches beyond the standard model. This article summarizes the science reach of ARIEL and describes the facility.

Keywords Rare isotope beam · Isotopic separation on-line · Photo-fission · Medical isotopes · Ion sources · Targets · Electron linear accelerator

1 Introduction

TRIUMF has recently embarked on the construction of ARIEL, the Advanced Rare Isotope Laboratory, with the goal to significantly expand the Rare Isotope Beam (RIB) program for Nuclear Physics and Astrophysics, Nuclear Medicine and Materials Science. At the heart of ARIEL there is a 500 kW, 50 MeV electron accelerator (e-linac) for isotope production via photo-production and photo-fission as well as a second proton beam line from TRIUMF's 500 MeV cyclotron for isotope production via proton-induced spallation, fragmentation, and fission. Also included in ARIEL are two production targets and related infrastructure, mass-separators and ion beam transport to ISAC, and a electron-beam ion source (EBIS) for charge breeding.

ISAC and ARIEL: The TRIUMF Radioactive Beam Facilities and the Scientific Program.

J. Dilling (✉) · R. Krücken · L. Merminga
TRIUMF, 4004 Wesbrook Mall, Vancouver, BC, V6T2A3, Canada
e-mail: JDilling@triumf.ca

The ARIEL scientific program will be implemented in stages beginning with direct ^{8}Li photo-production in ^{9}Be targets for materials science using β-NMR and β-NQR, which is used to investigate depth-dependent magnetic properties at surfaces, interfaces and in nano structures. In the next stage, we will employ actinide targets for the production of neutron-rich fission fragments with design goal of up to 10^{14} fissions per second.

The ARIEL facility also includes the implementation of a new proton beamline (BL4N) from the TRIUMF's main cyclotron delivering up to 100 μA proton beam onto an additional production target. Combined with ISAC, ARIEL will support delivery of three simultaneous RIBs, up to two accelerated, to experimental stations at the existing ISAC facility, as well as new beam species and increased beam development capabilities.

This new and worldwide unique capability will allow for a much better exploitation of the available forefront experimental facilities at ISAC. Aside from the tremendous gain in available time for β-NMR also other experimental programs that need extended periods of beam time (4–6 weeks at a time) will be enabled by ARIEL's multi-user capability. In addition medical isotopes will be harvested for a research program on alpha emitters that might provide a promising avenue for effective alpha-tumor therapy. In the following sections we will briefly summarize the scientific reach of the main thrusts enabled through ARIEL. The following science programs will be enabled or enhanced due to ARIEL:

- solid state interface studies and new material with β-NMR and NQR.
- nuclear astrophysics
- physics beyond the standard model
- fundamental nuclear physics
- medical isotope research.

2 Magnetic functionality at surfaces, interfaces and in nanostructures

The e-linac, through conversion of its electrons to bremstrahlung photons, will allow for the production of copious amounts of ^{8}Li (via the $^{9}Be\ (\gamma,p)^{8}Li$ reaction) that are needed for a full exploitation of TRIUMF's unique β-NMR facility. As microelectronic devices shrink in size, the role of interfaces becomes increasingly important or even integral to their function. Giant magneto-resistance (Nobel prize 2007) in magnetic multi-layers is a prominent example, which was only discovered 15 years ago but is now in wide spread use in read heads for hard-disks. Interfaces are a new frontier in condensed matter physics since the behavior of electrons at an interface are distinct from those in the bulk and have unpredictable but potentially useful properties. This has implications for battery material, understanding high-temperature superconductivity and the functionality of nano-devices, as well as for the development of quantum computers.

Beta-NMR is one of the most informative ways to study local magnetic properties of surfaces and interfaces as a function of depth. Depth controlled β-NMR, to the level of a few nanometers, is only possible at the β-NMR facility at TRIUMF. Due to the limited beam time currently available for β-NMR ($\approx$5 weeks per year) the facility cannot realize its full scientific potential and only a selected user program can currently be carried out. The e-linac will enable more than 12 weeks of beam

Table 1 Predicted in-target production rates (nuclei/sec) of key nuclei with 500 kW electron beam (50 MeV, 10 mA) on liquid lead converter and 20 gr/cm^2 UCx target (1.2e13 fissions/s)

Nuclei	Production rates (nuclei/sec)
Ni-72	2.0E + 08
Zn-78	3.4E + 09
Kr-91	2.3E + 11
Kr-94	1.3E + 11
Rb-97	1.1E + 11
Sn-132	2.5E + 10
Sn-134	2.4E + 09
Xe-142	5.2E + 10
Xe-144	7.9E + 09
Cs-144	6.0E + 10
Cs-146	9.2E + 08

time for the β-NMR facility, which is critical for a viable experimental program in a fast moving field that requires the capability to carry out characterizations of new materials quickly.

3 Nuclear structure along the astrophysical r-process

The e-linac will produce unprecedented intensities of rare isotope beams of neutron-rich nuclei through photo-fission induced in actinide targets, e.g. ^{238}U. The production intensities of very neutron-rich nuclei are much higher than currently available at ISAC with its 10 μA proton beam. At the same time the photo-fission will avoid the production of high intensity contaminations that are produced via proton induced spallation and limit access to the most neutron-rich nuclei produced. Table 1 summarizes FLUKA calculations of in-target production rates of several neutron rich nuclei expected from the ARIEL e-linac.

In nature these very neutron-rich nuclei are only created for fractions of a second in violent star explosions (supernovae) or mergers of neutron stars during the so-called astrophysical r-process. However, they hold the key to our understanding of where in the universe the chemical elements heavier than iron were produced. With the e-linac we will bring the nuclear physics of supernova explosions to the laboratory and enable the identification and study of the nuclei involved in the r-process. In conjunction with precision astronomy and sophisticated astrophysical modeling using large scale computing facilities these studies will allow us to identify the astrophysical site of the r-process.

Measuring the properties of these nuclei is also critical for the quest to develop the long-sought unified theory for nuclei that will provide a predictive theoretical treatment of all nuclei based on the underlying fundamental force at work inside protons and neutrons, quantum chromo-dynamics (QCD). In particular experimental investigations of the evolution of shell structure and collective behavior in very neutron-rich nuclei will be essential to test the accuracy of modern theoretical approaches which predict substantial deviations from the well known nuclear structure near stability. The beams provided by ARIEL in combination with the available state-of-the-art experimental facilities at ISAC will enable the characterization of the important properties of these very neutron-rich nuclei and thus facilitate the development of this unified theoretical framework for nuclear physics.

4 Isotopes as laboratory for physics beyond the standard model

The multi-user capability facilitated by ARIEL will be particularly beneficial for those experiments that use rare isotopes of elements such as francium and radon as laboratories to search for physics beyond the standard model. The new proton target with its high resolution mass-separator and innovative ion-sources will provide intense, clean beams of heavy elements for long beam times of precision experiments that are searching for signs of new forces and broken fundamental symmetries. Important examples in this context are the recently initiated francium parity non-conservation program and the search for non-vanishing electric dipole moments using radon, radium or francium isotopes. These experiments need to run for hundreds of days per year to achieve the relevant sensitivities where tiny effects may become visible. In particular cases these experiments can reach sensitivities similar to those achieved in direct searches for such particles at the LHC. TRIUMF has a history as a precision laboratory and these new RIB experiments as well as the construction of the new ultra cold neutron (UCN) source and the associated neutron electric dipole moment experiment (nEDM) at TRIUMF will ensure that possible discoveries of new physics beyond the standard model of particle physics can be made at TRIUMF.

5 Medical isotopes

The production of radioisotopes via proton induced spallation in a uranium or thorium target is also very attractive from the point of nuclear medicine. In particular several alpha-emitting radioisotopes or their parent nuclei can be produced, which have great potential to be used in tumor therapy. The production intensities available using a 10 μA 500 MeV proton beam are sufficient for initiating a research program into developing radiopharmaceuticals for tumor therapy. The ion beams of the radioisotopes of interest would be collected following the high-resolution mass separator of the ARIEL facility in a dedicated new implantation station for medical isotopes.

In particular the isotope ^{211}At offers the possibility of selectively delivering a lethal dose of radiation to tumor cells by means of targeted radionuclide therapy, such as labeled anti-bodies. A number of studies have concluded that ^{211}At needs to be made available in clinically relevant amounts. Rather than preparing ^{211}At directly, at ARIEL the precursor, ^{211}Rn, would be produced, captured and purified for shipment to the appropriate collaborator. In this generator concept the ^{211}At is generated in transit through the decay of ^{211}Rn with a 14.7 h half-life.

6 The ARIEL facility

The ARIEL facility at TRIUMF presents a major new addition to the accelerator complex. A site overview of TRIUMF, including the existing and planned infrastucture is shown in Fig. 1. To fulfill its scientific mission, the ARIEL facility comprises five major elements:

- Superconducting 50 MeV, 10 mA cw linear electron accelerator (e-linac) and a beamline to transport high current electron beam to target stations

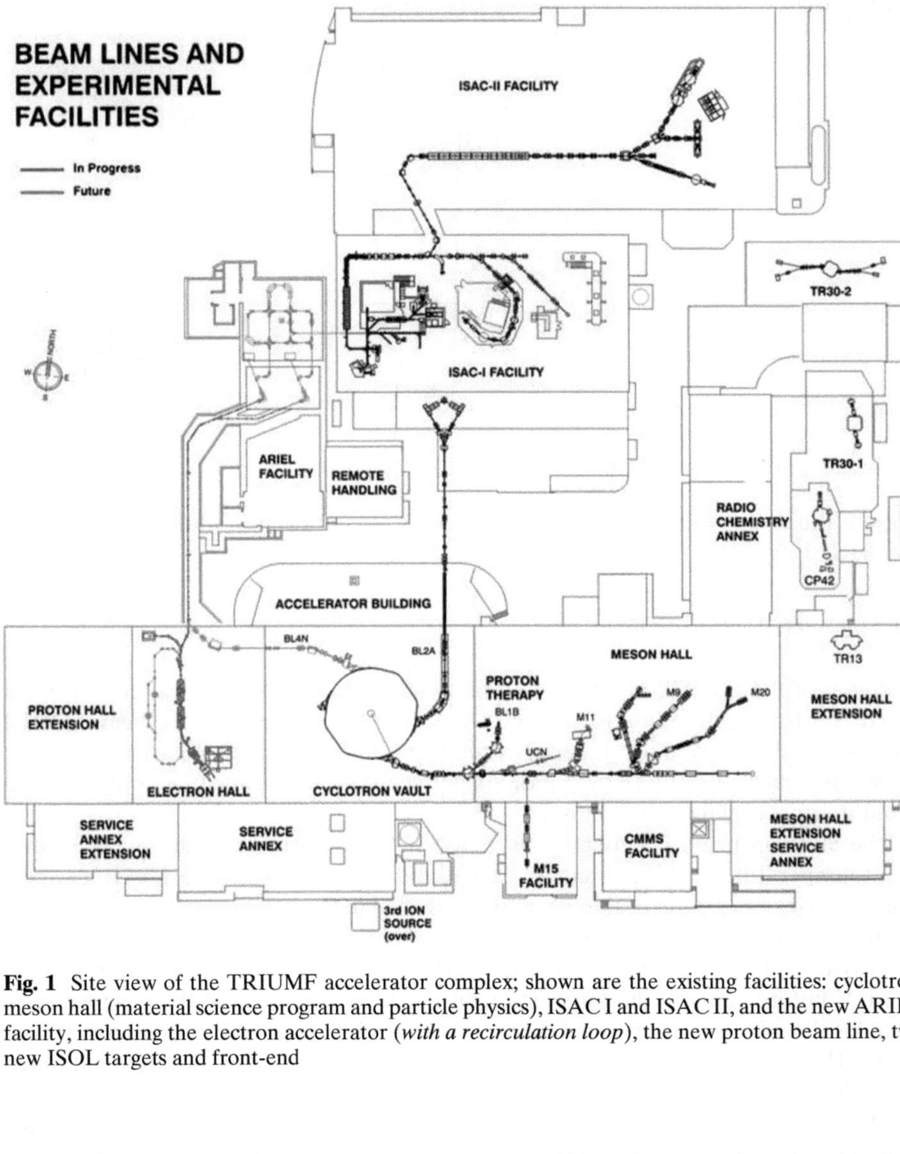

Fig. 1 Site view of the TRIUMF accelerator complex; shown are the existing facilities: cyclotron, meson hall (material science program and particle physics), ISAC I and ISAC II, and the new ARIEL facility, including the electron accelerator (*with a recirculation loop*), the new proton beam line, two new ISOL targets and front-end

- A new proton beamline (BL4N) capable of transporting up to 100 μA from the cyclotron to the target stations
- Two new high-power ISOL (Isotope Separation On-Line) target stations
- Mass separators and ion transport to the existing ISAC-I and ISAC-II accelerator complexes
- A new building to house the two target stations, remote handling infrastructure, chemistry labs, mass separators and front-end (Fig. 2); a tunnel for the proton and electron beamlines; renovations to the existing proton hall for its conversion to the electron hall; associated services.

Funded up to now are the 25 MeV, 100 kW electron linac and civil construction to encompass the objectives of the entire ARIEL program. In the following sections we

Fig. 2 Architects' rendition of the ARIEL building

present the design criteria, design choices and status of each of the five elements of the ARIEL facility.

6.1 The ARIEL electron linac

The goal of the ARIEL e-linac is to deliver 50–75 MeV, 10 mA cw electron beam as a driver for photo-fission of actinide targets to produce rare isotope beams (RIB) for nuclear physics, and for materials science research. The e-linac parameters were chosen to reach rates up to 10^{14} fissions per sec. The beam must be continuous to avoid thermal cycling on the target.

The electron beam is generated in a 300 kV DC thermionic gun, bunched in a room temperature 1.3 GHz buncher cavity, and accelerated by five 1.3 GHz superconducting cavities. One of these cavities is housed in the injector cryomodule whose energy gain is 10 MV, and the others are housed in two accelerator cryomodules, with two cavities each for an energy gain of 20 MV. The final electron beam energy is 50–75 MeV before going to the target stations. A future extension of the electron linac includes the addition of a recirculation arc that can be tuned either for energy recovery (ERL) operation or for energy doubling (RLA). When tuned as an RLA, it transports the beam through the linac for a second pass, thus reaching RIB energy up to 75 MeV with only one accelerator cryomodule installed. In the ERL scenario, a high brightness electron beam is interleaved with the single-pass RIB beam and accelerated in the same linac [1]. This beam could be used to drive an IR or THz free electron laser (FEL) or a Compton back-scattering X-ray source [2]. Below we present an overview of the major e-linac systems. A detailed description can be found in [3].

6.1.1 Electron source

The electron source is based on a 300 kV DC thermionic gridded cathode assembly that is modulated at 650 MHz. The RF amplitude and the grid bias are adjusted to provide ±16° (at 650 MHz) for 16 pC per bunch. To minimize field emission within the source, the design has an inverted electrode profile compared to the classical electron gun design. This reduces the surface area of the high voltage electrode, reducing the likelihood of field emission. To reduce the required length of the ceramic, the gun vessel is mounted in a vessel of pressurised SF_6 insulating gas.

For the purposes of emittance characterization and the implementation of a 650 MHz modulation scheme, a 100 keV DC gun was acquired from Jefferson Laboratory. Evaluation of the cathode on this electron gun has been crucial in developing the rf modulation requirements and has provided confidence in simulations. The 300 kV electron source has been assembled and HV testing is imminent.

6.1.2 Superconducting RF and high power RF systems

The e-linac accelerating cavities are TTF-type 9-cell niobium cavities with modified end cells and asymmetric beam pipes to ensure adequate damping of Higher Order Modes. The nominal operating gradient is 10 MV/m at $Q_0 = 1 \times 10^{10}$. HOM cavity studies in the range of 1.5–3.25 GHz have been completed and it was determined that damping of dipole modes with rf absorber material yields loaded shunt impedance at least 2 times below the BBU threshold, in energy recovery mode. The cavities are being fabricated by a Canadian manufacturer, PAVAC, Inc., and two of the three cavities required for the first phase of the e-linac have been received and undergoing testing. To reach the ultimate 0.5 MW of beam power, each cavity must provide 100 kW to the beam, and is equipped with two Cornell/CPI 50 kW cw couplers. The cavity tuner is a CEBAF style scissor tuner with room temperature motor.

The injector and main linac cryomodules borrow heavily from the ISAC-II heavy ion linac top-loading cryomodule design. The cryomodule vacuum vessel is a stainless steel vessel in a rectangular box shape. The cavities are mounted on a strongback slung from the top assembly from struts. All components are installed and supported from the top plate and the top assembly is loaded into the cryomodule from above.

The e-linac cryogenic system distribution replicates the ISAC-II system, which is based on a parallel feed of atmospheric LHe from a main trunk line to each of the cryomodules. The LHe is drawn from a main supply dewar supplied from a 4 K cold box. The 2 K liquid is produced in each cryomodule.

The construction and beam test of the Injector Cryomodule is the subject of collaboration between TRIUMF and the Variable Energy Cyclotron Centre (VECC) in Kolkata [4]. Beam tests of the injector cryomodule are expected in the fall 2013.

The first accelerator cryomodule is presently under construction, consists of two 9-cell cavities and requires four power couplers operating at 50 kW cw. A 300 kW klystron will be adequate to provide this power. 3 dB hybrids and phase shifters will be used to divide power equally and maintain required phase of the rf voltage in the accelerating cavities with respect to the beam. The second accelerator cryomodule will be installed as part of TRIUMF's next 5 year plan, between 2015–2020.

In a joint venture with Helmholtz Zentrum Berlin, two 300 kW cw klystrons have been purchased from CPI to power the injector and first accelerator cryomodule. These klystrons are designed to deliver at least 270 kW (rated for 290 kW) RF output power, at expected efficiency of minimum 52 %. In the long run the injector klystron will be used to power the second accelerator cryomodule, and a dedicated injector power source will have to be sourced.

6.1.3 Beam physics and engineering challenges

The ARIEL e-linac is a state of the art accelerator. At beam power of 0.5 MW, several design and operational challenges must be overcome, including: Careful control of beam losses (few Watts per meter; even less in superconducting cavities); development of a failsafe rastering scheme (>~1 kHz) to reduce the power density on target; development of non-intercepting diagnostics and a pulsing scheme to safely raise the beam power without de-stabilizing the rf; clean 650 MHz-modulation of the electron gun cathode grid; control of beam alignment in spite of large (~3 Gauss) stray ambient magnetic field from the neighboring cyclotron main magnet.

6.2 New proton beamline

The new proton beamline, BL4N, begins at the 4th extraction port of the cyclotron, and replaces beamline 4 that used to provide beams for the proton physics program which ceased operations about 10 years ago. Beamline 4N will be capable of transporting proton beam of energy between 470 and 500 MeV, and intensity up to 100 μA from the cyclotron to the target stations. Additional specifications include control of beam losses to below 1 nA/m, and beam size at the target within 2–10 mm (FWHM), and uniform beam density distribution on the target.

The beamline transport design and features are described in detail in references [5] and [6]. The optics design utilizes the experience gained from the operation of two high energy beamlines in the cyclotron, 1A and 2A, and has resulted in transverse beam sizes less than 15 mm rms, and a number of novel and improved features.

These include the compensation of the cyclotron's dispersion and foil move; collimation of large angle scattered particles from the stripper foil inside the cyclotron vault which has adequate shielding, and collimation of beam halo due to energy dispersion. These latter two features result in low loss transport and the ability to use smaller aperture magnets (e.g. 71 mm aperture quads would be adequate), thus reducing cost. There is flexibility in the magnetic lattice, so that different optics configurations can easily be setup. A dedicated matching section to control the beam size on the target is envisioned, and beam rastering on the target with AC steering magnets operating at frequency greater than 500 Hz. Lower frequencies would cause fatigue failure of the target. These features will result in BL4N being low-loss transport, more stable and more tunable than any of the existing primary cyclotron beamlines.

An upgrade program aiming to enhance the cyclotron capabilities and ensure its higher level performance consistent with the addition of BL4N is ongoing [7]. Elements of this program include a new extraction probe, an upgrade to the RF distribution system for 15–20 % RF power increase, upgrade of the ion source to 1 mA high brightness beam, centre region diagnostics upgrade, the simultaneous extraction of four high intensity beams, increase of the total extracted beam intensity towards ~400 μA, and intensity stabilization for multiple extracted beams.

6.3 ARIEL target hall

The ARIEL target hall comprises two ISOL target stations, east and west, actinide and conventional laboratories, target assembly laboratories, and a dedicated hot-cell facility for target diagnosis. The two target stations will be compatible with actinide target operation, and up to 500 kW electron beam and up to 50 kW proton beam power. The west ARIEL target station will be initially operated with electrons. Once the new proton beamline becomes operational, it will receive proton beam only, and electrons will be confined to the east target station, where RIB delivery to users will take place simultaneously with development activities, primarily of the photo-converter. During the operation of both targets with electrons, the sharing of the beam current on the targets will be flexible, to maximize both the user and the development programs.

In departure from the ISAC paradigm, the two ARIEL target stations are designed with completely independent services (e.g. cooling water, vacuum system, electrical, and nuclear ventilation), and adequate shielding such that personnel access to one target station will be permitted while the other is producing and delivering beam. In addition, the target stations will be equiped with remote connection and disconnection of services for rapid target exchange (2–3 days). The ARIEL target stations design principles are described in ref. [8].

6.4 Mass separators and ion transport to ISAC

The purpose of the ARIEL front end is to deliver beams from the ARIEL target stations to the existing ISAC accelerators and experimental facilities with the required purity, quality and flexibility.

The ARIEL front end connects the two ARIEL target stations to the existing ISAC facility and will expand delivery to eventually three simultaneous RIB beams with up to two simultaneous accelerated beams. The design goal of the low energy beam transport (LEBT) lines and mass separators is to provide sufficient flexibility that any of the target stations, existing or new, can deliver beams to any of the three experimental areas, low energy, medium energy or high energy, so that the RIB beams from each target can be optimized for a given experiment. Another goal is to provide a second path to the low energy experimental area since this is the area with the largest inventory of experimental infrastructure.

The ARIEL front end [9] consists of mass separators, with low, medium and high resolution, and a LEBT switchyard to deliver two simultaneous radioactive ion beams to the existing ISAC beamlines. The LEBT switchyard is designed such that beam from any of the two targets can be transported through the High Resolution Spectrometer (HRS) while the beam from the other target is sent to the Medium Resolution Spectrometer (MRS). In addition a bypass line is available such that a beam can be sent to the experimental area directly after the pre-separator.

Each of the two target/ion source units has an identical pre-separator with a mass resolution of 500 for 10 $\pi\,\mu$m through a 1 mm object slit. The low resolution separators and selection slits are located inside the target hall and they are engineered for remote handling. The pre-selected beams are then transported to the adjacent mass separator room via the LEBT transport lines.

The MRS is designed to produce a resolution of ~2500 for 10 $\pi\,\mu$m. The HRS leg uses two dipoles in tandem and includes an RFQ cooler to reduce the transverse emittance and energy spread of the beam before separation to achieve an effective resolution of ~20000 for 2.5 $\pi\,\mu$m. Both separators are planned to be installed on a high voltage platform to allow acceleration of the beam before separation.

The low energy beam transport (LEBT) consists of electrostatic transport sections delivering beams at source potential (up to $\leq$60 kV). The LEBT periodic and bending sections are copies of the existing ISAC low energy transport beam lines that have been demonstrated to be extremely reliable.

The LEBT comprises of an evacuated beam pipe with electrostatic elements inside including quadrupoles and steerers, joined to box-like chambers housed with electrostatic dipoles or diagnostic elements including profile monitors, faraday cups and slits.

To deliver simultaneously two accelerated beams, one to the medium energy the other to the high energy area, a new accelerator path with a new RFQ (RFQ2) is necessary. An Electron Beam Ion Source (EBIS) is envisioned to boost the 1+ ions from the on-line source. The EBIS will efficiently produce charge states compatible with $4 \leq A/q \leq 6$ for all masses so the RFQ2 specification for the new accelerator leg is conservatively set to $A/q \leq 7$, and the injection energy for the RFQ is ~8 keV/u, given that the target-ion source is rated for operation at ≤60 kV.

A new beamline running north of the existing RFQ1 will provide separate paths for ISAC-I and ISAC-II accelerated beams. A new medium energy transport section incorporates a switchyard that allows sending either of the RFQ1 or RFQ2 beams to either of ISAC-I or ISAC-II simultaneously. The new beamline to ISAC-II will include a room temperature drift tube linac (DTL2) to boost the energy from 0.15 MeV/u to 1.5 MeV/u for beams up to $A/q = 7$ to match the beam into the ISAC-II superconducting linac.

6.5 ARIEL conventional facilities

The ARIEL conventional construction includes the electron hall renovation, a new compressor building, and the ARIEL building (Fig. 2). The ARIEL main construction tender was issued December 2011 and awarded February 2012. Both the electron hall and the compressor building were completed in late 2012. The ARIEL building is expected to be complete in early August 2013.

References

1. Chao, Y., et al.: RF separator and septum layout concepts for simultaneous beams to RIB and FEL users at ARIEL. In: Proceedings of IPAC'11 Conference (2011)
2. Merminga, L.: ARIEL: TRIUMF's advanced rare isotope laboratory. In: Proceedings of IPAC'11 Conference (2011)
3. Koscielniak, S.: ARIEL e-linac (2013). doi:10.1007/s10751-013-0907-5
4. Laxdal, R., et al.: TRIUMF/VECC e-linac injector beam test. In: Proceedings of Linac 2012 Conference (2012)
5. Baartman, R.: http://lin12.triumf.ca/text/Talks/2009AAC/BL4N.pdf (2009)
6. Rao, Y.-N., Baartman, R.: Beam Line 4 North (BL4N) optics design. TRIUMF Design Note, TRI-DN-13-13 (2013)
7. Bylinski, Y.: Cyclotron upgrade. In: Cyclotrons 2010 Conf. Proceedings (2010)
8. Bricault, P.G., Ames, F., Dombsky, M.: Kunz, P., Lassen, J.: Rare Isotope Beams at ISAC - Target & Ion Source Systems (2013). doi:10.1007/s10751-013-0880-z
9. Marchetto, M., Baartman, R., Laxdal, R.E.: ARIEL front end (2013). doi:10.1007/s10751-013-0908-4

DOI 10.1007/978-94-007-7963-1_31
Reprinted from *Hyperfine Interactions* Journal, DOI 10.1007/s10751-013-0907-5

ARIEL e-linac

Electron linear accelerator for photo-fission

Shane Koscielniak

Published online: 15 October 2013

Abstract The design and implementation of a 1/2 MW beam power electron linear accelerator (e-linac) for the production of rare isotope beams (RIB) via photo-fission in the context of the Advanced Rare IsotopE Laboratory, ARIEL (Koscielniak et al. 2008; Merminga et al. 2011; Dilling et al., Hyperfine Interact, 2013), is described. The 100 % duty factor e-linac is based on super-conducting radiofrequency (SRF) technology at 1.3 GHz and has a nominal energy of 50 MeV. This paper provides an overview of the accelerator major components including the gun, cryomodules and cryoplant, high power RF sources, and machine layout including beam lines. Design features to facilitate operation of the linac as a Recirculating Linear Accelerator (RLA) for various applications, including Free Electron Lasers, are also noted.

Keywords Linear accelerator · Superconducting · Electron

1 Introduction

The TRIUMF ten year plan (2010–2020) seeks to triple the laboratory nuclear physics scientific output by the addition of two new rare isotope beam (RIB) sources to complement the three exising experimental areas at the ISAC facility which previously have shared a single "driver" proton beam line (BL2A) for RIB production—resulting in under-utilization of experimental resources. The plan foresees two more

TRIUMF receives federal funding via a contribution agreement through the National Research Council of Canada.

ISAC and ARIEL: The TRIUMF Radioactive Beam Facilities and the Scientific Program.

S. Koscielniak (✉)
TRIUMF, 4004 Wesbrook Mall, Vancouver B.C., Canada
e-mail: shane@triumf.ca

drive beams: a 50 MeV 10 mA electron linac followed (after 2015) by installation of a 100 μA 500 MeV proton beamline (BL4N) from port 4 of the existing H- cyclotron. In June 2010 that plan moved from dream to first phase of implementation. TRIUMF and the University of Victoria B.C., and twelve collaborating Canadian universities, have been awarded funds by the Canadian federal and British Columbia provincial governments for the construction of the Advanced Rare IsotopE Laboratory (ARIEL) in support of expanding the rare isotope program that serves nuclear structure and astrophysics studies as well as materials science and nuclear medicine. At the centre of this new facility will be a 1/2 MW beam power electron linear accelerator located in an existing vault. The e-linac is to be followed by two new high-power target stations and mass separators in the new ARIEL building, and ion transport to the ISAC-I and ISAC-II accelerator complexes. At the time of writing, civil construction to encompass the objectives of the entire ARIEL program and the linac first stage (see below) are funded, but not the ISOL target stations and ion transport, nor the BL4N proton beamline.

Initially, the electron beam (e-beam) will be shared between the targets. Later, the west target will be dedicated to protons, and the east to electrons; the west terminus of BL4N minimizes the number of strong bending magnets for the high-momentum proton beam. Beyond this plan, it is foreseen to add a recirculation path to the linac and run an IR or THz FEL in the back leg while simultaneously delivering e-beam to the RIB users; or to perform energy doubling for RIB users alone by two accelerating passes through the linac. The ARIEL civil construction comprises a new target building, a tunnel connecting with the e-hall, a helium compressor building for the attendant cryogenic plant, and substantial renovation of the former p-hall to house the e-linac. TRIUMF has embarked on construction of a 300 keV thermionic gun, a 10 MeV injector cryomodule (EINJ) and two 20 MeV accelerator cryomodules (named EACA and EACB), and e-beam transfer lines; and these are the focus of this paper. The e-linac construction is staged: initially there will be the EINJ and EACA, yielding an e-beam energy of up to 30 MeV; later EACB is added, raising the energy to 50 MeV or more.

1.1 Photo-Fission

Neutron-rich heavy nuclides will be produced from photo-fission of actinide targets. The e-beam falls on a convertor to produce gamma rays, which in turn fall on a production target, followed by isotope separation on line. The substitution of a ^{9}Be target and the γ,p reaction allows the copious production of ^{8}Li for materials science research. There are good reasons to consider photo-fission over fission/spallation at a proton driver. Although the range and breadth of products is smaller, the fractional yield of some neutron-rich species is higher. Moreover, there is lower isobaric contamination, simpler nuclide separation, less activation of the target area and easier remote handling. Although the number of fissions per electron is much less than per proton, this is easily compensated by source electron flux. Finally, the e-beam travels at light speed from the start; thus a single RF structure may be used throughout, leading to cost savings as compared with a proton linac which typically has a complicated front end accelerator.

2 E-linac overview

The e-linac 1/2 MW beam parameters (50 MeV, 10 mA) derive from the requirement of in-target fission rates up to 5×10^{13}/s, and the production efficiency versus energy which falls steeply below 20 MeV and saturates above 60 MeV; after that it is a better investment to increase the electron flux. Beam must be continuous to avoid thermal cycling of the target.

The 50 MeV e-linac major components are a 300 keV electron gun, 10 MeV injector cryomodule (EINJ) with one 9-cell Nb cavity, and two 20 MeV accelerator cryomodules (EAC A,B) each containing two 9-cell Nb elliptical cavities. The EAC is conceived as the concatenation of two mirror symmetric EINJ. The linac is divided into three cryomodules for several reasons: (i) the EINJ, which is constructed first, prototypes most features of the EAC; (ii) one EAC can be taken off-line for repair without taking the facility off-line; and (iii) the e-beam transport between EINJ and first EAC provides the location for a beam merger optics for a future recirculation ring. The e-linac is linked to the ARIEL target halls, to the north, by an 80 m beamline in a new tunnel.

The bunch repetition rate of 650 MHz is chosen different from the cavity resonance frequency, 1.3 GHz, in order to: (i) simplify the beam position measurement system, impedance matching is easier and signal dispersion is less; and (ii) create the opportunity to interleave a high-brightness bunch train (using the empty RF buckets) for the FEL application.

The cavity frequency (1.3 GHz), type (9-cell elliptical) and operating temperature (2K) have been chosen to benefit from the two decades of development, at the TESLA Test Facility (TTF [4]), in the design and manufacture of high gradient, non-multipactor, low field emission, $\beta = v/c = 1$ bulk-niobium cavities for a TeV-scale linear collider for high-energy physics studies. Note, if normal-conducting copper cavity technology were adopted for e-linac, the wall-plug power consumption would have been ~6 MW, rather than the 1.5 MW for the SRF implementation, and c.w. operation would have been unimaginable at these gradients.

Unlike the 1 % duty factor in the TTF application, the e-linac cavity considerations are dominated by the high c.w. beam current and power. The nominal electric field accelerating gradient, $E = 10$ MV/m, chosen for e-linac is modest, as can be achieved with chemical etch alone. The operating quality factor, Q, of 10^{10} is more challenging and may require cavity baking at 650 ^{0}C. If the cavities were to attain $Q = 10^{10}$ while operating at $E = 14$ MV/m, there is sufficient cryogenic capacity to operate the linac at 75 MeV; and the beam lines are compatible with this energy.

In the following sections we shall describe the major e-linac systems: (i) the vault, (ii) electron gun, (iii) buncher cavity, (iv) beam lines, (v) cryomodules, (vi) superconducting RF cavities, (vii) cryoplant, and (viii) high-power RF systems. Figure 1 shows the major components in the electron hall.

2.1 Electron hall

The e-linac is housed in an existing heavily shielded vault formerly used for a nuclear structure experimental program using a 1/2 kW proton beam. The vault, formerly

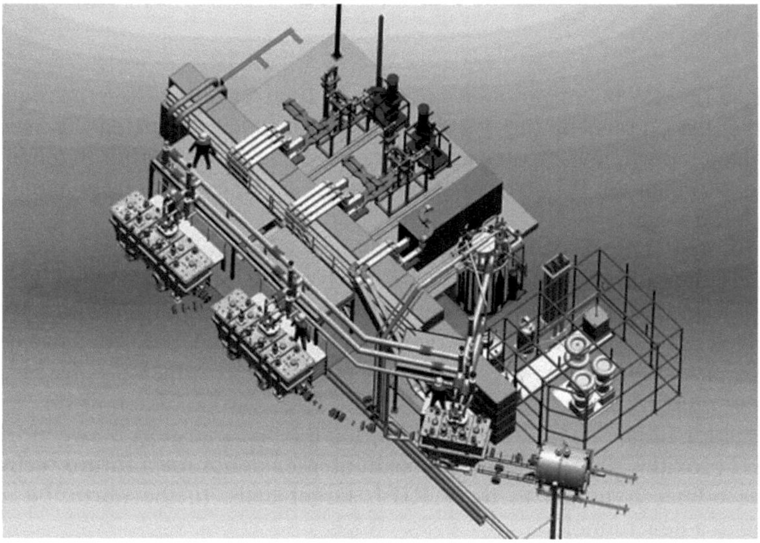

Fig. 1 Isometric view of the e-linac gun and HV cage, cryomodules, beamlines, klystrons and IOT, and coldbox in the electron hall. The north-south shield wall and access platform are also shown

called the proton hall (p-hall) is re-purposed and renamed the electron hall (e-hall). The e-hall floor at, the B2 sub-basement level, is divided into three areas: (i) linac to the west, (ii) equipment (klystrons and 4K coldbox) to the east, and (iii) p-beamline to the north. Between these areas, a half-height (10 ft) concrete wall runs N-S to shield equipment from e-beam losses, and a full-height wall runs E-W to shield equipment from future BL4N p-beam losses. Klystron HV and magnet DC power supplies along with controls and beam diagnostic electronics racks will be located above the linac on concrete roof beams, spanning the vault E-W, at ground level. 12.5 kV switchgear is brought here for connection of the HV power supplies to the AC mains. A 10-ton full-coverage crane at the underside of the roof beams allows transport of cryomodules, etc., in an otherwise congested area.

Large equipment shall be delivered to the B2 level by a vertical hatchway at the NW corner of the hall, and small items from a B2 level opening at the south wall. Personnel access is from the existing staircase at the SE corner of the hall, and emergency egress via a new stairway at the NE corner. During e-linac operations, no personnel are admitted to the e-hall. Throughout the e-hall, shielding for personnel radiation protection shall provide a safe working environment outside the hall in the event of a full catastrophic beam loss sustained for up to one hour. For example, concrete shielding up to 5 ft thickness is added to the south wall to protect the existing service annexe in the event of beam loss occurrence during recirculation.

2.2 Electron gun

Due to modest brilliance requirement (e-beam is spent on target) and its low maintenance characteristic, a 10 mA DC thermionic gridded gun is chosen for the source with emittance 5μm/plane (1σ, normalized). The gun 300 kV high-voltage (HV) supply, purchased from Glassman, resides in air-insulated Faraday cage. The gun's bias voltage and heater power supplies are mounted on a HV platform in air, in the same cage, and fed by an isolation transformer.

The gun energy, 300 keV, is the lowest at which the following 9-cell SRF cavity can provide adequate longitudinal acceptance. In deference to 300 kV operation, the gun adopts a novel layout in which the electrode profiles are inverted resulting in a long anode tube. This minimizes the surface area of the high voltage electrode (cathode), reducing the likelihood of field emitted electrons. Use of the long anode tube results in the need for a strong and compact solenoid to capture the e-beam. The anode tube is fitted with saddle-type steering coils to compensate ambient magnetic fields. To reduce the length of the gun HV isolation ceramic and to provide consistent environmental conditions, the gun is operated in an SF6 insulating gas filled vessel pressurized at up to 1 atmosphere gauge. The gas management system provides means to evacuate the vessel, purge it with N2 or fill it with SF6, and recover the SF6 gas to a twice-volume storage bag at 1 atm absolute.

The gun is modulated at 650 MHz, producing bunches $\pm 16°$ FW, obviating the need for a chopper and beam dump; an important consideration with 3kW e-beam power. The EIMAC Y-845 gridded dispenser type cathode has been chosen for this purpose. The gun modulation scheme is quite novel. First, to avoid placing the 650 MHz amplifier and macro-pulser on a high voltage platform along side the gun, RF power is transmitted via a dielectric (ceramic) waveguide between the grounded vessel and the gun; i.e. the amplifier is at ground potential. Second, to avoid reflected RF power from the cathode/grid, a coaxial transmission line matching section is inserted between the 50Ω RF cables and the EIMAC cathode. The line has a series of impedance steps, and a 1/4 lambda stub for the heater cable.

2.3 Buncher cavity

The gun is followed by a buncher cavity to improve the longitudinal match to the 9-cell cavity in the injector (EINJ). The 1.3 GHz copper cavity is of the Daresbury-EMMA design, and was procured from Niowave, USA. The measured Q and shunt resistance of the buncher are 20,000 ad 3.3 MΩ [$V^2/(2P)$ definition] respectively. With these parameters, gap voltage 30 kV and 10 mA e-beam, the generator power required for phase angles of -90 and $-90 \pm 15°$ are 160 and 290 W. A solid state amplifier, operating in class AB providing 400 W, has been purchased from Bruker BioSpin to meet this need.

2.4 Beam lines

The gun and cryomodules are linked by short beam line sections. The Electron Low energy Beam Transport (ELBT) contains 3 solenoids in-line to the injector cryomodule, a variety of beam diagnostics and a spectrometer stub. The injector is followed by a merger section (EMBT) consisting of two small-angle dipole magnets that symmetrically sandwich a combined function magnet, and quadrupole optics. The merger anticipates the energy recovery ring for the FEL, and is achromatic. The 10 MeV merger section provides the opportunity for clipping off momentum tails, produced during the gun modulation, and is equipped with a single-sided collimation jaw. Transport between the accelerator cryomodules is performed by quadrupoles; and the first such transport, at 30 MeV, is equipped with a spectrometer stub and low-power (250 W) Faraday cup.

At the 50 MeV exit of the second accelerator cryomodule, the EHAT optics leads to several branchings of the beamline. A reversible polarity dipole switches the e-beam between the EHBT or EHDT. The Electron High energy Dump Transport (EHDT) leads to a 100 kW c.w. capable beam dump inside the linac vault for tuning purposes. The "main line", the Electron High energy Beam Transport (EHBT), leads northward to the target stations 80 metres distant. This line begins with a matching section, a dogleg that bends in horizontal and vertical planes, and a quadrupole matching system to the following periodic 8-cell FODO lattice. The dogleg implements a change in beam line height between the linac vault and tunnel that accommodates the future necessity of running the electron beam below the 500 MeV proton beamline magnets. At the end of the tunnel, a vertical dogleg brings the e-beam into the plane of the targets. Thus far, the transport is achromatic. Finally the line divides into two branches with bends and matching quadrupoles before each target, east and west. As noted earlier, although the nominal beam energy is 50 MeV, the energy could be raised to approximately 75 MeV either by a recirculation arc, or by operating the cavities at higher gradient if the quality factor permits. For that reason all the beamline magnets are designed compatible with 75 MeV energy.

Throughout the high energy transport the intrinsic beam spot size is of order 1 mm r.m.s. radius leading to a power density capable of melting metals in hundred of micro-seconds. Evidently, the beam spot must be enlarged before it is applied to the dump or targets. Optical magnification alone is not the answer, since it will also magnify any beam halos present. Instead the beam is swept in a two-dimensional raster over a square of area 4×4 cm^2 by AC dipoles excited at kHz frequencies—according to the scheme adopted at Jefferson laboratory [5].

2.4.1 Quadrupole magnets

The majority of EHBT quadrupoles have integrated strength (gradient times effective length) of typically 0.4 T, but extending upto 0.8 T. EMBT quads are typically no more than 0.1 T strength. At these levels, even short quads with aspect ratio (effective length over aperture diameter) of 1 will have poletip fields 2 kGauss and 500 Gauss, respectively; well below the saturation fields of steel. As space is limited, it is advantageous to make the quads as short as practical; this has the benefit of making the poletip field as large as possible, minimizing hysteresis effects. In the short quad limit [6], e-beam aberrations are minimized by taking the poles to be locally spherical with radius $4/\pi$ times the aperture radius. Areas with compact optics, such as the doglegs, have quadrupole strength up to 1.2 T and these approximate to the long-magnet limit in which entrance/exit field effects do not dominate; their poles are locally cylindrical with radius 1.145 times the aperture radius. The e-beam transport magnetic quadrupoles are divided in three classes: (i) weak, < 0.2 T; (ii) medium, < 0.5 T; (iii) strong, < 1.3 T. The weak and medium designs retain the same (spherical) poles and yolks, but differ in their coils and cooling: air-cooled or chill-plates, respectively. The strong quads follow a conventional water-cooled design, and have cylindrical poles.

2.4.2 Magnetic field compensation

An unusual feature of the beam lines is the necessity for magnetic field compensation and shielding. Within the e-hall, the nearby proton cyclotron leads to a roughly vertical field of up to 5 Gauss. In the ARIEL tunnel, the e-beam must be shielded from the stray field of the proton beamline 3 ft above. In the e-hall, a combination of Helmholtz coils and mu-metal shields is used: large, segmented coils above and below the e-beam plane null-out the average field, while layered mu-metal shields the residual magnetic field. Adopting a $\sqrt{3}:1$ ratio between coil width and separation of the rectangular coils results in a leading order dipole field with quartic corrections in the field expansion about the beam line. In the tunnel, mu-metal shields alone are sufficient.

2.4.3 Recirculation arc

A recirculation arc is foreseen, but presently unfunded, that would connect the output of the linac with the input of the first accelerator cryomodule (EACA). A magnetic chicane in the back leg permits the path length to be tuned either for Energy Recovery Linac (ERL) operation or for energy doubling (Recirculating Linear Accelerator, RLA).

When tuned as an RLA, the recirculated beam is further accelerated on the second pass through the linac, thus reaching RIB energy up to 75 MeV with only the EACA installed. When tuned as an ERL, the e-beam could be used to drive an IR or THz free electron laser (FEL) or a Compton back-scattering X-ray source in the back leg of the recirculation loop. After recirculation the beam is decelerated in the linac and dumped at 5–10 MeV.

The primary function of the e-linac is to deliver RIB to nuclear physics experimenters; and this is not compromised by ERL operation. In the foreseen scenario, two beams are delivered simultaneously to two different users: a high-brilliance electron beam is interleaved with the single-pass RIB e-beam bunches and accelerated/decelerated in the same linac. This is achieved by running both beams through EINJ and time-separating them (after EACB) in the EHAT transport section. In this scheme, a 650 MHz RF separator directs alternating bunches into the two apertures (left or right) of a septum magnet, from whence one train is directed into the recirculation loop and the other to the EHBT transport to the RIB production targets.

In the RLA mode, there is a single beam and single user (RIB); separation of the first pass e-beam into the arc and the second pass out to the EHBT is by pure magnetic separation of two beams of disparate energy.

All modes, RIB or RLA alone and simultaneous ERL/RIB, would share some of the e-beam transport in common. The modes have been studied, mutually consistent parameters found, and footprints assigned in the e-hall floor plan to a recirculation arc and the modified EHAT optics with beams separation.

2.5 Cryomodules

Cryomodule designs come in two varieties, distinguished by how the cold mass is inserted to (and suspended from) the cryostat: either end-loading cylinders (as at the TTF) or top-loading boxes (as in ISAC-II Heavy Ion Linac). Adopting the box concept allows the e-linac to benefit from the crane and clean room infrastructure at ISAC-II, as well as the experience with design for c.w. operation. Thus, the e-linac borrows heavily from the ISAC-II heavy ion linac top-loading cryomodule design. The cryostat is a stainless steel vacuum vessel in a rectangular box shape with exterior ribs to strengthen against the vacuum load. The cold mass (cavity string and 2-phase He pipe) is supported from a strong back which, in turn, is held by 2-point and 3-point suspension rods from the lid. All components are installed and supported from the top plate and the assembly is loaded into the cryomodule from above. Unlike ISAC, the e-linac modules have separate beam line and insulation vacuums. The entrance and exit beam tube regions are angled to allow the installation of the cold mass as a hermetically sealed unit with warm isolation valves at either end. The box design has sufficient head room for the incorporation of a dedicated 4K/2K cryo-insert in each module for the production of 2K LHe in situ.

Because of the c.w. requirement, several design features are driven by the large dynamic, and relatively small static, heat loads; a situation very different from the 1 % duty factor TTF. For example, the cryogen plumbing must be resized, the TTF 5K thermal shield may be omitted and tuners activated by warm motors. The modules have a single thermal shield at 77K, two layers of magnetic shielding (room temp and 4K) and three shells of multi-layer insulation (MLI). The 150 mm ID He return pipe (2-phase pipe) and large chimney (ID=90 mm) per cavity are sized for c.w. operation compatible with a doubling of the gradient and allowing a heat removal of 0.5W/cm^2.

Warm actuators for the tuners were chosen because of the small number of cryomodules and the vulnerability this presents to high-availability operation. The cold tuners themselves are of the CEBAF scissor type and are modified for the different cavity frequency and number of cells. The two 60 kW c.w. RF power input couplers per cavity, developed by Cornell and manufactured by "Communication and Power Industries" (CPI) are diametrically opposed about the beam pipe to minimize transverse kicks. Unlike the Cornell application, which uses G10 composite suspension, movement of the e-linac strong backs during cool down is greater than can be accepted by the coupler, and so they are mounted in a TRIUMF-designed gimbal.

2.5.1 Cryo-insert

A 4K reservoir on board each cryomodule acts as phase separator and the 4K return vapour is sent back to the cold box. The 2K liquid is produced in each cryomodule by passing the 4K liquid through a heat exchanger in counterflow with the returning exhaust gas from the 2K phase separator and expanding the gas to 3.2 kPa through a Joule-Thomson (JT) expansion valve. The two-phase header pipe above the cavity string acts as a 2K phase separator and delivers cold gas back through the 4K/2K heat exchanger to the sub-atmospheric pumping system.

2.6 Superconducting RF cavities

The e-linac accelerating cavities are TTF-style 9-cell niobium cavities with modified end cells and asymmetric beam pipes to ensure adequate damping of Higher Order Modes (HOM). The 7 interior cells have 35 mm radius iris, while the coupler-side cell has a 48 mm iris, and the damper side iris is 39 mm radius. Even if there was no HOM issue, the Cornell/CPI couplers dictate an asymmetric beam pipe.

The HOM damping target is set by regenerative (2-pass) Beam Break Up (BBU) as may occur during beam recirculation through the linac. The BBU threshold is estimated in excess of 20 mA if loaded shunt resistance of all modes satisfy $(R/Q) \times Q_L < 10^7$ ohm. HOM cavity studies [7] in the range of 1.5–3.25 GHz show that the combination of a stainless steel damping ring on the coupler end, and CESIC ring on the other, damp all HOMs to $(R/Q) \times Q_L < 2 \times 10^6$ ohm. The fundamental input couplers alone damp many modes, but the beamline absorbers are essential to reaching this value.

2.7 Cryoplant

With only 3 cryomodules to support, the cryosystem is medium-sized; and it is not cost effective to install a central 2K plant. Instead, 2K He is produced where it is needed: inside the modules. Thus the purpose of the cryo-system is to supply liquid helium (LHe) first at 4K in a closed re-liquefaction and refrigeration loop; and second, produced at 2K in the cryomodules by sub-atmospheric (SA) pumping, and then returned to the closed loop. This function is supported by several (connected) geographically distributed sub-systems. In the e-hall: the 4K cold box and LHe dewar, LHe distribution trunk, 4K/2K cryo-inserts, and 4K gaseous He return with vaporizers. In the compressor building: the SA pumping, the main and recovery/purification compressors, an oil removal and gas management (ORGM) package and helium purifier. In the open air: two room-temperature, pressurized gaseous He (GHe) storage tanks (one for clean, and one for dirty He) 113 m^3 each, GHe trunking between compressor building and e-hall. The SA pumps are located distant from the e-hall; this choice eliminates acoustic noise and significant vibration at the e-linac, and avoids irradiation of the pumps thus facilitating their refurbishment at a vendor location without nuclear workers.

The HELIAL cold-box (refrigerator) and ORGM is manufactured by "Air Liquide Advanced Technologies", and the compressors by "Kaeser". The cold box is normally operated in mixed liquefaction/refrigeration mode: 210 l/hr and 130 W, respectively. The GHe circuit design places a strong emphasis on monitoring and maintaining He purity free of oil, moisture or other contaminants. In addition, there is a liquid nitrogen (LN2) supply at 77K delivered from an external supply, and exhausted to the atmosphere. The LN2 is used to pre-cool the GHe in the cold box, and to cool the 77K thermal shield.

The basic concept of the e-linac cryogenic system replicates the ISAC-II delivery. GHe is delivered from the storage tank to the cold box which produces 4K liquid at 1.3 Bar to a supply dewar that feeds LHe to a common LHe delivery trunk with parallel 4K feeds to each cryomodule. The 2K liquid is produced in the 4K/2K cryo-insert in each cryomodule, as noted above, in response to sub-atmospheric pumping. Warmed SA gas is returned to the main compressor. In the event of a power failure, the recovery compressor (on emergency power supply) returns the gas to the storage tank; if the gas becomes contaminated the same compressor acts in purification mode and recirculates the He through carbon filters and molecular sieves.

2.8 High power RF sources

The injector has one 9-cell cavity, while the two accelerator cryomodules each have two 9-cell cavities; for a total of five 1.3 GHz SRF cavities. The energy of e-linac is 50 MeV with an e-beam current of 10 mA. To reach the ultimate 1/2 MW beam power, each cavity must provide 100 kW, and each is equipped with two Cornell/CPI 50 kW c.w. input couplers; for a total of ten couplers.

The RF power concept chosen for the three cryomodules is shaped to a large degree by the narrow scope of available c.w. RF sources at 1.3 GHz. Originally it was conceived that each cavity would be driven by a 100 kW RF source, but the vendor, “e2V”, ceased production. The present implementation has the injector fed by up to two 30 kW c.w. Inductive Output Tubes (IOT). This is not ideal, since a 60 kW injector does not meet the 100 kW requirement for the first SRF cavity. As vendor products evolve, the IOTs may be replaced or upgraded. Contrastingly, the accelerator is powered by two c.w. 270 kW klystrons, one per cryomodule. This is a cost effective arrangement, with a minor disadvantage that the RF controls must work with the vector sum of the cavity field levels rather than the individual values.

2.8.1 30 kW IOT and transmitter

For the first stage of installation, the 30 kW output power of the IOT is split in two to provide rf power for 2 couplers in the EINJ. The 30 kW RF source consists of an IOT, CPI model VKL9130, housed in a transmitter from Bruker BioSpin. The latter comprises a 1.3 GHz driver amplifier, power supplies for the IOT, and an IOT power output section. The solid state driver operates in class AB and provides 400 W RF power. The supply includes the switch-mode HV power supply for the IOT beam voltage; and filament, focus, grid-bias and ion pump power supplies. The WR650 waveguide output section includes a 30 kW capable circulator and 50 ohm load equipped with direction couplers for output power measurements. A power coupler conditioning station utilizes the same IOT.

2.8.2 290kW klystron

In a joint venture with Helmholtz Zentrum Berlin, the c.w. klystron for the first accelerator cryomodule has been procured from CPI, USA. This klystron is designed to deliver RF output power to at least 270 kW with incremental gain ≥ 0.5 dB/dB, and saturation at 290 kW. The klystron has a specified perveance of 0.55 $\mu A/V^{1.5}$ and beam voltage of 65 kV; efficiency is expected to be at least 52 %. With power loss of 20 % in the waveguide distribution system, 108 kW will be available at each cavity.

Three dB hybrids and phase shifters will be used to divide power equally and maintain required phase of the RF voltage in the accelerating cavities with respect to the beam. Two 30 dB variable attenuators are employed for RF conditioning the cavities, one cavity at a time.

The klystron has a variety of power supplies: a switching mode HV supply for the 65 kV beam voltage, and gun heater/filament, focusing coils, ion pump power supplies. The HV supply is located on the roof beams, approx 40 m cable length from the klystron on the e-hall floor (B2 level). The low voltage supplies may be located with the klystron.

2.9 Conclusions

The construction of the ARIEL facility marks the beginning of a new era at TRIUMF in which there will be three simultaneous drive beams for RIB production—enabling concurrent operation of three experimental areas in the ISAC facility that will serve the needs of a broad scientific user community for many years to come. This community includes nuclear physics and nuclear medicine and materials science. A particular feature of the facility will be the delivery of unprecedented intensities of rare, short-lived exotic isotopes, and in particular those with extreme neutron excess.

At the heart of this facility expansion are a MW-class electron linear accelerator under construction, and a future proton beamline from the TRIUMF H- cyclotron. The 50 MeV e-linac is as innovative as the original 500 MeV cyclotron or the light-ion linacs in the ISAC complex. The e-linac not only introduces a new particle, the electron, into the mainstay of TRIUMF operations but also introduces many new technologies to the laboratory. Foremost are the 1.3 GHz elliptical RF cavities operating at 2K with superconducting niobium and superfluid helium; in a supporting role are the innovative 300 kV electron source and the klystron high-power RF sources. Moreover, TRIUMF has adopted the 1.3 GHz SRF technology originally intended for a future low duty factor high-energy-physics linear collider and taken it into the pioneering realm of high average current continuous operation.

To conclude, the vastly augmented scientific user facility that is "ARIEL" and the newly introduced accelerator technologies that are "e-linac" combine to make their completion a transformative "moment" in the history of the laboratory that equals the beginning of the ISAC era and is only surpassed by the founding of the cyclotron.

References

1. Koscielniak, S., et al.: Proposal for a 1/2-MW electron linac for rare isotope and materials science. In: Proc. of EPAC08, p. 985, Genoa, Italy (2008)
2. Merminga, L., et al.: ARIEL: TRIUMF's Advanced IsotopE laboratory. In: Proc. of IPAC2011, p. 1917, San Sebastian, Spain (2011)
3. Dilling, J., Kruecken, R., Merminga, N.: ARIEL overview. Hyperfine Interact. (this issue, 2013)
4. Richard, F., Schneider, J., Trines, D., Wagner, A.: TESLA Technical Design Report. http://tesla.desy.de/new-pages/TDR-CD/start.html (2001)
5. Yan, C., Sinkine, N., Wojcik, R.: Linear beam raster for cryogenic targets. Nucl. Inst. Methods Phys. Res. A **539**, 1–15 (2005)
6. Baartman, R.: Quadrupole shapes. Phys. Rev. ST Accel. Beams **15**, 074002 (2012)
7. Kolb, P., et al.: HOM Cavity Design for the TRIUMF E-linac. In: Proc. of SRF2011, p. 203, Chicago, IL USA (2011)

DOI 10.1007/978-94-007-7963-1_32
Reprinted from *Hyperfine Interactions* Journal, DOI 10.1007/s10751-013-0908-4

ARIEL front end

M. Marchetto · R. A. Baartman · R. E. Laxdal

Published online: 23 October 2013

Abstract The ARIEL project at TRIUMF will greatly expand the variety and availability of radioactive ion beams (RIBs) (Laxdal, Nucl Inst Methods Phys Res B 204:400–409, 2003). The ARIEL front end connects the two ARIEL target stations to the existing ISAC facility to expand delivery to two and eventually three simultaneous RIB beams with up to two simultaneous accelerated beams (Laxdal et al. 2008). The low-energy beam transport lines and mass separators are designed for maximum flexibility to allow a variety of operational modes in order to optimize the radioactive ion beam delivery. A new accelerator path is conceived for high mass delivery from an EBIS charge state breeder. The front-end design utilizes the experience gained in 15 years of ISAC beam delivery.

Keywords ARIEL · Front · End · Separator · Charge breeder · LASER · Post accelerator

1 Introduction

In the ISAC-I facility, 500 MeV protons from the cyclotron [3] at up to 100 μA impinge on one of two production targets [4] to produce radioactive isotopes. The isotopes are ionized and the resulting beam is mass-separated ($\Delta M/M \sim 1/2500$ for $\pi\epsilon = 10\pi$ μm) and transported in the low-energy beam transport (LEBT) electrostatic beamline to either the low-energy experimental area or through a series of room temperature accelerating structures (RFQ [5], DTL [6]) to the ISAC-I

ISAC and ARIEL: The TRIUMF Radioactive Beam Facilities and the Scientific Program.

M. Marchetto (✉) · R. A. Baartman · R. E. Laxdal
TRIUMF, 4004 Wesbrook Mall,
Vancouver, BC, V6T 2A3, Canada
e-mail: marco@triumf.ca

medium energy experimental area. The 35 MHz RFQ accelerates ions with $A/q \leq 30$ to 150 keV/u and the post-stripper (106 MHz) DTL accelerates ions with $A/q \leq 7$ to energies fully variable from 117 keV/u to 1.8 MeV/u. The accelerated beam can also be transported to the ISAC-II 40 MV Superconducting Linear Accelerator [7] (SC-linac) for acceleration above the Coulomb barrier and delivered to the ISAC-II high energy area. The ISAC ECR charge breeder [8], CSB1, installed in the ISAC mass-separator room, is used to boost the charge state of masses with $A > 30$ to allow acceleration in the RFQ.

Even though the ISAC facility offers intense RIBs with world class accelerator and end stations some restrictions exist. The present facility has ten major experimental end stations but only one RIB beam available at any given time. Since the cyclotron is typically shut down for four months of the year there is less than one month of radioactive beam time per experimental facility in a year. The beam time is further reduced due to target/source development since there is no on-line target test area. The present target design and hotcell arrangement coupled with hall access restrictions demands a lengthy (weeks) turn around for target replacement and periodic vault access is required during the switchover reducing the beam delivery from the other target. Even though two target areas exist the proton driver beam cannot be time shared since there is only one mass separator and LEBT.

The ARIEL facility (see Fig. 1) is a major addition to the TRIUMF complex that will greatly expand the scientific potential of the TRIUMF radioactive ion beam facility.

Fundamental to the ARIEL proposal is the goal to deliver three simultaneous RIBs to three experimental areas. The upgrade involves the addition of a high power electron linac (50 MeV/10 mA – 0.5 MW) [9], to produce RIBs through photo-fission. A second beamline from the existing cyclotron will deliver a new 500 MeV/100 μA – 50 kW proton-driver [10]. The new driver beams are coupled with two new production areas with independent separators and a flexible beam delivery system allowing simultaneous operation. The new target stations are designed for actinide target use with hermetically sealed target volumes for self containment. They are engineered for a turn around of only a few days. A new accelerator front end including a second charge state booster provides the possibility of a second accelerated beam. Three simultaneous beams will at least triple the beam hours on target. Additionally, the electron driver beam could be available during the cyclotron shutdown periods enhancing the beam time even further or augmenting development time.

2 Front-end overview

The purpose of the ARIEL front end is to deliver beams from the ARIEL target stations to the existing ISAC beam lines with sufficient purity, quality and flexibility. The front end consists of mass separators and a LEBT switchyard to deliver two simultaneous radioactive ion beams (RIBs) to the existing ISAC beamlines. Combined with the existing RIB beam delivered from the ISAC target area the ARIEL/ISAC facility will allow the delivery of three simulataneous RIB beams to three ISAC experimental areas. The areas include two low energy areas, the ISAC medium energy area and the ISAC-II high energy area. The ARIEL front-end installation

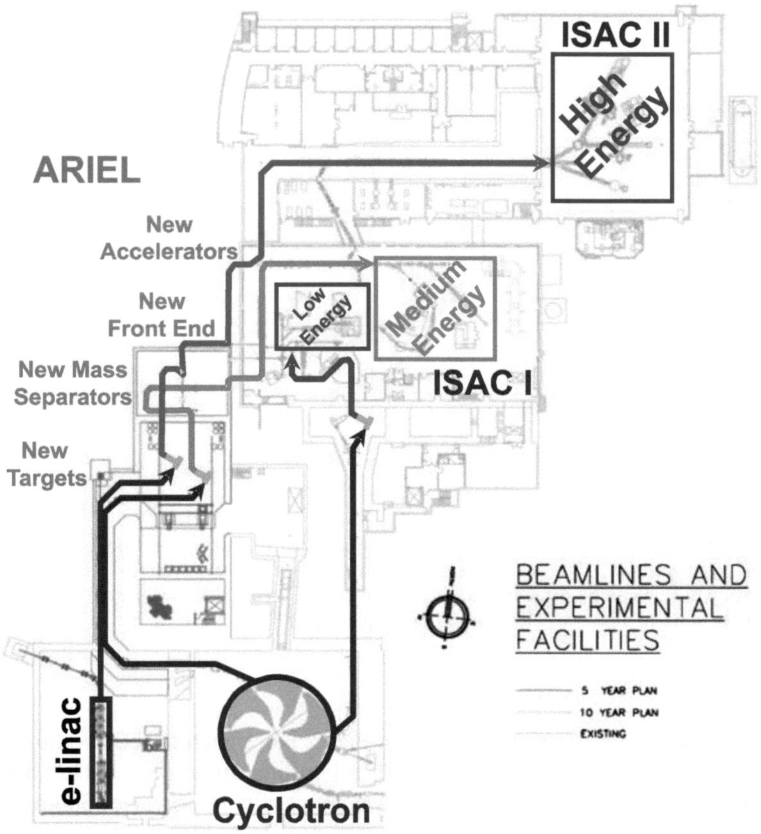

Fig. 1 ARIEL and ISAC facilities layout. The ARIEL facility gives ISAC a multi-users capability by allowing the delivery of three simulataneous radioactive ion beams to the three different experimental areas (low, medium and high energy)

is planned in a staged way to allow early use of the facility and extensions as capital and personnel allow.

3 Mass separators

It is useful to consider a switchyard that is capable of either low resolution, medium resolution or high resolution separation schemes depending on the experiment. The LEBT switchyard is designed with sufficient flexibility that each of the targets can pass a beam through the High Resolution Spectrometer (HRS) while the beam from the other target is sent to the Medium Resolution Spectrometer (MRS). In addition a bypass line is available so that if not required a beam can be sent to the experimental area directly after the pre-separator. In general, reduced resolution schemes reduce the tuning time required in order to achieve a suitable transmission. In this way the operators can optimize the separation required for each ion while minimizing the tuning time.

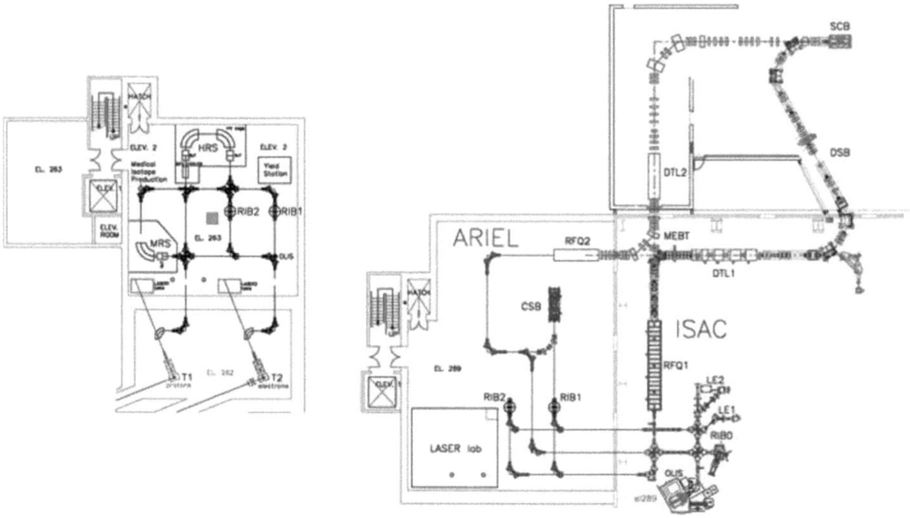

Fig. 2 ARIEL front end layout. The *left picture* represents the footprint of the underground target hall and separator room. The beam is transported through vertical section (RIB1 and RIB2) to ground level where it can be direceted to the existing ISAC beam line and post accelerators (RFQ1 and DTL1) or send to a new accelerator path (RFQ2 and DTL2)

The proposed configuration is shown in Fig. 2. Here each of the two target/source units have an identical pre-separator with a bending angle of $\theta = 112°$ and radius of curvature of $\rho = 0.5$ m that produces a mass resolving power of 500 for 10π μm through a 1 mm object slit. The bending angle is dictated by the target orientation. The low resolution separators and object and image (selection) slits are located inside the target hall (see Fig. 2 left picture) to contain most of the unwanted radioactive elements. The key elements are engineered for remote handling. The pre-selected beams are then transported to the adjacent mass separator room via the LEBT transport lines.

The MRS leg includes a single dipole with $\theta = 90°$ and radius of curvature of $\rho = 1.2$ m and is designed to produce a resolving power of ~2500 for 10π μm. The HRS leg uses two $\theta = 90°$ $\rho = 1.2$ m dipoles in tandem and includes an RFQ cooler to reduce the transverse emittance and energy spread of the beam before separation to achieve an effective resolving power of ~20000 for 2.5π μm. Both separators are planned to be installed on a high voltage platform to allow acceleration of the beam before separation.

4 Low-energy beam transport

The low-energy beam transport (LEBT) consists of electrostatic transport sections delivering beams at source potential (up to ≤60 kV). The LEBT periodic and bending sections are carbon copies of the existing ISAC low energy transport beam lines [11] that has been demonstrated to be extremely reliable. Electrostatic components are cheaper than magnetic, but also simpler to tune since the settings depend only upon beam energy, and not on mass. One can go from one mass to another by simply changing the separator dipole, without retuning the electrostatic elements. The voltages are not high due to the low beam energy.

The LEBT is comprised of an evacuated beam pipe with electrostatic elements inside including quadrupoles and steerers, joined to box-like chambers housed with electrostatic dipoles or diagnostic elements including profile monitors, Faraday cups and slits. Substantial LEBT transport has already been installed in ISAC. The ARIEL front-end design is based on the standard building blocks that comprise the present ISAC installation with the exception of the line downstream of the new charge state booster, CSB2 (see below). Here the vacuum must be improved from the standard $2 \times 10^{-7}\,Torr$ to $< 5 \times 10^{-8}\,Torr$ to facilitate transportation of the higher charge states. The ARIEL LEBT has a design acceptance of 150 μm.

The LEBT quadrupoles have a bore radius of 25 mm with a nominal length of 50 mm. For the typical focal length $f = 0.3$ m, the required electrode voltage is 2.5 kV at the maximum beam energy of 60 keV. The electrodes are cylindrical in section, with radius 1.145 times the bore radius and are cut from an aluminum extrusion.

The bends have an aperture of 38 mm, with a radius of 254 mm and spherical shape. Spherical electrodes give equal focusing in the bend and non-bend directions. The electrode potentials are 9 kV at the full energy of 60 keV. The outer electrode can swing open on a pivot to allow switching between straight-through (bend off) and 45° bend.

Besides special sections for matching to the separators, CSB and accelerators there are only three different modules; the periodic cell, a 90° achromatic bend, and an insertion for creating small waists, e.g. for the buncher. The backbone of the transport is a FODO periodic cell with a phase advance of 90° degrees per cell. Each cell is 1 m in length, and the quads are arranged in doublets, 0.31 m between quad centres. These parameters are of course a compromise between good optical properties and minimum cost. The bend module consists of two 45° bends with a symmetric triplet between. The triplet is tuned to make the module achromatic. Since it is symmetric as well, this also zeroes the effect of transverse motion on particle arrival time.

The design goal of the new installation is to provide sufficient flexibility that any of the target stations, existing or new, can deliver beams to any of the three experimental areas, low-energy, medium-energy or high-energy, so that the RIB beams from each target can be optimized for a given experiment. Another goal is to provide a second path to the low-energy experimental area since this is where there is the largest inventory of experimental infrastructure.

On the mass separator floor the switchyard delivers beams to two vertical beamlines RIB1 and RIB2 (being RIB0 the existing ISAC vertical section) that bring the beam to the ARIEL ground floor. From here beamlines join the existing ISAC-LEBT infrastructure. A new connecting line is added to the ISAC low-energy experimental area splitting the area into two halves; with each line serving three experimental areas allowing two simultaneous low-energy experiments. The ARIEL EBIS charge state booster, CSB2, will be installed on the ground floor of the ARIEL building. LEBT beamline will connect RIB1 and the CSB2 as well as from the ISAC LEBT to CSB2 so that beams from the existing target area can also be boosted in CSB2. A low-energy beam transport also connects CSB2 to a new RFQ2.

5 New accelerator leg

A new accelerator path is necessary to deliver simultaneously two accelerated beams one to the medium energy the other to the high energy area. The technique of accelerating a 1+ beam through a series of stripping and acceleration steps was considered but abandoned due to the high total accelerating voltages required. Instead an EBIS charge breeder (CSB2) will be added to boost the 1+ ions from the on-line source. The choice of EBIS technology is based on the following. Firstly, the performance of the CSB1 source (Phoenix 14.5 GHz ECRIS) provides only a modest ionization efficiency (2–5 %) with a rather large stable contamination coming from the background gas. This demands that a second stripping stage be employed to help clean the impurities from the RIB [12]. Secondly, an EBIS source would give us a new technology that would make the combined ISAC/ARIEL facility more flexible. TRIUMF has a working EBIS with the TITAN experimental installation [13]. Other facilities in North America, such as MSU and ANL, are also pursuing EBIS technology so that development can be streamlined through collaboration. Higher charge states, better efficiencies and lower background contamination are all cited as reasons for pursuing the EBIS technology. A slight disadvantage for experimenters is that the beam would be delivered in a pulsed mode.

It is expected that the EBIS will efficiently produce charge states compatible with $4 \leq A/q \leq 6$ for all masses so the RFQ2 specification for the new accelerator leg was conservatively set to $A/q \leq 7$. Since the target ion source is rated for operation at ≤ 60 kV, this sets the injection energy for the RFQ at ~8 keV/u. As in ISAC-I a single gap, multi-harmonic pre-buncher at 11.8 MHz is used upstream of the RFQ2. This produces reasonable space between bunches of 85 ns for the experimenters and also reduces the longitudinal emittance from the RFQ. An RF frequency of 70.7 MHz, six times the bunch period, is chosen for RFQ2 giving a reasonable opening cell length of 8.7 mm. The exit of the RFQ2 is injected into a beamline cross downstream of RFQ1 so that the exit energy of RFQ2 and RFQ1 are chosen to be the same at 150 keV/u. This sets the RFQ maximum accelerating voltage to 1 MV. A 11.8 MHz chopper is positioned downstream of RFQ2 to remove beam from the satellite buckets between the main bunches.

A new beamline running north of the existing RFQ1 will provide separate paths for ISAC-I and ISAC-II accelerated beams. A new medium energy transport section incorporates a switchyard that allows sending either of the RFQ1 or RFQ2 beams to either of ISAC-I or ISAC-II simultaneously. The new beamline to ISAC-II will include a room temperature drift tube linac (DTL2) at 106 MHz to boost the energy from 0.15 MeV/u to 1.5 MeV/u for beams up to $A/q = 7$ to match the beam into the ISAC-II SC-linac. The drift tube linac would require a maximum of 9.5 MV of accelerating potential and would consist of five IH tanks and four sets of quarupole triplets between tanks for periodic transverse focussing. An optional stripper could be employed between DTL2 and the SC-linac to boost the charge state to achieve a higher final energy.

Table 1 ARIEL front-end staged installation

ARIEL front-end component	Stage 1	Stage 2	Stage 3	Stage 4	Stage 5	Stage 6	Stage 7
West target station	√	√	√	√	√	√	√
West pre-separator	√	√	√	√	√	√	√
East target station					√	√	√
East pre-separator					√	√	√
LEBT bypass	√	√	√	√	√	√	√
MRS				√	√	√	√
HRS and RFQ cooler		√	√	√	√	√	√
RIB1	√	√	√	√	√	√	√
RIB2					√	√	√
LEBT to LE1/LE2	√	√	√	√	√	√	√
LEBT to RFQ1			√	√	√	√	√
CSB2			√	√	√	√	√
RFQ2						√	√
LEBT to RFQ2						√	√
DTL2							√

The table shows the installed ARIEL components at each stage. The target stations are not part of the front-end but they define the starting point

6 Front-end staging installation

The front-end is planned to be installed in different stages. This approach allows to balance funds and manpower, and at the same time it gives the possibility of delivering ARIEL science early. Table 1 summarizes the staging.

The stage 1 installation will deliver light mass RIB's using only the pre-separator. The light RIB beams will be sent directly to the low energy experimental area. The front end components to be installed are the pre-separator of the west target station and ~50 m of electrostatic beam line to ISAC LEBT.

The following two stages (2 and 3) are respectively meant to increase the resolution capability with the installation of the HRS and the mass range to be delivered to medium and high energy experimental stations using the CSB2 through the existing ISAC accelerator chain. The installation of ~30 m of electrostatic beam lines is required besides the high resolution spectrometer, RFQ cooler and charge state booster. Most of the components for stage 2 and 3 can be installed without interruption of the beam delivered to the low energy area.

Stage 4 and 5 are going to increase the RIB production capability and delivery. The major installation of the east target is accompanied by the installation of the MRS and the completion of the LEBT switch-yards consisting of adding $\sim 50\ m$ of electrostatic beam lines. At these stages ARIEL and ISAC will have the capability of delivering three simultaneous RIB's: two beams to the low energy experimental area and one to either the medium or the high energy experimental stations.

The last two stages (6 and 7) are going to give full capability to the combined ARIEL-ISAC facility. Still three simultaneous RIB's can be delivered but now with two post accelerated beams. A major investment is required on the accelerator side with the installation of the new RFQ2 and DTL2. In this final configuration any of the three experimental areas (low, medium and high) can have a RIB beam at the same time.

References

1. Laxdal, R.E.: Acceleration of radioactive ions. Nucl. Inst. Methods Phys. Res. B **204**, 400–409 (2003)
2. Laxdal, R.E., et al.: The Proposed ISAC-III at TRIUMF, EPAC2008, p. 3635. Genoa, Italy (2008)
3. Bylinskii, I., et al.: TRIUMF 500 MeV Cyclotron Refurbishment, Cyclotrons and Their Application, p. 143. Giardini Naxos, Italy (2007)
4. Bricault, P., et al.: Progress Towards New RIB and Higher Intensities at TRIUMF, CYCLOTRONS 2010, p. 365. Lanzhou, China (2010)
5. Poirier, R.L., et al.: CW Performance of the TRIUMF 8 Meter Long RFQ for Exotic Ions, LINAC2000, p. 1023. Monterey, California, USA (2000)
6. Laxdal, R.E., et al.: Beam Commissioning and First Operation of the ISAC DTL at TRIUMF, PAC2001, p. 3942. Chicago, Illinoois, USA (2001)
7. Laxdal, R.E.: Commissioning and Early Experiments with ISACII, PAC2007, p. 2593. Albuquerque, New Mexico, USA (2007)
8. Ames, F., et al.: Commissioning of the ECRIS Charge State Breeder at TRIUMF, ECRIS2010, p. 178. Grenoble, France (2010)
9. Koscielnik, S.R., et al.: ARIEL and the TRIUMF e-Linac Initiative, a 1/2 MW Electron Linac for the Rare Isotope Beam Production, LINAC2008. Victoria, British Columbia, CANADA, September 2008, p. 383.
10. Dutto, G., et al.: Simultaneous Extraction of Two Stable Beams for ISAC, EPAC2008, p. 3505. Genova, Italy (2008)
11. Baartman, R., Welz, J.: 60 keV Beam Transport Line and Switch-Yard for ISAC, PAC7, p. 2778. Vancouver, British Columbia, Canada (1997)
12. Marchetto, M., et al.: In Flight Ion Separation Using a Linac Chain, LINAC2012. Tel-Aviv, Israel (2012)
13. Lapierre, A., et al.: The TITAN EBIT charge breeder for mass measurements on highly charged short-lived isotopes—first online operation. Nucl. Inst. Methods Phys. Res. A **624**, 54–64 (2010)

Author Index to Volume 225

MIX
Papier aus verantwortungsvollen Quellen
Paper from responsible sources
FSC® C105338

If you have any concerns about our products,
you can contact us on
ProductSafety@springernature.com

In case Publisher is established outside the EU,
the EU authorized representative is:
Springer Nature Customer Service Center GmbH
Europaplatz 3, 69115 Heidelberg, Germany

Printed by Libri Plureos GmbH
in Hamburg, Germany